DEVELOPING SKILLS IN ALGEBRA

A LECTURE WORKTEXT

FIFTH EDITION

A LECTURE WORKTEXT

J. LOUIS NANNEY
MIAMI-DADE COMMUNITY COLLEGE

JOHN L. CABLE
MIAMI-DADE COMMUNITY COLLEGE

WCB Wm. C. Brown Publishers

Book Team

Editor *Earl McPeek*
Developmental Editor *Theresa Grutz*
Production Editor *Eugenia M. Collins*
Designer *K. Wayne Harms*
Art Editor *Mary E. Swift*
Visuals Processor *Andréa Lopez-Meyer*

Wm. C. Brown Publishers

President *G. Franklin Lewis*
Vice President, Publisher *George Wm. Bergquist*
Vice President, Operations and Production *Beverly Kolz*
National Sales Manager *Virginia S. Moffat*
Group Sales Manager *Vincent R. Di Blasi*
Vice President, Editor in Chief *Edward G. Jaffe*
Marketing Manager *Elizabeth Robbins*
Advertising Manager *Amy Schmitz*
Managing Editor, Production *Colleen A. Yonda*
Manager of Visuals and Design *Faye M. Schilling*
Production Editorial Manager *Julie A. Kennedy*
Production Editorial Manager *Ann Fuerste*
Publishing Services Manager *Karen J. Slaght*

WCB Group

President and Chief Executive Officer *Mark C. Falb*
Chairman of the Board *Wm. C. Brown*

Library of Congress Catalog Card Number: 91–55594

ISBN 0–697–08585–6

Printed in the United States of America by Wm. C. Brown Publishers, 2460 Kerper Boulevard, Dubuque, IA 52001

10 9 8 7 6 5 4 3 2 1

Contents

PRODUCTS AND FACTORING

ALGEBRAIC FRACTIONS

EXPONENTS AND RADICALS

QUADRATIC EQUATIONS

SYSTEMS OF EQUATIONS AND INEQUALITIES

Preface

Purpose

Developing Skills in Algebra is designed for the student who needs a comprehensive review of the topics from elementary and intermediate algebra. The text contains the algebraic skills necessary to prepare the student for courses in college algebra and trigonometry. As the title implies, this text develops and maintains student proficiency and confidence in the basic algebraic skills. Students will benefit from the clear, concise explanations that we have developed during our many years as classroom teachers. The wide use of this text through its first four editions testifies to its popularity with students and instructors.

Changes to the Fifth Edition

In this fifth edition we have responded to the valuable suggestions from the many users and reviewers of the text.

- Explanations in many sections have been rewritten for greater clarity. Improvements have been made to sections 1–4, 1–7, 1–8, 2–1, 2–2, 2–6, 3–1, 3–2, 3–4, 3–5, 4–2, 4–5, 6–3, 6–7, 7–1, 7–3, 7–6, 7–7, 7–8, 7–9, 8–1, 8–4, and 8–7.
- Extra steps to enhance understanding in the worked out examples occur in every section.
- Additional worked out examples have been included in many sections. New examples have been added to sections 1–1, 1–4, 1–7, 1–8, 2–3, 3–1, 3–2, 4–5, 6–2, 7–3, and 7–9.
- Interesting applications have been added to many exercise sets. For instance, sections 1–1–2, 1–8–1, 2–3–1, 2–4–1, 3–2–1, 5–7–1, 7–9–1, 8–7–2, and 8–8–1 now include more motivating applications of mathematics to real-world situations.
- The scope of the word problems has been expanded to include work problems in sections 5–7 and 7–9.
- The topic of scientific notation has been included in section 6–2.
- A new format is used for the chapter summaries that keys concepts to sections within each chapter to make review much easier for students.

Prerequisites

It is assumed that the student is proficient in the arithmetic of rational numbers. If this is not the case, then it is advised that a course in prealgebra be required as a prerequisite.

Format

Each section of the text contains clear and concise explanations or rules followed by detailed examples and numerous exercises for each problem type. The student learns and maintains skills by being involved in "doing" the mathematics. Answers to the odd-numbered section exercises are given in the answer section at the end of the text to help students check their mastery of the skills associated with each problem type as they work the exercise.

Text Features

Chapter Survey

The Chapter Survey helps students and instructors identify concepts the student has already mastered and areas the student should concentrate on. Instruction can be tailored to the particular strengths identified by the Chapter Survey results. All answers are provided in the answer appendix and each is coded to the appropriate section.

Introductory Paragraph

The introductory paragraph overviews the topics to be covered, explains why it will be useful to master the particular skills and concepts, and ties the new material to previously learned topics where appropriate.

Objectives

Each section of the text begins with a boxed statement of the objectives. Thus, a clearly stated purpose for each section is available to both student and instructor.

Boxed Rules, Laws, and Definitions

Boxes emphasize important rules, laws, and definitions. The important ideas for each section are identified for the student in clear, concise language.

Numbered Examples

Each concept and skill is slowly developed using numerous examples that build on each other and increase slowly in level. The examples model the complete range of exercises in the text.

Cautions

We have drawn from our many years of teaching experience to identify common student errors. These "Cautions" are set apart and highlighted in the text to point out and correct student errors. These cautions should be of significant help to the student.

Section Exercises

The Section Exercises provide comprehensive practice on every skill and concept presented in the section. All exercise sets are graded in level of difficulty from easy to more challenging, and paired, with the "evens" reflecting the "odds" in content and level.

Chapter Summary

Each chapter is followed by a Summary that lists key words as well as rules and procedures covered in the chapter. These are referenced to the sections that discuss them to help students effectively and efficiently review the chapter.

Chapter Review

After the chapter summary, each chapter contains an exhaustive review of all of the skills and concepts presented in the chapter. This gives the student ample practice with the topics. Odd-numbered answers are given in the answer appendix. Each answer is coded to the section number so students can easily locate the correct area for review if they encounter difficulties.

Practice Test

A Practice Test, with answers keyed to the sections within the chapter, is given at the end of each chapter. This test is designed to help the student review and to prepare for examinations. Each test covers all of the key concepts presented in the chapter and is similar in content and length to an actual chapter test.

Cumulative Test

Following chapter 4 is a Cumulative Test designed to help students maintain mastery of previously learned concepts. All of the answers are given in the answer appendix and are coded to the appropriate chapter and section so that concepts needing reinforcement can be easily located.

End-of-Book Test

A final examination test, giving an overview of the entire text, follows the last chapter. Answers to this test are keyed to chapter and section for easy review. Both the Cumulative Test and the End-of-Book Test emphasize the building-block nature of mathematics and help students maintain previously learned skills.

Instructional Techniques

We are indeed gratified to learn of the variety of teaching situations in which this text has been used. Some schools have successfully used the text in the traditional classroom-lecture arrangement followed by outside assignments. Others have successfully used an "individualized approach" where the students work through the text on their own and use the instructor as a consultant. The text, together with the supplementary materials available from Wm. C. Brown Publishers, is especially suited for a laboratory situation.

Combinations, variations, and modifications of these instructional techniques are encouraged when they fit instructor preferences and specific student needs.

Supplementary Materials

For the Instructor

The *Instructor's Manual* portion of the combined ***Instructor's Manual/Test Item File*** includes all answers to problems in the textbook.

The *Test Item File/Quiz Item File* portion is a printed version of the computerized *TestPak* and *QuizPak* that allows you to choose test items based on chapter, section, or objective. The objectives are taken directly from the section objectives in *Developing Skills in Algebra: A Lecture Worktext,* Fifth Edition.

wcb ***TestPak,*** our computerized testing service, provides you with a mail-in/call-in testing program and the complete *Test Item File* on diskette for your use with IBM® PC, Apple® or Macintosh® computers. **wcb** *TestPak* requires no programming experience. Tests can be generated randomly, by selecting specific test items or objectives. In addition, new test items can be added and existing test items can be edited.

wcb ***QuizPak,*** a part of *TestPak 3.0,* provides students with true/false, multiple choice, and matching questions from the *Quiz Item File* for each chapter in the text. Using this portion of the program will help your students prepare for examinations. Also included with the **wcb** *QuizPak* is an on-line testing option that allows professors to prepare tests for students to take using the computer. The computer will automatically grade the test and update the gradebook file.

wcb ***GradePak,*** also a part of *TestPak 3.0,* is a computerized grade management system for instructors. This program tracks student performance on examinations and assignments. It will compute each student's percentage and corresponding letter grade, as well as the class average. Printouts can be made utilizing both text and graphics.

For the Student

The ***Student's Solutions Manual*** contains overviews of every chapter of the text, solutions to every-other, odd-numbered exercise problem, and chapter self-tests with solutions to help students check their mastery of key concepts. It is available for student purchase.

Audiotapes developed specifically for *Developing Skills in Algebra: A Lecture Worktext* can be used in a self-paced laboratory, or as a boost for the student who needs extra help after a lecture. In these tapes topics are reinforced with careful explanations of specific examples and reminders of key concepts. Students have the opportunity to listen to the reasoning behind the solutions and then practice solving specific problems in the text's exercise sets. Then students can listen to a carefully developed, complete explanation of the solution to the particular problem.

Videotapes reinforce key concepts with fully developed examples, additional explanations, and useful hints.

Tutorial drill and practice **Software** is available to give your students additional practice with and reinforcement of the basic concepts of algebra.

Acknowledgments

A text of this quality cannot be produced without extensive aid from many sources. We are indebted to the many users of the previous editions for their helpful suggestions. Special thanks go to the reviewers of the fifth edition and the previous editions, which include the following people:

Helen W. Baril
Quinnipiac College

Sandra A. Cameron
Central Ohio Technical College

Lucile Danker
Quincy College

Daniel Drucker
Wayne State University

Carol Elrod
Gainesville, GA

Nina R. Girard
University of Pittsburgh at Johnstown

John Grossman
St. Paul Technical Vocational Institute

Cheryl Hobneck
Illinois Valley Community College

Jack Howard
Cumberland University

Sally Ann Low
Angelo State University

Katherine J. McDonald
Wayne State University

Lois McMullan
East Central Junior College

Dr. Jan Melancon
Loyola University

Mark W. Neumann
UWC Rock County

Ned Schillow
Lehigh County Community College

Theresa M. Shustrick
University of Pittsburgh at Johnstown

Sally C. Webster
Husson College

We would also like to thank Earl McPeek (editor), Theresa Grutz (developmental editor), Gene Collins (production editor), Wayne Harms (designer), Mary Swift (art editor), and Andréa Lopez-Meyer (visuals processor) for their work on this edition.

SURVEY

The following questions refer to material discussed in this chapter. Work as many problems as you can and check your answers with the answer section in the back of the book. The results will direct you to the sections of the chapter in which you need to work. If you answer all questions correctly, you have a good understanding of the material contained in this chapter.

1. Give the negative or opposite of (-31). **1.** ____________

2. Add: $(-94) + (+47)$ **2.** ____________

3. Evaluate: $(-20) - (-14)$ **3.** ____________

4. Evaluate: $(+8) + (-35) - (+4)$ **4.** ____________

5. Evaluate: $(-14)(-3)$ **5.** ____________

6. Evaluate: $\dfrac{-36}{+6}$ **6.** ____________

7. Evaluate: $(-3)(+5)(-2)(-1)$ **7.** ____________

8. Simplify: x^4x^3 **8.** ____________

9. Multiply: $(7a^2b)(-5a^3b^4)$

9. ______

10. Divide: $\dfrac{27x^8y}{-3x^2y^3}$

10. ______

11. Combine similar terms:
$5x^2y - 2xy + 7xy^2 - 4x^2y - 3xy$

11. ______

12. Multiply: $-3a(a^2 - 2a + 1)$

12. ______

13. Evaluate: $7 + 4 \cdot 2 - 5$

13. ______

14. Simplify: $3x^2 + 4x[2x - (x - y)]$

14. ______

15. Evaluate: $3x^2 - 4x + 2$ when $x = -2$

15. ______

Fundamental Concepts

Certain fundamental ideas form the foundations of the study of algebra and of algebraic manipulations. It is therefore important that the student of algebra be proficient in the skills of arithmetic.

The topics to be covered in this chapter include the operations on signed numbers and simple operations on monomials and polynomials, including the technique of evaluating expressions. Mastering these skills is necessary to the further study of algebra.

1-1 ADDITION OF SIGNED NUMBERS

Upon completion of this section you should be able to:

1. Give the negative or opposite of a number.
2. Add positive and negative numbers.

The numbers first encountered in elementary arithmetic are those used in counting, or the set of **counting numbers,** $\{1, 2, 3, 4, \ldots\}$. This set is later extended to include zero and the negatives of the counting numbers giving us the set of **integers.**

$$\{\ldots, -4, -3, -2, -1, 0, 1, 2, 3, 4, \ldots\}$$

The set of **rational numbers**—those that can be expressed as a ratio of two integers—are generally referred to as **fractions.** This set includes the integers since, for example, 3 can be expressed as $\frac{3}{1}, \frac{6}{2}$, and so on. In fact, this set includes all number expressions that involve only the operations of addition, subtraction, multiplication, and division.

Next a set of numbers, called the **irrational numbers,** that cannot be expressed as the ratio of integers is developed. This set includes such numbers as π, $\sqrt{5}$, $\sqrt[3]{7}$, and so on.

The set of rational and irrational numbers together make up a set called the **real numbers.** Elementary algebra is a study of the real numbers and their properties.

The basic set of properties (axioms) of the real numbers can be used to justify all of the manipulations used in both arithmetic and elementary algebra. If this set of properties is accepted as being true, then all other properties can be proved as theorems. Although an intuitive approach rather than one of rigorous proofs is used in this text, mention is often made of these basic properties. They are listed on the inside front cover for easy reference.

Members of the set of real numbers can be classified as positive, negative, or zero. Numbers such as $+7$, $+\frac{3}{4}$, $+10$, and so on, are positive, while -3, $-\frac{4}{5}$, -5, and so on, are negative. Positive numbers are designated by a $(+)$ sign and negative numbers are designated by a (−) sign. Such numbers are often referred to as **signed numbers.** If a number is not preceded by a sign, it is understood that the number is positive. Numbers such as 9 and 35 are positive numbers.

We think of all numbers greater than zero as **positive** and all numbers less than zero as **negative.** Zero is neither positive nor negative.

It is important to distinguish between a negative number and the negative (sometimes called opposite or additive inverse) of a number. The **negative** or **opposite** of a given number is a number that when added to the given number yields zero. See Property 5 for real numbers.

Example 1 The opposite of $+3$ is -3 because $+3 + (-3) = 0$.

Example 2 The opposite of -3 is $+3$ because $-3 + (+3) = 0$.

EXERCISE 1–1–1

Give the negative or opposite of each.

1. $+5$ **2.** $+7$ **3.** $+\frac{3}{4}$ **4.** -2

5. x **6.** $(3 + 2)$ **7.** $-(7 + 2)$ **8.** $-a$

9. 0 **10.** $-\frac{7}{8}$

Since all real numbers are classified as positive, negative, or zero, this set can be represented on a number line. By using a straight line from plane geometry (a straight line has infinite length) and choosing a point to be represented by 0 and another point to be represented by 1, we can, with certain agreements, establish a correspondence between the real numbers and the points of the line. We must first agree to place the positive numbers to the right of 0 and the negative numbers to the left of 0. We must also agree that the length of the line segment from 0 to 1 will be used as a unit measurement between all real numbers that differ by 1, and that real numbers between 0 and 1 will be represented by proportional parts of this unit. Such a **real number line** is represented below.

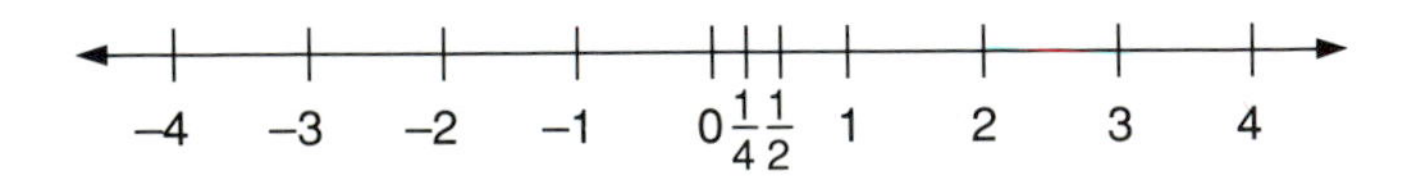

The arrows indicate that the line extends infinitely in each direction.

The operation of addition can be thought of as finding the end result when numbers are combined. We can illustrate addition of signed numbers by using the number line if we think of a positive number as movement along the line to the right and a negative number as movement along the line to the left.

Example 3 $+3$ can be illustrated by starting at zero on the number line and moving three units to the right.

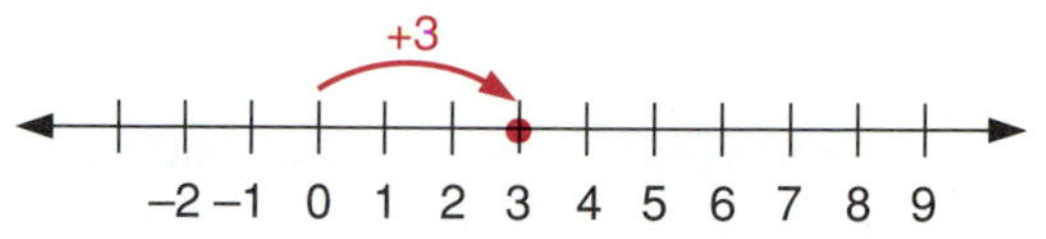

Example 4 Use the number line to find the sum of $+3$ and $+4$.

Solution Starting at 0 on the number line, we move three units to the right. From the point $+3$ we then move four units to the right.

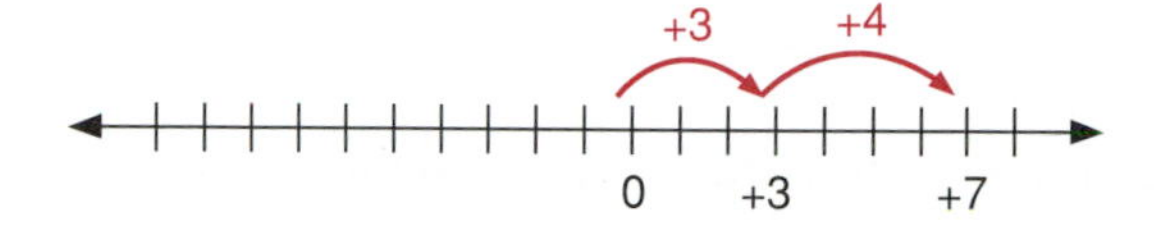

Notice that the result of these two movements is $+7$. Therefore, we can say that $(+3) + (+4) = +7$.

Example 5 Find $(-3) + (-4)$.

Solution

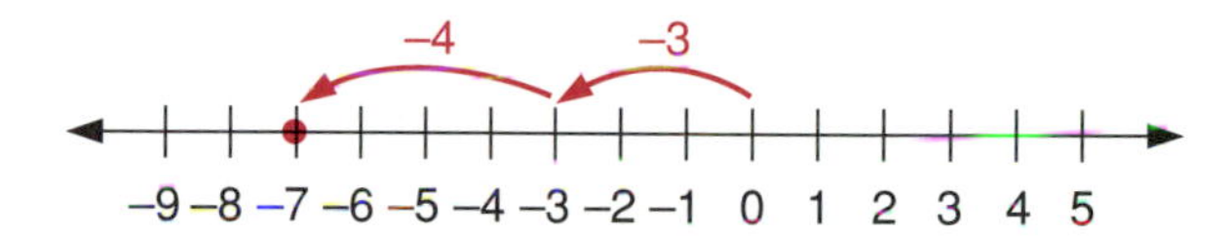

Since the end result is -7, we have $(-3) + (-4) = -7$.

If we move in a certain direction along the line and then move even further in that same direction, the end result will be the combination of the two movements in that same direction. We state this fact in a formal rule.

RULE
To add signed numbers having like signs add the numbers (without regard to sign) and use the common sign.

Example 6 $(+12) + (+5) = +17$

Example 7 $(-8) + (-7) = -15$

Example 8 $(-3) + (-7) = -10$

Example 9 $(+4) + (+5) = +9$

We now look at the other possibility, adding signed numbers when the signs are not alike.

Example 10 Add $+7$ and -2 using the number line.

Solution Starting at 0, we move seven units to the right to the point $+7$. From this point we then move two units to the left.

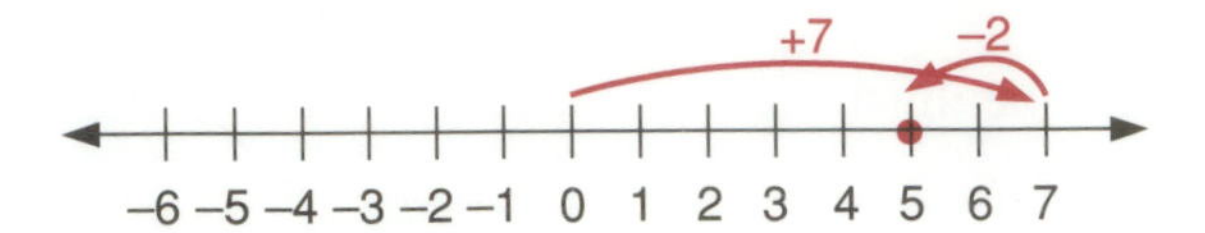

We see the end result is $+5$, so we conclude $(+7) + (-2) = +5$. Notice that we could have illustrated this problem by first moving to the left.

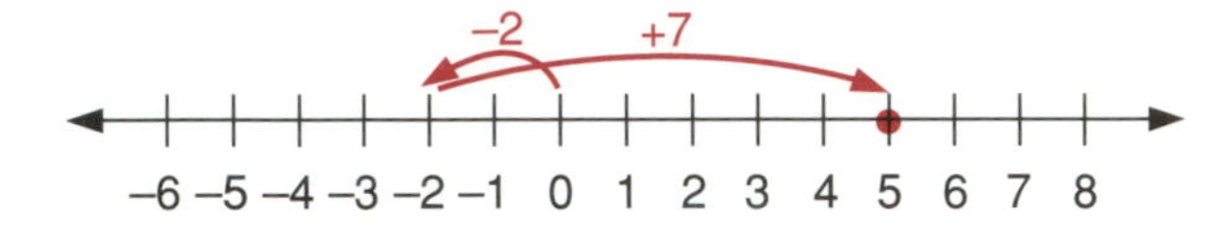

Our end result is still $+5$.

Example 10 illustrates the **commutative property of addition.**

$$(+7) + (-2) = (-2) + (+7) = +5$$

Example 11 Use the number line to find $(-8) + (+3)$.

Solution Starting at 0, we move eight units to the left to the point -8. From there we move three units to the right.

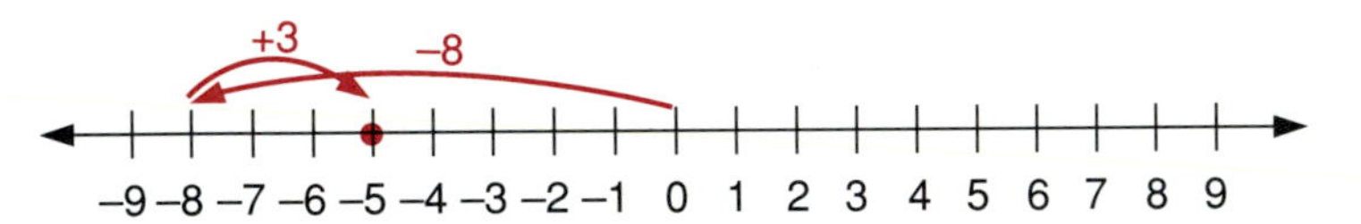

The end result is -5, so $(-8) + (+3) = -5$.

If we move along the number line in one direction and follow by a movement in the opposite direction, if the greatest movement is to the right, the end result will be to the right of zero. Conversely, if the greatest movement is to the left, the end result will be to the left of zero. We state this formally as a rule.

> **RULE**
> To add signed numbers with unlike signs subtract the smaller number from the larger number (without regard to sign) and use the sign of the larger number (without regard to sign).

Example 12 $(+6) + (-5) = +1$ (Ignoring signs, 6 is larger than 5.)

Example 13 $(-6) + (+5) = -1$ (Ignoring signs, 6 is larger than 5.)

Example 14 $(-50) + (+17) = -33$ (Ignoring signs, 50 is larger than 17.)

Example 15 $\left(+\frac{7}{8}\right) + \left(-\frac{2}{8}\right) = +\frac{5}{8}$ $\left(\text{Ignoring signs, } \frac{7}{8} \text{ is larger than } \frac{2}{8}.\right)$

You will encounter many problems that are stated in words. It is important that you read the problem carefully.

Example 16 From 6:00 P.M. until midnight the temperature falls four degrees, and from midnight until 4:00 A.M. it falls eight more degrees. What is the total fall in temperature from 6:00 P.M. to 4:00 A.M.?

Solution If a fall in temperature is represented by a minus sign, then a fall of 4° followed by a fall of 8° can be written as $(-4°) + (-8°) = -12°$.

EXERCISE 1-1-2

Find the sums.

1. $(+7) + (+10)$

2. $(+16) + (+28)$

3. $(-6) + (-9)$

4. $(-3) + (-21)$

5. $(-11) + (+33)$

6. $(-13) + (+27)$

7. $(-5) + (+1)$

8. $(-37) + (+13)$

9. $(+17) + (-43)$

10. $(+36) + (-19)$

11. $(+53) + (+71)$

12. $(-47) + (-25)$

13. $(+83) + (-92)$

14. $(+17) + (-18)$

15. $(-49) + (+51)$

16. $(+7) + (-7)$

17. $\left(+\frac{3}{7}\right) + \left(-\frac{5}{7}\right)$

18. $\left(-\frac{2}{5}\right) + \left(+\frac{4}{5}\right)$

19. $(-35) + (+35)$

20. $\left(+\frac{1}{2}\right) + \left(-\frac{1}{2}\right)$

21. Find the result of a rise in temperature of eight degrees followed by a fall of two degrees.

22. Find the result of a profit of \$30 combined with a loss of \$36.

23. The stock market gained ten points in the morning, then lost six points in the afternoon. What was the net gain or loss for the day?

24. A worker earned $25 in the morning and spent $18 in the afternoon. What was the net gain or loss for the day?

25. Find the result of a gain of two yards in a football game followed by a loss of six yards.

26. Find the result of a loss of two yards followed by a loss of five yards.

1–2 SUBTRACTION OF SIGNED NUMBERS

OBJECTIVES

Upon completion of this section you should be able to:

1. Subtract one signed number from another.
2. Add and subtract several signed numbers.

Now that we have learned to add signed numbers, the next basic operation we will study is subtraction. Subtraction is defined as adding the negative. This gives rise to a very simple rule for subtracting signed numbers.

RULE

To subtract one signed number from another change the sign of the number being subtracted and use the appropriate rule for addition.

Example 1 $(+5) - (+2) = (+5) + (-2) = +3$

Example 2 $(+5) - (-2) = (+5) + (+2) = +7$

Example 3 $(+8) - (+10) = (+8) + (-10) = -2$

Example 4 $(+8) - (-10) = (+8) + (+10) = +18$

Remember that a number written without a sign is always considered positive.

Example 5 $(+5) + (+3)$ can be written $5 + 3$.

Example 6 $(+5) - (+3)$ can be written $5 - 3$.

Note: In algebra no two signs are ever written together without the use of parentheses. Thus, $+5 + -3$ would not be a correct statement but $+5 + (-3)$ would have a specific meaning.

EXERCISE 1-2-1

Find the differences.

1. $(+9) - (+3)$

2. $(+8) - (+4)$

3. $(-5) - (+3)$

4. $(-7) - (+6)$

5. $(-11) - (-21)$

6. $(-13) - (-53)$

7. $(+8) - (-9)$

8. $(+21) - (-13)$

9. $(-14) - (+9)$

10. $(-21) - (+2)$

11. $29 - 14$

12. $73 - 57$

13. $25 - 36$

14. $17 - 54$

15. $(-12) - 15$

16. $\left(-\frac{5}{9}\right) - \frac{3}{9}$

17. $32 - (-5)$

18. $\frac{5}{8} - \left(-\frac{2}{8}\right)$

19. $(-14) - (-14)$

20. $\left(-\frac{3}{5}\right) - \left(-\frac{3}{5}\right)$

We have established the rules for adding and subtracting signed numbers. We now need to look at situations that involve both operations.

Consider the expression

$$3 - 2 + 5.$$

Since only two numbers may be combined at a time, this expression could have two meanings unless a rule is applied. For instance, if one person should decide to subtract 2 from 3 and add the result to 5, the answer would be 6. If another person should decide to add 2 and 5, then subtract the result from 3, the answer would then be -4. To avoid any confusion mathematicians have agreed to the following rule.

RULE
If an expression contains only additions and subtractions, these operations are performed in order from left to right.

The preceding expression, $3 - 2 + 5$, would therefore be evaluated as 6.

Example 7
$$\begin{aligned} 7 - 2 + 3 - 5 + 4 &= 5 + 3 - 5 + 4 \\ &= 8 - 5 + 4 \\ &= 3 + 4 \\ &= 7 \end{aligned}$$

Example 8 If at 9:00 A.M. the temperature is +20 degrees and from 9:00 A.M. until 1:00 P.M. it rises 14 degrees, then from 1:00 P.M. until 5:00 P.M. it falls 18 degrees, what is the temperature at 5:00 P.M.? Answer: +16 degrees.

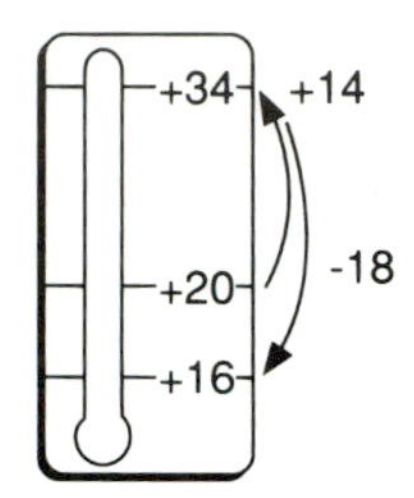

Starting at +20°, a rise of 14° takes the temperature to +34°. A fall of 18° takes it to +16°.

Examples 7 and 8 are examples of combining signed numbers. If "steps north" is represented as positive, then 6 steps north followed by 3 steps north equals 9 steps north could be written

$$+6 + 3 = +9.$$

If a fall in temperature is represented by a minus sign, then a fall of 3 degrees followed by a fall of 7 degrees can be written

$$-3° - 7° = -10°.$$

If a rise in temperature is plus and a fall in temperature is minus, then example 8 can be written

$$+20° + 14° - 18° = +16°.$$

EXERCISE 1–2–2

Evaluate each expression.

1. $3 + 5 + 7$ **2.** $16 + 4 + 2$ **3.** $-5 - 13 - 7$

4. $-7 - 5 - 13$ **5.** $5 - 6 + 4$ **6.** $9 - 7 - 6$

7. $-3 - 7 - 5$

8. $4 - 7 - (-6)$

9. $5 + 9 - 14$

10. $7 - 3 - 5$

11. $13 - 17 + 5 - 4$

12. $-4 + 5 - 12 + 6$

13. $-7 - (-5) + 7 - (-9)$

14. $15 + 37 - 15 + 3$

15. $27 - 59 - 27 + 50$

16. $14 - 27 - 13 - 8$

17. $8 - 7 + 2 - 4 + 9$

18. $-19 + 7 - 22 + 16$

19. $34 - 16 + 9 + 16 - 3$

20. $9 - 5 + 13 + 6 - 7 - 20$

21. If the temperature was 50 degrees at 6:00 A.M. and from 6:00 A.M. to 2:00 P.M. it rose 15 degrees, then from 2:00 P.M. to 7:00 P.M. it fell 9 degrees, what was the temperature at 7:00 P.M.?

22. A temperature starts at 10 degrees. It falls 15 degrees, then rises 8 degrees, then falls 3 degrees. What is the final temperature?

23. A weight-watcher lost five pounds the first week, gained two pounds the second week, lost three pounds the third week, and lost two pounds the fourth week. What was the net gain or loss for the four weeks?

24. In American football a team has four plays (downs) during which the ball must be moved ten yards forward or given up to the other team. On a series of four downs the Miami Dolphins did the following:

first down	gained five yards
second down	lost eight yards
third down	gained twelve yards
fourth down	gained two yards

Did the team gain enough yards to keep the ball? How many yards did they gain in the four downs?

1-3 MULTIPLICATION AND DIVISION OF SIGNED NUMBERS

OBJECTIVES

Upon completion of this section you should be able to:

1. Multiply and divide signed numbers.
2. Find the product of several signed numbers.

We now wish to consider the two remaining basic operations, multiplication and division. 2×3, $2 \cdot 3$, $(2) \times (3)$, $(2)(3)$, $(2) \cdot (3)$, and $2(3)$ are all ways of indicating the product of 2 and 3. If letters are used to represent numbers, multiplication can be indicated by writing the letters together without a sign or parentheses. For instance, if x and y represent numbers, then xy represents the product of x and y.

$\frac{2}{3}$ and $2 \div 3$ are ways of indicating the quotient of 2 and 3.

$\dfrac{\frac{5}{8}}{\frac{2}{3}}$ and $\frac{5}{8} \div \frac{2}{3}$ are ways of indicating the quotient of $\frac{5}{8}$ and $\frac{2}{3}$.

Multiplication is often regarded as repeated addition. Thus,

$$(3)(5) = 5 + 5 + 5 = 15$$

and

$$(4)(-3) = (-3) + (-3) + (-3) + (-3) = -12.$$

The two preceding examples illustrate that the product of two positive numbers is positive and the product of two numbers having different signs is negative. We would now like to determine the product of two negative numbers.

A property of multiplication often used is that the product of any real number and zero is zero. This can be seen in the following example.

$$\begin{aligned} 3(3) + 3(0) &= 3(3 + 0) && \text{(distributive property)} \\ &= 3(3) && \text{(additive identity)} \end{aligned}$$

But the only way $3(3) + 3(0) = 3(3)$ is for $3(0) = 0$.

We now return to the question "What is the product of two negative numbers?" Let us find the product of $(-3)(-5)$. Consider the expression $(-3)(+5) + (-3)(-5)$.

$$\begin{aligned} (-3)(+5) + (-3)(-5) &= (-3)[(+5) + (-5)] && \text{(distributive property)} \\ &= (-3)(0) && \text{(additive inverse)} \\ &= 0 \end{aligned}$$

Now if $(-3)(+5) + (-3)(-5) = -15 + (-3)(-5) = 0$, then $(-3)(-5) = 15$ since $-15 + 15 = 0$.

Division can be defined as multiplication by the reciprocal. Dividing by 2 is the same as multiplying by $\frac{1}{2}$; dividing by $\frac{3}{4}$ is the same as multiplying by $\frac{4}{3}$; and so on. Since division is defined in this way, it is obvious that the rules for multiplication and division of signed numbers must be identical.

The following are the rules (which we learned in arithmetic) for multiplying and dividing fractions.

a. multiplying $\left(\frac{a}{b}\right)\left(\frac{c}{d}\right) = \frac{ac}{bd}$

b. dividing $\frac{a}{b} \div \frac{c}{d} = \left(\frac{a}{b}\right)\left(\frac{d}{c}\right) = \frac{ad}{bc}$

The preceding discussion brings us to the following rules for multiplying and dividing signed numbers.

RULE
The product or quotient of two signed numbers having like signs will be positive.

Example 1 $(+3)(+2) = +6$

Example 2 $(-3)(-2) = +6$

Example 3 $\frac{+6}{+2} = +3$

Example 4 $\frac{-6}{-2} = +3$

Example 5 $(-10) \div (-2) = +5$

Example 6 $\left(+\frac{2}{3}\right)\left(+\frac{5}{7}\right) = +\frac{10}{21}$

Example 7 $(-2)\left(-\frac{1}{5}\right) = +\frac{2}{5}$

RULE
The product or quotient of two numbers having unlike signs will be negative.

Example 8 $(+3)(-2) = -6$

Example 9 $(-3)(+2) = -6$

Example 10 $\frac{+6}{-2} = -3$

Example 11 $\frac{-6}{+2} = -3$

Example 12 $(10) \div (-2) = -5$

Example 13 $\left(-\frac{1}{6}\right)\left(+\frac{2}{7}\right) = -\frac{2}{42} = -\frac{1}{21}$

Example 14 $(+7)\left(-\frac{3}{5}\right) = -\frac{21}{5}$

The rules for multiplication and division of two numbers can be stated briefly as, "Like signs give positive, unlike signs give negative."

CAUTION Do not confuse the rules of signs for addition and subtraction with the rules for multiplication and division. All work from this point on will be easier if you pause now and make sure that you know the rules of signs.

EXERCISE 1-3-1

Find the products and quotients.

1. $(-3)(-4)$

2. $(-3)(4)$

3. $5(-2)$

4. $-7(3)$

5. $(6) \cdot (2)$

6. $(-7)\left(-\frac{1}{7}\right)$

7. $\left(\frac{3}{4}\right)\left(\frac{4}{3}\right)$

8. $\left(\frac{2}{3}\right)(-12)$

9. $\left(-\frac{2}{5}\right) \div \left(\frac{3}{7}\right)$

10. $\left(-\frac{1}{2}\right) \div \left(-\frac{1}{8}\right)$

11. $\frac{-15}{-3}$

12. $\frac{-18}{6}$

13. $(-20) \div (-5)$

14. $(20) \div (5)$

15. $\frac{18}{-6}$

16. $(-6)\left(\frac{2}{3}\right)$

17. $\frac{(-6)}{\frac{3}{2}}$

18. $(-10)\left(\frac{3}{4}\right)$

19. $(12)\left(-\frac{2}{5}\right)$

20. $\left(-\frac{7}{8}\right)\left(-\frac{4}{3}\right)$

Multiplication, like addition and subtraction, can be performed on only two numbers at a time. If the product of three numbers is indicated, the product of any two of them must be multiplied by the third.

Example 15 $(-2)(-3)(4) = (6)(4) = 24$

or $(-2)(-3)(4) = (-2)(-12) = 24$

or $(-2)(-3)(4) = (-8)(-3) = 24$

Notice that the order of multiplication does not change the answer. The previous examples are illustrations of use of the associative and commutative properties of multiplication.

If the product of more than three numbers is indicated, this procedure is simply repeated.

Example 16 $(-2)(3)(-4)(5) = (-6)(-4)(5) = (24)(5) = 120$

Example 17 $(-3)(2)\left(-\frac{1}{5}\right) = (-6)\left(-\frac{1}{5}\right) = +\frac{6}{5}$

EXERCISE 1-3-2

Find the products.

1. $(2)(5)(7)$

2. $(4)(-6)(3)$

3. $(10)(5)(-8)$

4. $(-6)(2)(-3)$

5. $(-4)(-6)(2)$

6. $(6)(-2)(-5)$

7. $(4)(-9)(-3)$

8. $(-2)(-8)(-5)$

9. $(-1)(-5)\left(-\frac{1}{5}\right)$

10. $(-7)(-14)(0)$

11. $(3)(-1)(-5)(10)$

12. $(-4)(3)(-11)(2)$

13. $(-1)(-5)(-2)(-3)$

14. $(3)(-2)(-1)(-7)$

15. $(-10)(10)(-1)(-3)$

16. $(-8)(4)(-2)(10)$

17. $(-2)(-1)(-10)(4)(1)$

18. $\left(-\frac{1}{2}\right)(-14)\left(-\frac{3}{7}\right)(-2)$

19. $(27)\left(-\frac{1}{9}\right)\left(-\frac{1}{3}\right)(-37)$

20. $(49)(-72)(-104)(0)(-23)$

1-4 POSITIVE WHOLE NUMBER EXPONENTS

OBJECTIVES

Upon completion of this section you should be able to:

1. Evaluate expressions such as 2^4.
2. Simplify expressions such as $x^3 \cdot x^4$.
3. Find the products and quotients of expressions involving exponents.

In the first three sections we have studied operations using only numbers of arithmetic. You may have heard the statement "Algebra is an extension of arithmetic." As we continue our study the truth of this statement should become more obvious. In algebra letters are used as "placeholders" for numbers. Letters of the alphabet, both uppercase and lowercase, are used to represent numbers as we perform various operations.

In algebra special mathematical symbols and notations also are often used to simplify certain expressions. One such symbol that we introduce here is the *exponent*.

Numbers that are multiplied together are called **factors.** In the expression x^2, where x holds the place for some number, the number 2 is called an **exponent.** The exponent indicates that the **base** x is used as a factor two times. x^5 indicates that x is used as a factor five times.

$$x^2 = (x)(x)$$
$$x^3 = (x)(x)(x)$$
$$x^5 = (x)(x)(x)(x)(x)$$

x^2 is read "x squared" or "x to the second power." x^3 is read "x cubed" or "x to the third power." When the exponent is four or more, such as x^5, then x^5 is read only "x to the fifth power." If no exponent is written, it is assumed to be 1.

$$x^1 = x$$

Example 1 Evaluate 5^2.

Solution $5^2 = (5)(5) = 25$.

Example 2 Evaluate $(-3)^3$.

Solution We are asked to cube -3. So $(-3)^3 = (-3)(-3)(-3) = -27$.

Example 3 Evaluate $-(2)^4$.

Solution We are asked to find the negative of 2^4. So $-(2)^4 = -(2)(2)(2)(2) = -16$.

From the definition of exponent we can establish laws for multiplying and dividing numbers with exponents.

For instance, suppose we wish to multiply $x^3 \cdot x^5$. From the definition we have

$$x^3 = x \cdot x \cdot x \text{ and } x^5 = x \cdot x \cdot x \cdot x \cdot x$$
$$\text{so } x^3 \cdot x^5 = (x \cdot x \cdot x) \cdot (x \cdot x \cdot x \cdot x \cdot x)$$
$$= x \cdot x \cdot x \cdot x \cdot x \cdot x \cdot x \cdot x$$
$$= x^8.$$

If we generalize the above example to the product of $x^a \cdot x^b$, we have

$$x^a = \underbrace{x \cdot x \cdot x \cdot \ . \ . \ . \ x}_{a \text{ factors}}$$
$$x^b = \underbrace{x \cdot x \cdot x \cdot \ . \ . \ . \ x}_{b \text{ factors}}$$
$$\text{and } x^a \cdot x^b = \underbrace{(x \cdot x \cdot x \cdot \ . \ . \ . \ )}_{a \text{ factors}}\underbrace{(x \cdot x \cdot x \cdot \ . \ . \ . \ x)}_{b \text{ factors}}$$
$$\text{so } x^a \cdot x^b = x^{a+b}.$$

This is the first law of exponents.

LAW I
$x^a \cdot x^b = x^{a+b}$ (To multiply add exponents.)

CAUTION Be careful to note that the just stated rule applies only when the *bases are the same.* Also note that only the exponents are added. The base *never* changes.

Example 4 $x^3 \cdot x^4 = x^{3+4} = x^7$

Example 5 $x^8 \cdot x^2 = x^{10}$

Example 6 $x^4 \cdot x = x^4 \cdot x^1 = x^5$

Example 7 $x^3 \cdot y^2 = x^3 \cdot y^2$ (Rule does not apply.)

Example 8 $2^3 \cdot 2^4 = 2^7$

Example 9 $3^4 \cdot 3^2 = 3^6$ (Notice that the base does not change.)

Example 10 $2 \cdot 2^3 = 2^4$

Example 11 $3^2 \cdot 2^4 = 3^2 \cdot 2^4$ (Rule does not apply.)

EXERCISE 1-4-1

Evaluate.

1. 3^2
2. $(-3)^2$
3. $(2)^3$
4. $(-2)^2$
5. $-(2)^3$
6. $-(-2)^3$

Find the following products in terms of exponents. If the rule does not apply, so state.

7. $x^3 \cdot x^5$
8. $y^4 \cdot y^2$
9. $3^4 \cdot 3^3$
10. $2^3 \cdot 2^5$
11. $(x^2)(x^3)(x^4)$
12. $(x^4)(y^3)$

We now need to consider expressions that contain both numbers and letters.

In an expression such as $3x^4$, which means $3(x)(x)(x)(x)$, 3 is called the **coefficient,** x is called the **base,** and 4 is called the **exponent.**

coefficient → $3x^4$ ← exponent
↖ base

Note that only the x and *not* the 3 is raised to the 4th power. If both the 3 and the x were raised to the 4th power, the expression would be written as $(3x)^4$. Notice the parentheses! $(3x)^4 = (3x)(3x)(3x)(3x)$.

An indicated product such as $(2x^3)(3x^4)$ involves two previously discussed ideas. The coefficients 2 and 3 can be multiplied since the order of multiplication has no effect on the answer (commutative property of multiplication). x^3 and x^4 can also be multiplied by adding the exponents since the bases are the same. Thus,

$$(2x^3)(3x^4) = (2)(3)(x^3)(x^4) = 6x^7.$$

Example 12 $(5x^2)(-2x^3) = (5)(-2)(x^2)(x^3) = -10x^5$

Notice that the rules for multiplying signed numbers must always be considered.

Example 13 $(3x^2)(2y^2)(4z^5) = (3)(2)(4)(x^2)(y^2)(z^5) = 24x^2y^2z^5$

Notice that the coefficients are multiplied even though the bases are different.

EXERCISE 1-4-2

Find the products.

1. $(2x^2)(5x^3)$

2. $(4x^3)(3x^5)$

3. $(-3x^2)(7x^3)$

4. $(11a^3)(-6a^7)$

5. $(-8a^2)(-4b^3)$

6. $(7xy)(5xy)$

7. $(-2x^2y)(3xy^3)$

8. $(6a^2b)(5a^3c)$

9. $(2x^2)(-5x^5)(3x^4)$

10. $(-7a^2b^2)(-2ab)(3ab)$

11. $(-6xy)(-2x^2)(-4y^5)$

12. $(11ab)(3b^2c)(-5a^2c^3)$

13. $(-3x^2y^3)(-7xy^5z^2)(5yz)$

14. $(6x^2)(-5y^2)(2z)$

15. $(-2a^2b)(-5b^2c^2)(6c^3d^2)$

We will now state and use the second law of exponents.

LAW II

$$\frac{x^a}{x^b} = x^{a-b} \quad \text{if } a \text{ is greater than } b.$$

$$\frac{x^a}{x^b} = \frac{1}{x^{b-a}} \quad \text{if } a \text{ is less than } b.$$

$$\frac{x^a}{x^b} = 1 \quad \text{if } a = b.$$

Example 14 $\dfrac{x^5}{x^3} = x^{5-3} = x^2$

Example 15 $\dfrac{x^5}{x^7} = \dfrac{1}{x^{7-5}} = \dfrac{1}{x^2}$

This rule comes directly from the definition of exponents. For instance,

$$\frac{x^5}{x^3} = \frac{(x)(x)(x)(x)(x)}{(x)(x)(x)}.$$

We can now use a rule from arithmetic that states a nonzero number divided by itself is equal to 1. Thus, we may divide three x's in the numerator by the three x's in the denominator obtaining

$$\frac{x^5}{x^3} = \frac{(\not{x})(\not{x})(\not{x})(x)(x)}{(\not{x})(\not{x})(\not{x})} = x^2.$$

Example 16 $\frac{x^{10}}{x^7} = x^3$

Example 17 $\frac{x^3}{x^8} = \frac{1}{x^5}$

Example 18 $\frac{2^4}{2^6} = \frac{1}{2^2}$ (Note that the base does not change.)

Example 19 $\frac{x^3}{y^2} = \frac{x^3}{y^2}$ (Rule does not apply since the bases are not the same.)

Example 20 $\frac{12x^5}{-2x^3} = -6x^2$ (Coefficients are divided, exponents are subtracted.)

EXERCISE 1–4–3

Find the quotients.

1. $\frac{x^4}{x^3}$ **2.** $\frac{x^8}{x^5}$ **3.** $\frac{x^2}{x^7}$

4. $\frac{3^6}{3^2}$ **5.** $\frac{5^4}{5^5}$ **6.** $\frac{x^3}{y^5}$

7. $\frac{-4x^5}{2x^3}$ **8.** $\frac{2^4}{2^4}$ **9.** $\frac{x^{10}}{4x^7}$

10. $\frac{-15x^4}{-3xy}$ **11.** $\frac{x^{12}y^5}{x^7y^2}$ **12.** $\frac{12x^5y^6}{-4x^2y}$

13. $\frac{24x^2y^7}{-6x^5y^3}$ **14.** $\frac{-3x^4y^3}{27x^7y^5}$ **15.** $\frac{-x^2y^2}{-x^2y^2}$

16. $\dfrac{12ab}{-24a^3b^2}$ **17.** $\dfrac{-15xy}{21yz}$ **18.** $\dfrac{7a^2bc}{84ab^2c}$

19. $\dfrac{27w^3x^2y^4}{-36w^2y^4}$ **20.** $\dfrac{-132a^3bc^4}{77a^5c}$

1–5 LITERAL EXPRESSIONS AND COMBINING SIMILAR TERMS

OBJECTIVES

Upon completion of this section you should be able to:

1. Distinguish between terms and factors.
2. Identify similar terms.
3. Combine similar terms.

Any expression in which letters are used to represent numbers is called a **literal expression.** For instance, if x and y represent numbers, then $2x + 3y^2$ is a literal expression. In an indicated product the numbers (or letters) being multiplied are called **factors.** In an indicated sum (or difference) the expressions being added (or subtracted) are called **terms.**

Example 1 $3xyz$ has four *factors:* 3, x, y, and z.

Example 2 $2x + 3y - 5z$ has three terms: $2x$, $3y$, and $-5z$. (Notice that each term has two factors.)

CAUTION It is important that we clearly distinguish between *terms* and *factors.* Take special note of the fact that only an indicated product has factors and *only* an indicated sum (or difference) has terms.

DEFINITION

Two terms are said to be **similar** (like) if the literal factors of one term are exactly the same as the literal factors of the other.

Example 3 $3a$ and $6a$ are similar.

Example 4 $3a$ and $3b$ are *not* similar.

Example 5 $2x^2$ and $3x^2$ are similar.

Example 6 $2x^2$ and $3x$ are *not* similar. (x^2 and x are not exactly the same.)

Example 7 $5abc^2$ and $-2abc^2$ are similar.

 **ULE**

Only similar terms can be added or subtracted. To add or subtract similar terms combine the coefficients and use this result as the coefficient of the common literal factors of the term. (This is an application of the distributive property.)

Example 8 $2xy + 3xy = (2 + 3)xy = 5xy$

Example 9 $2x^2 + 3x^3 = 2x^2 + 3x^3$ (cannot be combined because the literal factors x^2 and x^3 are not the same)

Example 10 $5x^2y - 3x^2y = (5 - 3)x^2y = 2x^2y$

Example 11 $-7xz - 2xz = -9xz$

Example 12 $-3xy + 2x = -3xy + 2x$ (cannot be combined)

Example 13 $2xy + 3xy - 4x = 5xy - 4x$

Note in this last example that the expression is simplified by combining the two terms that are similar. To simplify, all similar terms are combined. If more than one term occurs in the simplified form, the order is not important.

EXERCISE 1-5-1

Simplify by combining similar terms.

1. $5a + 7a$
2. $9x - 5x$
3. $3a - 10a$
4. $11x^2 + 7x^2$
5. $8xy - 12xy$
6. $15x - 5y$
7. $17x^3 + 3x^2$
8. $13xyz + 9xyz$
9. $-7ab^2c - 5abc$
10. $x^2y + 12x^2y$
11. $3x - 7x + 5x$
12. $20a + 13b - 11a$
13. $16a^2b - 31a^2b + 5ab^2$
14. $8x^2y - 3x^2y + 5x^2y$
15. $2ab + 5bc - 6ac$
16. $5xy^2 - xy^2 + 2x^2y^2$
17. $4xy + 13a - 7xy - 5a$
18. $16a^2b - 4a^2b + 5a^2b^2 - 7a^2b$

19. $21xyz + 15xy - 17xyz + 7xy^2$

20. $6xy - 5yz - 9xy + 7yz + 3xy - 2yz$

1-6 PRODUCT OF A MONOMIAL AND ANOTHER POLYNOMIAL

OBJECTIVES

Upon completion of this section you should be able to:

1. Identify monomials, binomials, and trinomials.
2. Multiply a polynomial by a monomial.

$a(b + c) = ab + ac$ is called the *distributive property* of multiplication over addition. It gives us a way of changing from a product of factors, (a) and $(b + c)$, to a sum of terms, ab and ac.

A numerical example will show that this property is valid.

$$5(2 + 7) = 5(9) = 45$$

also

$$(5)(2) + (5)(7) = 10 + 35 = 45$$

We obtain the same answer in each case and conclude that $5(2 + 7) = (5)(2) + (5)(7)$. This property is not limited to two terms. It is true for any number of terms.

$$a(b + c + d + \cdots) = ab + ac + ad + \cdots$$

This rule is called the *generalized distributive property.*

A literal expression having one or more terms that contain only nonnegative whole number exponents is called a **polynomial.** Special names are sometimes used for polynomials of one, two, or three terms. A polynomial having *one* term is called a **monomial.** A polynomial with *two* terms is called a **binomial.** A polynomial with *three* terms is called a **trinomial.**

Example 1 $5xy$ is a monomial.

Example 2 $5x + y$ is a binomial.

Example 3 $3x^2y^4z$ is a monomial.

Example 4 $3x + 2y - 4z$ is a trinomial.

Polynomials of more than three terms generally have no special names.

RULE

To multiply a polynomial by a monomial multiply *each term* of the polynomial by the monomial. (Note that this is an application of the distributive property.)

Example 5 $5x(2x - 3y) = (5x)(2x) - (5x)(3y) = 10x^2 - 15xy$

Example 6 $-3(2x^2 - 5x + 7) = -6x^2 + 15x - 21$

Note that the sign of every term changes when multiplying by a *negative* number.

Example 7 $2xyz(3x^2 + 3y^2 - 3z^2) = 6x^3yz + 6xy^3z - 6xyz^3$

A polynomial enclosed in parentheses and preceded by a positive sign is understood to be the product of $+1$ and the polynomial. For instance, $(3x - 2y)$ is the same as $(+1)(3x - 2y)$.

Example 8 $(2x^2 + 7x - 2) = 2x^2 + 7x - 2$

A polynomial enclosed in parentheses and preceded by a negative sign is understood to be the product of -1 and the polynomial. For instance, $-(3x - 2y)$ is the same as $(-1)(3x - 2y)$.

Example 9 $-(3x^2 - 2x + y) = -3x^2 + 2x - y$

Note that when removing parentheses preceded by a negative sign, all signs within the parentheses are changed.

EXERCISE 1-6-1

Multiply.

1. $3(x + 2y)$

2. $-2(2x + y)$

3. $x(x + y)$

4. $3x(2x + 5y)$

5. $-(2x - 7y)$

6. $2x(x - 5y)$

7. $5(2x + y - 3z)$

8. $-6x(x - 3y)$

9. $-x(2x - 4y)$

10. $-2x(3x - 2y - z)$

11. $-(3x + 5y - z)$

12. $3xy(2x - 3y - 5z)$

13. $-7y(x - 2y - 3z)$

14. $2x(-x^2 + 3x + 1)$

15. $3x(2 + 5x - 4x^2)$

16. $-5y(-4y^2 - 2y + 1)$

17. $2xy(3x^2y - 2xy^2 + xy)$

18. $-7xy(2x^2y^3 + 3x^2y - 5xy^2)$

19. $2xyz(x^2y - 5y^2z + 9xz)$ **20.** $-xyz(x^2y^2z - 10xz + 3y^2z^3)$

1-7 ORDER OF OPERATIONS AND GROUPING SYMBOLS

OBJECTIVES

Upon completion of this section you should be able to:

1. Evaluate expressions involving several operations such as $7 - 2^3 \times 5 + 3$.
2. Simplify expressions such as $2(x + y) - [3x - (x + y)]$.

As previously discussed, if a number expression contains only addition and subtraction, these operations must be performed, in order, from left to right. Now let us consider expressions that also contain other operations.

Should $2 \times 3 + 5$ be 2×8 or $6 + 5$?

Unless a rule is established, there could be no answer to this question. Therefore, mathematicians have agreed to the following rule.

RULE

If no parentheses occur in an expression to be evaluated, exponents (powers) must be worked first, then multiplication and division (left to right), then addition and subtraction (left to right).

A number expression enclosed in parentheses is treated as if it were a single number. Therefore when parentheses do occur in an expression, the operations within the parentheses are performed first. Listing them in order, we have

1. parentheses
2. exponents (powers)
3. multiplication and division (left to right)
4. addition and subtraction (left to right)

Example 1

$$\begin{aligned} 2^3 - 3 \times 4 + 6 &= 8 - 12 + 6 \\ &= 2 \end{aligned}$$

Example 2

$$\begin{aligned} 2^3 - 3 \times (4 + 6) &= 8 - 3(10) \\ &= 8 - 30 \\ &= -22 \end{aligned}$$

Note here that the parentheses change the normal order of operations so that addition is performed *before* multiplication.

Example 3

$$\begin{aligned} 5 - 4 \times 12 \div 3 \times 2 &= 5 - 48 \div 3 \times 2 \\ &= 5 - 16 \times 2 \\ &= 5 - 32 \\ &= -27 \end{aligned}$$

EXERCISE 1–7–1

Evaluate.

1. $4 \times 3 + 6$

2. $4 \times (3 + 6)$

3. $6 \div 2 + 1$

4. $6 \div (3 + 6)$

5. $5 \times 2 - 3 \times 2$

6. $5 \times (2 - 3) \times 2$

7. $12 \div 2 - 6 \div 3$

8. $12 \div (2 - 6) \div 3$

9. $5 - 2 \times 4$

10. $(5 - 2) \times 4$

11. $7 + 3 \times 5 - 2$

12. $(7 + 3) \times (5 - 2)$

13. $2^2 + 3 \times 2$

14. $3^2 + 2 \times 7 + 5$

15. $2 \times 5 + 7 \times 3 - 5 \times 2$

16. $2 \times 8 \div 4 \times 3 \div 2 - 6$

17. $3 - 8 \div 2 \times 5$

18. $4 \times (3 - 5)^3 \div 2$

19. $7 - 2^3 \times 5 + 3$

20. $4^2 \div 2 - 3 \times 2^3 + 6$

21. $16 + 4 \times 3 - 5^2$

22. $16 + 4 \times (3 - 5)^2$

23. $15 \times 6 \div 3 - 2^2$

24. $15 \times (6 \div 3 - 2)^2$

25. $7 - 3 + (5 - 1)^2 - 2$

26. $5 \times 3 - (6 + 2)^2 + 4$

27. $2 - (6 + 4) \times 8 \div 4$

28. $4 \times (3 - 9) \div (2 + 1) - 3$

29. $4 + 27 \div 3 \times 2 - 6$

30. $1 - 16 \times 2 \div 4 + 3$

Sometimes the meaning of an English sentence can be changed completely by the use of a comma. For example,

"Let's feed the lions, Jim."
"Let's feed the lions Jim."

Mathematical sentences may also need punctuation to clarify the meanings. The symbols for punctuation or grouping of numbers are parentheses ( ), brackets [ ], and braces { }. These three symbols are used in exactly the same way and all are used simply for clarification. For instance, $5 - [3 + (2 - 1) + 4]$ could be written using only parentheses, but $5 - (3 + (2 - 1) + 4)$ would not be as clear at first glance. To avoid any confusion we alternate the symbols.

RULE
When simplifying an expression having grouping symbols within grouping symbols, remove the *innermost* set of symbols first.

Example 4 $5x - 2[3x - (2x + 3)]$

We first remove the parentheses.

$$5x - 2[3x - 2x - 3]$$ (Notice the sign changes.)

We next remove the brackets by multiplying by -2.

$$5x - 6x + 4x + 6$$

Finally combining like terms, we have

$$3x + 6.$$

Example 5 $2x + [-4x - (3x - 6)] = 2x + [-4x - 3x + 6]$
$= 2x - 4x - 3x + 6$
$= -5x + 6$
$= 6 - 5x$

EXERCISE 1-7-2

Remove all grouping symbols and combine like terms.

1. $2x + [3x - (2x + 1)]$ **2.** $5a + [4a - (3a - 2)]$ **3.** $4x - [3x - (x + 1)]$

4. $10x + [x - (3x - 5)]$ **5.** $2x - [5x - (3x + 1)]$ **6.** $a - [4a - (4 - 2a)]$

7. $2x - 3[1 - (3x + 4)]$ **8.** $5x - 2[6 - (4x - 3)]$ **9.** $5 - [x + (8 - 3x)]$

10. $11 + [2a - (3a - 9)]$ **11.** $[3a - (4 - 2a)] - 3a$ **12.** $[4x - (3 - 2x)] - x$

13. $8x + 2[3x - 4 - (x + 5)]$ **14.** $4a + 3[3a - 5 - (2a + 1)]$ **15.** $5a - [-4 - 3(2a - 3)]$

16. $3x - [7 - 2(4x - 3)]$

17. $2x - 3 + [4x - (x - 7)]$

18. $5a - 2 + [3a - (a + 1)]$

19. $7a + 4 - 4[3a - (5 - 2a)]$

20. $x - 5 - 2[9x - (7 - x)]$

21. $6a + 3b + [a + 3(3a - 4b)]$

22. $3x - 2y + [2x - 5(x + y)]$

23. $10x - y - 2[3x - 6(2x - 5y)]$

24. $9a - 5b - 3[6b - 4(2a - b)]$

25. $a + b - [a - b - (a + b)]$

26. $2x + y - [y - x - (3x + 2y)]$

27. $6x + \{3x - [4x - 3 - (x - 1)]\}$

28. $a + b - \{a - b - [2a + (b - a)]\}$

29. $2x^2 - 3x + 3\{x - 3x^2 - 2[4x - (2x^2 - x)]\}$

30. $3x^2 - 1 - 2\{-5x^2 + 3 - 3[4x - (x^2 - 2x)]\}$

1–8 EVALUATING LITERAL EXPRESSIONS AND USING FORMULAS

OBJECTIVES

Upon completion of this section you should be able to:

1. Evaluate literal expressions by substituting numbers for letters.
2. Apply this technique in working with formulas.

In a literal expression such as $3x$ the letter x is holding the place for some number. When letters are used to represent various numbers they are often called **variables.** As we replace x with various numbers, we will obtain specific values for the expression. For instance, if we replace x with the number 5, we will have

$$3x = 3(5) = 15.$$

(5 ← value of x)

If we let $x = 8$, then

$$3x = 3(8) = 24.$$

If $x = -2$, then

$$3x = 3(-2) = -6.$$

The numerical value of a literal expression such as $2ab + c$ can be found if we know the values of a, b, and c. When given a specific value for a letter, we may substitute that value in the expression every time the letter occurs.

Example 1 Evaluate $2ab + c$, when $a = 3$, $b = -2$, $c = 15$.

Solution

$$2ab + c = 2(3)(-2) + (15)$$

(3 ← value of a; −2 ← value of b; 15 ← value of c)

$$= -12 + 15$$
$$= 3$$

In an expression such as $x^2 - 2xy + 3x$ the letter x must represent the same number every time it occurs.

Example 2 Evaluate $x^2 - 2xy + 3x$, when $x = 2$, $y = 3$.

Solution

$$x^2 - 2xy + 3x = (2)^2 - 2(2)(3) + 3(2)$$
$$= 4 - 12 + 6$$
$$= -2$$

Example 3 Evaluate $x^2 - 2xy + 3x$, when $x = 4$, $y = -1$.

Solution

$$\begin{aligned} x^2 - 2xy + 3x &= (4)^2 - 2(4)(-1) + 3(4) \\ &= 16 + 8 + 12 \\ &= 36 \end{aligned}$$

When evaluating a literal expression, we must be careful to remember the order of operations that we discussed in the previous section. For instance, the expressions $2x^2$ and $(2x)^2$ would have different values.

A good practice that may help avoid errors is to replace the variables with parentheses and then enter the proper number value within the parentheses.

Example 4 Evaluate $2x^2$, when $x = 4$.

Solution

$$\begin{aligned} 2x^2 &= 2(4)^2 \\ &= 2(16) \\ &= 32 \end{aligned}$$

Example 5 Evaluate $(2x)^2$, when $x = 4$.

Solution

$$\begin{aligned} (2x)^2 &= (2 \cdot 4)^2 \\ &= (8)^2 \\ &= 64 \end{aligned}$$

CAUTION Be careful with an expression such as $-x^2$. Realize that only the value of x is being squared and that the negative sign will precede the result. Recognize that $-x^2$ means $-1x^2$.

For example, if $x = 4$ then

$$-x^2 = -(4)^2 = -16.$$

If we want to square $-x$, we will write $(-x)^2$. For example, if $x = 4$, then

$$(-x)^2 = (-4)^2 = 16.$$

One of the most common uses of evaluating expressions is in working with **formulas.**

Example 6 The perimeter (distance around) a rectangle can be found by using the formula $P = 2\ell + 2w$, where P represents the perimeter, ℓ represents the length of the rectangle, and w represents its width. Find the perimeter of a rectangle if the length is 5.4 inches and the width is 3.6 inches.

Solution We first write the formula for the perimeter.

$$P = 2\ell + 2w$$

We next make the substitutions $\ell = 5.4$ in. and $w = 3.6$ in.

$$\begin{aligned} P &= 2(5.4 \text{ in.}) + 2(3.6 \text{ in.}) \\ &= 10.8 \text{ in.} + 7.2 \text{ in.} \\ &= 18.0 \text{ in.} \end{aligned}$$

Example 7 The formula for finding the area A of a trapezoid is

$$A = \frac{1}{2}h(b + c),$$

where h represents the height, and b and c represent the two bases.

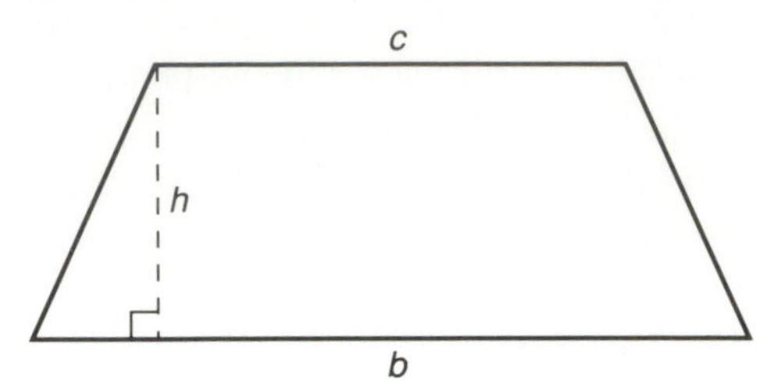

Evaluate $A = \frac{1}{2}h(b + c)$, when $h = 6$ meters, $b = 4$ meters, and $c = 1$ meter.

Solution

$$\begin{aligned} A &= \frac{1}{2}(6 \text{ m})(4 \text{ m} + 1 \text{ m}) \\ &= \frac{1}{2}(6 \text{ m})(5 \text{ m}) \\ &= 15 \text{ m} \end{aligned}$$

Example 8 The volume V of a right circular cone where the height is h and the radius of the base is r is found by the formula

$$V = \frac{1}{3}\pi r^2 h.$$

Find the volume, if $r = 3$ cm, $h = 14$ cm and $\pi = \frac{22}{7}$.

Solution

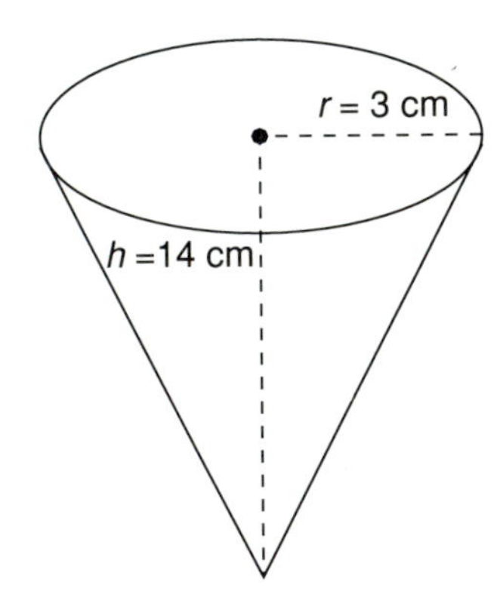

$$\begin{aligned} V &= \frac{1}{3}\left(\frac{22}{7}\right)(3 \text{ cm})^2(14 \text{ cm}) \\ &= \frac{1}{3}\left(\frac{22}{7}\right)(9 \text{ cm}^2)(14 \text{ cm}) \\ &= 132 \text{ cm}^3 \end{aligned}$$

Example 9 A baseball is thrown upward with an initial velocity v of 40 ft/sec. The distance s of the ball above the ground at t seconds is given by $s = vt + \frac{1}{2}gt^2$, where g is the acceleration due to gravity. Find the distance of the ball above the ground at the end of 2 seconds. Use $g = -32 \text{ ft/sec}^2$.

Solution Using the formula $s = vt + \frac{1}{2}gt^2$ and making the substitutions $v = 40$ ft/sec, $t = 2$ sec, and $g = -32$ ft/sec², we obtain

$$\begin{aligned} s &= vt + \frac{1}{2}gt^2 \\ &= \left(40\,\frac{\text{ft}}{\text{sec}}\right)(2\text{ sec}) + \frac{1}{2}\left(-32\,\frac{\text{ft}}{\text{sec}^2}\right)(4\text{ sec}^2) \\ &= 80\text{ ft} - 64\text{ ft} \\ &= 16\text{ ft.} \end{aligned}$$

EXERCISE 1-8-1

Evaluate.

1. $3x + 5$, when $x = -7$

2. $16 - 2x$, when $x = -3$

3. x^2, when $x = -5$

4. $-x^2$, when $x = -5$

5. $(-x)^2$, when $x = -5$

6. $5x^2$, when $x = 4$

7. $(5x)^2$, when $x = 4$

8. $x^2 + 3x - 1$, when $x = -2$

9. $3x^2 - 5x + 2$, when $x = 3$

10. $2x^2 - 3x + 1$, when $x = -3$

11. $3x^2 + xy - 6$, when $x = 5$, $y = -1$

12. $2x^3 + 3xyz - 2z^2$, when $x = 2$, $y = 5$, $z = -1$

13. The perimeter P of a rectangle is given by $P = 2\ell + 2w$, where ℓ and w represent the length and width. Find P when $\ell = 12$ cm and $w = 7$ cm.

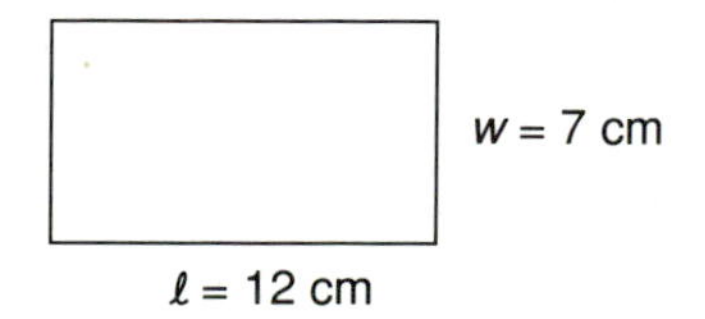

14. The area A of a rectangle is given by $A = bh$, where b represents the base and h represents the height. Find A when $b = 9$ ft and $h = 3$ ft.

$h = 3$ ft

$b = 9$ ft

15. The perimeter P of a square is given by $P = 4s$, where s represents the length of one side. Find P when $s = 8$ meters.

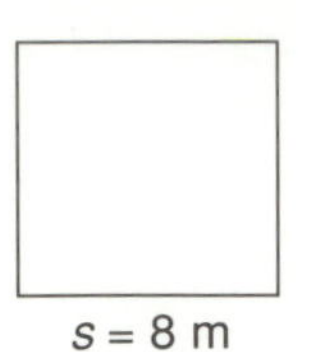

16. The area A of a square is given by $A = s^2$. Find A when $s = 3.5$ inches.

17. The perimeter P of a triangle is given by the formula $P = a + b + c$, where a, b, and c represent the three sides of the triangle. Find P when $a = 14.5$ ft, $b = 10.6$ ft, and $c = 17.8$ ft.

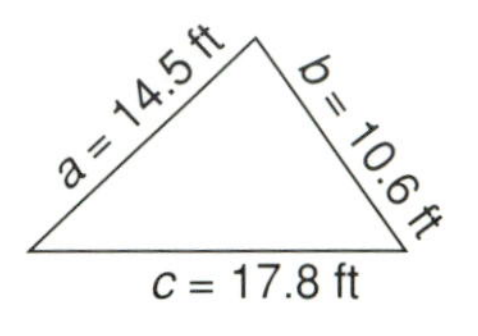

18. The area A of a triangle is given by $A = \frac{1}{2}bh$, where b represents the base and h represents the height. Find A if $b = 7$ in. and $h = 6$ in.

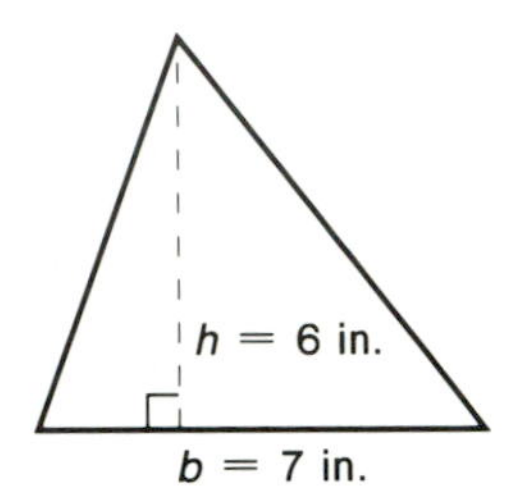

19. A distance formula from physics is $d = rt$, where r represents the rate and t represents the time. Find d when $r = 55$ mi/hr and $t = 4$ hours.

20. A force formula from physics is $F = ma$, where m represents the mass of an object and a represents its acceleration. Find F (the correct unit will be pounds) when $m = 120$ slugs and $a = 32$ ft/sec^2.

21. A formula from business for finding interest is $I = Prt$, where P represents the principal amount invested, r represents the rate of interest, and t the time invested. Find I when $P =$ \$8,000, $r = 8\%$ per year, and $t = 3$ years.

22. The circumference C of a circle is given by $C = \pi d$, where d represents the diameter of the circle and π is a constant number that is approximately equal to $\frac{22}{7}$. Find C when $d = 14$ meters.

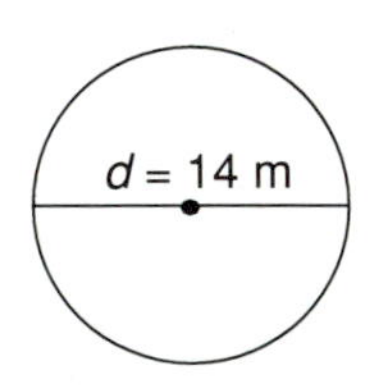

23. The area A of a circle is given by $A = \pi r^2$, where r represents the radius of the circle. Find A when $\pi = \frac{22}{7}$ and $r = 7$ inches.

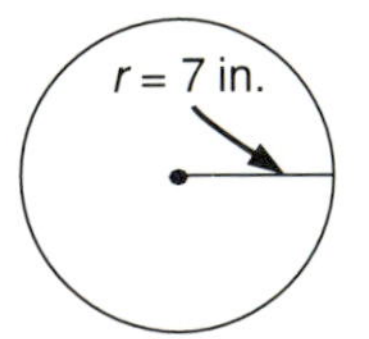

24. A formula for changing Fahrenheit temperature to Celsius is given by $C = \frac{5}{9}(F - 32)$. Find C when $F = 68$.

25. A formula for changing Celsius temperature to Fahrenheit is $F = \frac{9}{5}C + 32$. Find F when $C = 30$.

26. The volume of a rectangular solid is given by $V = xyz$. Find V when $x = 3$ cm, $y = 1.5$ cm, $z = 4$ cm.

$z = 4$ cm
$y = 1.5$ cm
$x = 3$ cm

27. The volume of a cube is given by $V = s^3$. Find V when $s = 3.5$ meters.

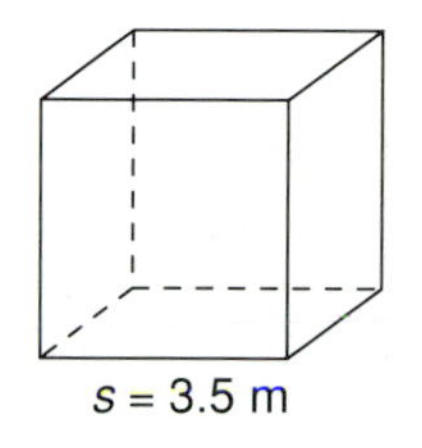

28. The volume of a cylinder of height h having a circular base with radius r is given by $V = \pi r^2 h$. Find V when $\pi = \frac{22}{7}$, $r = 1.4$ ft, and $h = 3$ ft.

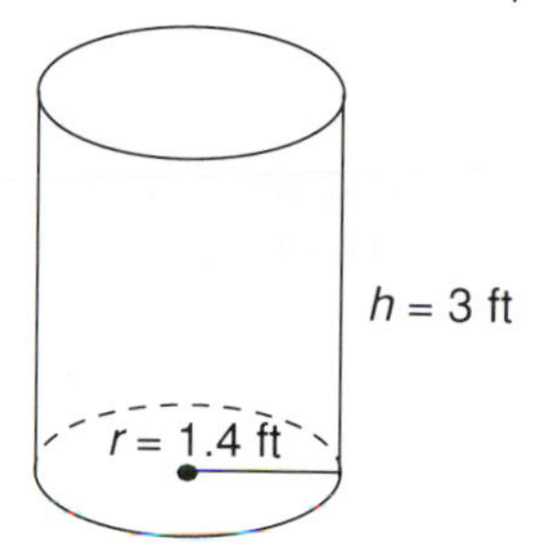

29. The area of a trapezoid is given by $A = \frac{1}{2}h(b + c)$. Find A when $h = 9$ in., $b = 13$ in., and $c = 7$ in.

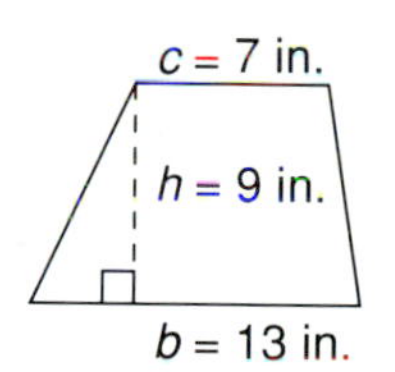

30. The volume of a right circular cone is given by $V = \frac{1}{3}\pi r^2 h$. Find V when $\pi = \frac{22}{7}$, $r = 7$ meters, and $h = 9$ meters.

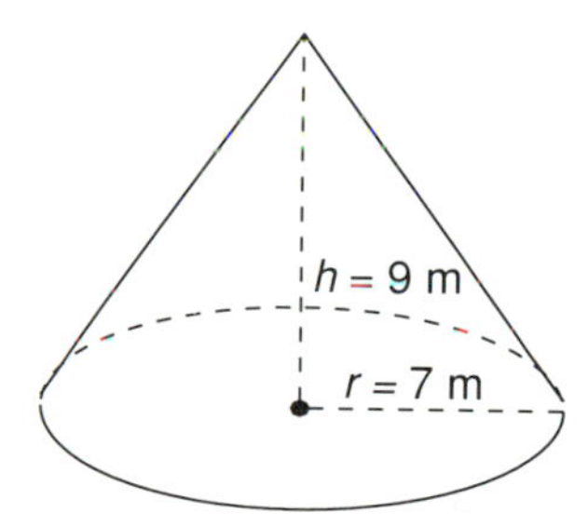

31. A distance formula for a free-falling object is given by $s = \frac{1}{2}gt^2$, where g represents acceleration due to gravity and t represents time. Find s when $g = 32$ ft/sec^2 and $t = 5$ seconds.

32. The volume of a sphere is given by $V = \frac{4}{3}\pi r^3$. Find V when $\pi = \frac{22}{7}$, $r = 2.1$ meters.

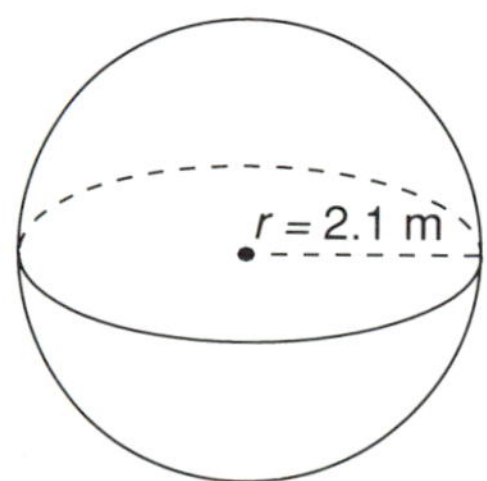

33. If an object is thrown downward from a height h with an initial velocity v, then the distance s of the object above the ground at any time t can be found using the formula $s = h - vt - \frac{1}{2}gt^2$, where g is the acceleration due to gravity. Find s when $h = 5{,}500$ cm, $v = 100$ cm/sec, $t = 3$ seconds, and $g = 980$ cm/sec^2.

34. A formula for determining the total amount T in an account when a principal P is invested at a rate of interest r for a time t is $T = P(1 + rt)$. Find the total amount in the account at the end of 5 years if $12,000 was invested at 8.3% annual interest.

35. In mathematics a formula for determining the last term L in an arithmetic progression is $L = a + (n - 1)d$, where a is the first term, n represents the number of terms in the progression, and d is the common difference between the terms. Find the last term in a progression having 9 terms if the first term is 3 and the common difference is $\frac{1}{2}$.

36. A formula from physics for finding the average velocity v of an object that travels a distance d in a given time t is $v = \frac{d}{t}$. Find the average velocity of an object that travels 1,200 feet in 2.5 seconds.

37. The sum S of the terms of a geometric progression can be found by using the formula $S = \frac{a}{1 - r}$, where a is the first term of the progression and r is the common ratio with a value between -1 and 1. Find S when $a = 12$ and $r = \frac{1}{4}$.

38. The tax-free yield formula $F = T(1 - B)$ is used to find the equivalent tax-free yield F of a taxable rate T on an investment by an individual whose tax bracket is B. If a person is in the 28% tax bracket, find the equivalent tax-free yield of an 8% taxable investment.

39. A formula from physics for the kinetic energy K of an object having a mass m and velocity v is $K = \frac{1}{2}mv^2$. Find the kinetic energy of a mass of 90 slugs moving at 40 ft/sec. (The correct unit for kinetic energy will be foot-pounds.)

40. The volume V of a prismatoid is found by using the formula $V = \frac{h}{6}(B + 4M + b)$, where h is the altitude, B is the area of the lower base, M is the area of the midsection, and b is the area of the upper base. Find V when $h = 15$ cm, $B = 12$ cm^2, $M = 8$ cm^2, and $b = 4$ cm^2.

41. The total current I in a circuit containing two resistors R and r in parallel with a voltage V across them is given by the formula $I = V\left(\frac{1}{R} + \frac{1}{r}\right)$. Find I (in amps) when $V = 30$ volts, $R = 2$ ohms, and $r = 5$ ohms.

42. A formula from mathematics for the sum S of the terms of a geometric progression is $S = \frac{a - r\ell}{1 - r}$, where a is the first term, ℓ is the last term, and r is the common ratio. Find the sum of a geometric progression having a first term of 4, a last term of 6, and a common ratio of $\frac{1}{3}$.

CHAPTER 1 SUMMARY

The number in brackets refers to the section of the chapter that discusses the concept.

Terminology

- Elementary algebra is a study of the **real numbers** and their properties. [1–1]
- Positive and negative real numbers are referred to as **signed numbers.** [1–1]
- The **negative** or **opposite** of a given number is a number that when added to the given number yields a sum of zero. [1–1]
- The **real number line** is used to show a correspondence between the real numbers and the points of the line. [1–1]
- **Factors** are numbers that are multiplied. [1–4]
- An **exponent** indicates the number of times a base is used as a factor. [1–4]
- A numerical factor is sometimes referred to as the **coefficient.** [1–4]
- A **literal expression** is an expression in which letters are used to represent numbers. [1–5]
- In an indicated sum or difference, the expressions being added or subtracted are called **terms.** [1–5]
- **Similar terms** have exactly the same literal factors. [1–5]
- A **polynomial** is a literal expression having one or more terms that contain only nonnegative whole number exponents. [1–6]
- A **monomial** is a polynomial having one term. [1–6]
- A **binomial** is a polynomial having two terms. [1–6]
- A **trinomial** is a polynomial having three terms. [1–6]
- Parentheses, brackets, and braces are used as **grouping symbols.** [1–7]
- **Variables** are letters that are used to represent numbers. [1–8]

Rules and Procedures

Signed Numbers

- To add signed numbers having like signs add the numbers (without regard to sign) and use the common sign. [1–1]
- To add signed numbers with unlike signs subtract the smaller number from the larger number (without regard to sign) and use the sign of the larger number (without regard to sign). [1–1]
- To subtract one signed number from another change the sign of the number being subtracted and use the appropriate rule for addition. [1–2]
- The product or quotient of two signed numbers having like signs will be positive. [1–3]
- The product or quotient of two numbers having unlike signs will be negative. [1–3]

Laws of Exponents

- $x^a \cdot x^b = x^{a+b}$ [1–4]
- $\frac{x^a}{x^b} = x^{a-b}$ if a is *greater* than b. $\frac{x^a}{x^b} = \frac{1}{x^{b-a}}$ if a is *less* than b. [1–4]

Similar Terms

- Only similar terms can be added or subtracted. [1–5]
- To add or subtract similar terms combine the coefficients and use this result as the coefficient of the common literal factors of the term. [1–5]

Multiplying Polynomials by Monomials

- To multiply a polynomial by a monomial multiply each term of the polynomial by the monomial. [1–6]

Order of Operations

- If an expression contains only additions and subtractions, these operations are performed in order from left to right. [1–2]
- If no parentheses occur in an expression to be evaluated, exponents must be worked first, then multiplication and division (left to right), then addition and subtraction (left to right). [1–7]
- When simplifying an expression having grouping symbols within grouping symbols, remove the innermost set of symbols first. [1–7]

Evaluating Literal Expressions

- To evaluate a literal expression substitute the correct number for each letter and then evaluate the resulting numerical expression. [1–8]

CHAPTER 1 REVIEW

Add.

1. $(+7) + (-9)$

2. $(-17) + (-5)$

3. $\left(+\frac{5}{9}\right) + \left(-\frac{3}{9}\right)$

4. $(-5) + (+8) + (-10)$

5. $(+6) + (-15) + (+4)$

Combine.

6. $5 - 8 + 2$

7. $4 - (-6) - 1$

8. $-3 + (-4) + 9$

9. $-13 + 5 - (-16) - 2$

10. $-4 + 9 + 2 - 8 - 7 + 6$

Evaluate.

11. $(8)(-5)(-7)$

12. $(-4)(3)(10)$

13. $\left(\frac{4}{9}\right) \div \left(-\frac{2}{3}\right)$

14. $\frac{-32}{-8}$

15. $(-3)(-1)(5)(-10)\left(-\frac{2}{5}\right)(3)$

Simplify.

16. $(x^3)(x^7)$

17. $(-2x)(4x^3)(x)(-3x^2)$

18. $(3x^2y)(5xy^3z)(-2y^2z)$

19. $\frac{9x^7y^3}{-3x^2y}$

20. $\frac{8x^4y^4}{2xy^6}$

Combine.

21. $9a - 3a$

22. $5x - 8x - x$

23. $3x^2 + 2x - 2x^2 + x^2$

24. $6xy^2 + 7x^2y - 3xy^2$

25. $-13ab + 5a - 7ab + 2$

Multiply.

26. $5(2x - 4y)$

27. $-(3x^2 + 2x - 5)$

28. $6x(2x^2 - 3x - 4)$

29. $3xy(2x^2y - 5xy^3 + 2)$

30. $-2abc(3a^2b - 5abc^2 - bc)$

Simplify.

31. $2 \times 3 + 6 \times 4$

32. $3 \times 9 \div 3 \times 2$

33. $5x^2 + 3(2x - 1) - 4x(2x^2 - 3x - 1)$

34. $2x[3x - 2(x + 3) + x]$

35. $3x^2 - 2x\{2x - 4[3 - 4(x - 5)]\}$

Evaluate.

36. $3x^2$ when $x = -5$

37. $-x^3$ when $x = -2$

38. $2x^2 - 3x + 1$ when $x = -4$

39. $5x^3 - 2xy^2 - 3y$ when $x = -3, y = 4$

40. $V = \frac{1}{3}\pi r^2 h$. Find V when $\pi = \frac{22}{7}$, $r = 4$, $h = 21$

CHAPTER

PRACTICE TEST

1. Give the negative or opposite of $-\frac{3}{5}$.

1. ______________________

2. Add: $(-24) + (-16)$

2. ______________________

3. Add: $\left(-\frac{7}{11}\right) + \left(+\frac{5}{11}\right)$

3. ______________________

4. Combine: $(+18) - (-23)$

4. ______________________

5. Combine: $15 + (-5) - (-31)$

5. ______________________

6. Combine: $3 - 11 + 21 + 9 - 25 - 7$

6. ______________________

7. Multiply: $\left(\frac{2}{3}\right)(-42)$

7. ______________________

8. Divide: $\left(-\frac{1}{3}\right) \div \left(-\frac{5}{6}\right)$

8. ______________________

9. Multiply: $-11\left(\frac{1}{3}\right)(-42)\left(-\frac{2}{7}\right)$

9. ______________________

10. Evaluate: $(-3)^5$

10. ______________________

11. Multiply: $(3x^2y^3)(-2xy^2)(4x^3y^2)$

11. ______________________

12. Divide: $\dfrac{99a^6}{33a^4}$

12. ______________________

13. Divide: $\dfrac{12x^5y^3}{-3xy^4}$

13. ______________________

14. Simplify: $13x - 5y - 8y$

14. ______________________

15. Simplify:
$3a^2b - 4ab^2 + 6a^2b + 10ab^2 - 8a^2b + 7ab$

15. ______________________

16. Multiply: $4a^2b^2(5a - 2a^2b - 5b^2)$

16. ______________________

17. Simplify: $25x - (4x + 3y - 2)$

17. ______________________

18. Simplify: $3x^2 - [5x - 4x(x - 3)]$

18. ______________________

19. If $x = -2$, evaluate $3x^2 - 15x + 4$

19. ______________________

20. $s = h - \dfrac{1}{2}gt^2$. Find s when $h = 1{,}996$, $g = 980$, and $t = 2$

20. ______________________

CHAPTER

2 SURVEY

The following questions refer to material discussed in this chapter. Work as many problems as you can and check your answers with the answer section in the back of the book. The results will direct you to the sections of the chapter in which you need to work. If you answer all questions correctly, you have a good understanding of the material contained in this chapter.

1. Are $x + 4 = 2x + 1$, and $x = 3$ equivalent equations?

1. ______________________

Solve for x.

2. $x + 5 = 3$

2. ______________________

3. $-3x = 12$

3. ______________________

4. $\frac{1}{2}x = 6$

4. ______________________

5. $\frac{x}{3} = \frac{3}{7}$

5. ______________________

6. $x - 7 = 4x + 5$

6. ______________________

7. $2x - 3(x - 2) = \frac{1}{2}(x + 1)$

7. ______________________

8. $3(2a - 3x) = 4(a + 2x)$

8. ______________________

9. $\frac{2}{5}x + 2 \geq \frac{x}{2} + 1$

9. ______________________

10. $|x - 3| > 5$

10. ______________________

First-Degree Equations and Inequalities: One Unknown

The manipulative skills of algebra ultimately lead to the solution of equations. Equations give answers to problems, and the solving of problems is the central theme of algebra.

In this chapter we will study the techniques of solving first-degree equations in one unknown. The fundamental concepts that you learned in the previous chapter supply the necessary tools for this chapter. Other types of equations will be studied once we have mastered the further techniques necessary for solving them.

2–1 CONDITIONAL EQUATIONS

OBJECTIVES

Upon completion of this section you should be able to:

1. Identify equations as true, false, or conditional.
2. Identify equivalent equations.

An **equation** in mathematics states that two number expressions are equal. For example,

1. $5 + 2 = 8$
2. $5 + 3 = 8$
3. $5 + x = 8$

These three statements are all equations. The first is false, the second is true, and the third is neither true nor false. Equations such as number 3 are called **conditional** equations. In this chapter we will work with conditional equations.

In an equation such as $5 + x = 8$ the letter x holds the place for a numerical value. Since x can take on various values, we often refer to this letter as a **variable.** The variable is often simply referred to as the **unknown.**

Depending on the numerical value we substitute for x, we will obtain either a true statement or a false statement. To **solve** a conditional equation we must find a replacement for x that will make the equation a true statement. Such a replacement is called a **solution** to the equation.

EXERCISE 2-1-1

Identify the following equations as true, false, or conditional:

1. $5 + 6 = 12 - 1$

2. $2 + 7 = 8 - 1$

3. $x - 3 = 5$

4. $27 - 21 = 17 - 23$

5. $x = 8$

Replace the variable x with the number 3 and determine if it is a solution.

6. $x + 17 = 20$

7. $x - 8 = -11 + 6$

8. $x + 4 = 2x - 1$

9. $2x - 9 = -6 + x$

10. $x - 6 = 2x - 3$

Two or more equations are **equivalent** if their solutions are identical. For instance, $2x + 1 = 5$ and $3x + 1 = 7$ both have a solution of 2 and are therefore equivalent equations. $5x - 1 = 14$ and $x = 3$ both have a solution of 3 and are therefore equivalent.

Example 1 Show that $3x - 3 = x + 7$ and $x = 5$ are equivalent equations.

Solution The second equation states $x = 5$. Thus if we substitute 5 for x in the first equation, we obtain

$$3(5) - 3 = 5 + 7$$
$$\text{or} \quad 12 = 12,$$

which is an identity. Therefore $x = 5$ is a solution to both equations, and they are thus equivalent.

EXERCISE 2-1-2

Determine which of the following pairs of equations are equivalent.

1. $x + 7 = 2x + 6$ and $x = 1$

2. $3x - 5 = 2x + 1$ and $x = 5$

3. $2x + 1 = x$ and $x = 1$

4. $4x - 3 = x + 6$ and $x = 3$

5. $5x - 4 = 3x + 2$ and $x = -3$

6. $3 - 2x = 5 - x$ and $x = -2$

7. $2x - 2 + x = 6 + x$ and $x = -8$

8. $3x + 1 = x - 5 + 4x$ and $x = 6$

9. $3x + 1 = x + 5$ and $x = 2$

10. $2x + 17 = 10 - 5x$ and $x = -1$

2–2 RULES FOR SOLVING FIRST-DEGREE EQUATIONS

OBJECTIVES

Upon completion of this section you should be able to:

1. Solve simple equations such as

   $x + 7 = 2$

   $3x = 12$

   $\frac{1}{5}x = 2.$

2. Solve proportions.

The **degree** of a polynomial equation in an unknown is the highest power of the unknown in any term. For instance $x^2 = 9$ is a *second-degree equation* in x since 2 is the highest power of x. $x^3 + 2x^2 - x = 14$ is a *third-degree equation* in x since 3 is the highest power of x. Equations in which the highest power of the unknown is 1 are **first degree-equations,** which we will study in this chapter.

$x = 5$ is an example of the simplest type of first-degree equation. Clearly, this equation is conditional since it is true only if x is replaced by 5.

The methods for solving equations involve the changing of a more complicated equation, such as $2x - 4 = x + 1$, to an equivalent equation, such as $x = 5$. The question now becomes, "How may an equation be changed without changing the value of its solution?"

To answer this question we must refer to the basic properties of equals. If two numbers are equal, such as $4 = 4$, we realize that we could add the same number, such as 3, to each side and the resulting numbers would still be equal. That is, $4 + 3 = 4 + 3$ or $7 = 7$. We could also subtract the same number from each side. For example, $4 - 3 = 4 - 3$ or $1 = 1$. We could multiply each side by the same number, such as 5, getting $4(5) = 4(5)$ or $20 = 20$. Finally we could divide each side by the same number, such as 2, getting $4 \div 2 = 4 \div 2$ or $2 = 2$. In each case the resulting quantities are equal.

Since an equation is a statement that two quantities are equal, we will apply these properties to them.

RULE

If any number is added to or subtracted from both sides of an equation, the resulting equation will be equivalent to the original equation.

Example 1 $3x + 7 = x - 9$ is equivalent to $3x + 7 + \boxed{10} = x - 9 + \boxed{10}$. (10 has been added to both sides of the equation.)

Of course, our object is to reduce an equation to the simplest form of $x =$ some number. We will therefore carefully choose the number we wish to add to or subtract from both sides.

Example 2 Solve $x - 7 = 3$ for x.

Solution Since we wish to arrive at an equation having only the unknown x on one side, we will add 7 to both sides.

$$x - 7 + \boxed{7} = 3 + \boxed{7}$$

$$\text{or} \quad x = 10$$

Example 3 Solve for x: $3x + 5 = 2x - 3$

Solution Subtract 5 from both sides.

$$3x + 5 - \boxed{5} = 2x - 3 - \boxed{5}$$
$$\text{or} \quad 3x = 2x - 8$$

Subtract $2x$ from both sides.

$$3x - \boxed{2x} = 2x - 8 - \boxed{2x}$$
$$\text{or} \quad x = -8$$

EXERCISE 2–2–1

Solve for x.

1. $x + 3 = 5$
2. $x + 7 = 2$
3. $x - 4 = 6$
4. $x - 8 = -11$
5. $x - 1 = 1$
6. $x + 5 = 5$
7. $x + 7 = -7$
8. $3x + 2 = 2x + 7$
9. $2x - 9 = 4 + x$
10. $5x - 2 = 4x - 5$

We will now give two more rules for changing an equation to an equivalent equation. These rules will involve the coefficients of the variable in the equation.

ULE

If both sides of an equation are divided by the same nonzero number, the resulting equation is equivalent to the original equation.

Note that the rule states we must divide by a nonzero number. This condition exists since division by zero is meaningless.

Example 4 Solve for x: $2x = 10$

Solution Divide each side by 2.

$$\frac{2x}{\boxed{2}} = \frac{10}{\boxed{2}}$$

or $$x = 5$$

Example 5 Solve for x: $-3x = 15$

Solution Divide each side by -3.

$$\frac{-3x}{-3} = \frac{15}{-3}$$

or $$x = -5$$

EXERCISE 2-2-2

Solve for x.

1. $2x = 24$

2. $3x = 12$

3. $5x = 15$

4. $3x = 1$

5. $4x = 2$

6. $7x = -14$

7. $9x = 3$ **8.** $-2x = 6$ **9.** $-x = -5$

10. $-12x = 4$

RULE
If both sides of an equation are multiplied by the same nonzero number, the resulting equation is equivalent to the original equation.

Note that the rule states we must multiply by a nonzero number. Multiplying both sides of an equation by zero yields $0 = 0$, which is true but useless.

Example 6 Solve for x: $\frac{1}{3}x = 5$

Solution Multiply each side by 3 $\left(\text{reciprocal of } \frac{1}{3}\right)$.

$$\boxed{3}\left(\frac{1}{3}x\right) = \boxed{3}(5)$$

or $\quad x = 15$

EXERCISE 2-2-3

Solve for x.

1. $\frac{1}{4}x = 3$ **2.** $\frac{1}{5}x = 2$ **3.** $\frac{1}{6}x = -1$

4. $\frac{x}{3} = 9$

5. $\frac{x}{5} = 25$

6. $\frac{1}{9}x = -3$

7. $\frac{x}{4} = 16$

8. $-\frac{1}{2}x = 7$

9. $-x = 5$

10. $-\frac{1}{3}x = 27$

If an equation contains more than one fraction, we multiply both sides of the equation by the least common denominator of all fractions involved. This process will eliminate the fractions and give an equivalent equation.

Example 7 Solve for x: $\frac{x}{3} = \frac{2}{5}$

Solution Multiply both sides by 15.

$$\boxed{15}\left(\frac{x}{3}\right) = \boxed{15}\left(\frac{2}{5}\right)$$

$$\text{or} \qquad 5x = 6$$

Divide each side by 5.

$$\frac{5x}{\boxed{5}} = \frac{6}{\boxed{5}}$$

$$\text{or} \qquad x = \frac{6}{5}$$

Solving simple equations by multiplying both sides by the same number occurs frequently in the study of **ratio** and **proportion.**

The **ratio** of a number x to a number y can be written as $x:y$, or $\frac{x}{y}$. In general, the fractional form is more meaningful and useful. Thus we will write the ratio of 3 to 4 as $\frac{3}{4}$. A **proportion** is a statement that two ratios are equal.

Example 8 What number x has the same ratio to 3 as 6 has to 9?

Solution To solve for x we first write the proportion.

$$\frac{x}{3} = \frac{6}{9}$$

Next we multiply each side of the equation by 9.

$$\boxed{9}\left(\frac{x}{3}\right) = \boxed{9}\left(\frac{6}{9}\right)$$
$$3x = 6$$
$$x = 2$$

Example 9 $\frac{2}{5}$ of a meter is how many tenths of a meter?

Solution We set up the equation.

$$\frac{x}{10} = \frac{2}{5}$$

Multiply each side by 10.

$$\boxed{10}\left(\frac{x}{10}\right) = \boxed{10}\left(\frac{2}{5}\right)$$
$$\text{or} \quad x = 4$$

EXERCISE 2-2-4

Solve for x.

1. $\frac{x}{7} = \frac{1}{2}$

2. $\frac{x}{6} = \frac{5}{3}$

3. $\frac{2x}{3} = \frac{7}{8}$

4. $\frac{2x}{5} = \frac{2}{3}$

5. $\frac{3x}{5} = \frac{1}{2}$

6. What number has the same ratio to 15 as 2 has to 3?

7. What number has the same ratio to 5 as 6 has to 30?

8. What number has the same ratio to 49 as 2 has to 7?

9. What number has the same ratio to 8 as 2 has to 3?

10. The ratio of miles to kilometers is 5 to 8. How many miles are there in 100 kilometers?

11. The ratio of teeth on a gear A to those on gear B is 5 to 7. If gear B has 35 teeth, how many teeth does gear A have?

12. The ratio of kilograms to pounds is 5 to 11. Find your weight in kilograms.

2-3 COMBINING RULES TO SOLVE FIRST-DEGREE EQUATIONS

OBJECTIVES

Upon completion of this section you should be able to:

1. Solve equations that require more than one operation.
2. Apply a step-by-step procedure to solve first-degree equations.

Most equations you will encounter will involve some combination of the operations of addition, subtraction, multiplication, and division. They may also involve parentheses, combining like terms, and most of the other skills developed in chapter 1.

Example 1 Solve for x: $12 + 3(x - 1) = x + 3$

Solution We observe that it will take more than one operation to find the solution. Where do we begin? The parentheses seem to make the equation more involved so we will begin by multiplying the numbers within the parentheses by the coefficient 3. Thus,

$$12 + \boxed{3(x - 1)} = x + 3$$

becomes $$12 + \boxed{3x - 3} = x + 3.$$

Next we subtract 3 from 12.

$$\boxed{12} + 3x - \boxed{3} = x + 3$$

obtaining $$9 + 3x = x + 3$$

We then subtract x from each side.

$$9 + 3x - \boxed{x} = x + 3 - \boxed{x}$$

or $$9 + 2x = 3$$

Now we subtract 9 from each side.

$$9 + 2x - \boxed{9} = 3 - \boxed{9}$$

or $$2x = -6$$

Finally we divide both sides by 2

$$\frac{2x}{\boxed{2}} = \frac{-6}{\boxed{2}}$$

obtaining the result

$$x = -3.$$

Example 2 Solve for x: $\frac{1}{3}x - 2 = \frac{1}{2}(1 - x)$

Solution Here we have the additional complication of fractions. The first step, however, is the same as in the solution of the previous equation—remove the parentheses. Thus we obtain

$$\frac{x}{3} - 2 = \frac{1}{2} - \frac{x}{2}.$$

Next we clear fractions by multiplying each side of the equation by the LCD (lowest common denominator), which in this case is 6.

$$\boxed{6}\left(\frac{x}{3} - 2\right) = \boxed{6}\left(\frac{1}{2} - \frac{x}{2}\right)$$

$$\boxed{6}\left(\frac{x}{3}\right) - \boxed{6}(2) = \boxed{6}\left(\frac{1}{2}\right) - \boxed{6}\left(\frac{x}{2}\right)$$

$$\text{or} \qquad 2x - 12 = 3 - 3x$$

Adding $3x$ to each side

$$2x - 12 + \boxed{3x} = 3 - 3x + \boxed{3x}$$

we obtain

$$5x - 12 = 3.$$

Next we add 12 to both sides.

$$5x - 12 + \boxed{12} = 3 + \boxed{12}$$

$$\text{or} \qquad 5x = 15$$

Dividing each side by 5

$$\frac{5x}{\boxed{5}} = \frac{15}{\boxed{5}}$$

we obtain

$$x = 3$$

for our solution.

You will find that the following step-by-step procedure for solving equations will help you avoid mistakes as you find a solution.

A STEP-BY-STEP PROCEDURE FOR SOLVING EQUATIONS THAT REDUCE TO FIRST-DEGREE EQUATIONS

1. Remove any parentheses.
2. Multiply both sides by the least common denominator of all fractions appearing in the equation.
3. Combine similar terms on each side of the equation.
4. Add or subtract terms on both sides of the equation to get the unknown on one side and everything else on the other.
5. Divide both sides of the equation by the coefficient of the unknown.
6. Simplify the solution.

Example 3 Solve for x: $\frac{5}{7}x - x - \frac{5}{3} = \frac{1}{21}(3x - 38)$

Solution Remove parentheses.

$$\frac{5}{7}x - x - \frac{5}{3} = \frac{3x}{21} - \frac{38}{21}$$

Multiply both sides by 21.

$$\boxed{21}\left(\frac{5}{7}x - x - \frac{5}{3}\right) = \boxed{21}\left(\frac{3x}{21} - \frac{38}{21}\right)$$

$$\boxed{21}\left(\frac{5}{7}x\right) - \boxed{21}(x) - \boxed{21}\left(\frac{5}{3}\right) = \boxed{21}\left(\frac{3x}{21}\right) - \boxed{21}\left(\frac{38}{21}\right)$$

This becomes

$$15x - 21x - 35 = 3x - 38.$$

Combine similar terms.

$$-6x - 35 = 3x - 38$$

Subtract $3x$ from both sides.

$$-6x - 35 - \boxed{3x} = 3x - 38 - \boxed{3x}$$

$$\text{or} \quad -9x - 35 = -38$$

Add 35 to both sides.

$$-9x - 35 + \boxed{35} = -38 + \boxed{35}$$

$$\text{or} \quad -9x = -3$$

Divide both sides by -9.

$$\frac{-9x}{\boxed{-9}} = \frac{-3}{\boxed{-9}}$$

Simplify.

$$x = \frac{1}{3}$$

CAUTION In an equation involving fractions make sure you multiply *every* term by the common denominator and not just the terms containing fractions. Thus to solve the equation

$$x + \frac{1}{5} = 1 - \frac{x}{2}$$

we must multiply every term by 10. Therefore,

$$\boxed{10}(x) + \boxed{10}\frac{1}{5} = \boxed{10}(1) - \boxed{10}\left(\frac{x}{2}\right).$$

Example 4 Solve for x: $\frac{2}{3} - \frac{x-2}{2} = \frac{1}{6}$

Solution Multiply both sides by 6.

$$6\left(\frac{2}{3} - \frac{x-2}{2}\right) = 6\left(\frac{1}{6}\right)$$

$$6\left(\frac{2}{3}\right) - 6\left(\frac{x-2}{2}\right) = 6\left(\frac{1}{6}\right) \qquad \left(\text{Note that } \frac{x-2}{2} \text{ is } \textit{one} \text{ term.}\right)$$

$$4 - 3(x-2) = 1$$

$$4 - 3x + 6 = 1$$

$$-3x + 10 = 1$$

$$-3x = -9$$

$$x = 3$$

Example 5 Mrs. Jones collected \$960.00 principal and interest from an investment. If the interest was one-half the original investment (principal), what was the amount of the original investment?

Solution To form an equation for solving this problem we find from the first statement that the

(original investment) + (interest) = \$960.

From the second statement we see that

$$(\text{interest}) = \frac{1}{2}(\text{original investment}).$$

If we let x = the original investment, then $\frac{1}{2}x$ = the interest.

Thus we have

$$x + \frac{1}{2}x = 960$$

$$2x + x = 1{,}920 \qquad \text{(multiplication rule)}$$

$$3x = 1{,}920$$

$$x = \$640$$

The original investment was \$640.

Check: Does $640 + \frac{1}{2}(640) = 960$?

640 + 320	960
960	

EXERCISE 2–3–1

Solve.

1. $3x + 5 = x + 7$

2. $4x - 3 = x - 9$

3. $3x - 1 = 2(x - 5)$

4. $3x + 2 = 5x - 8$

5. $x - 7 = 4x + 5$

6. $2x + 3 = 3x + 5$

7. $x - 6 = 5x - 14$

8. $3(x - 4) = 5 + 2(x + 1)$

9. $5x - 4 = 2x + 6$

10. $3x + 7 = 5x - 4$

11. $3x + 2(x - 5) = 7 - (x + 3)$

12. $5x - 3(x + 1) = 5$

13. $7x + 5 - 2(x - 1) = 21$

14. $5 + 6x - 3 = 2 + 4x$

15. $3(y + 1) + 3 = 7(y - 2)$

16. $5x - (2x - 3) = 4(x + 9)$

17. $\frac{x}{2} = \frac{1}{5} - x$

18. $4 - \frac{y + 2}{3} = 1$

19. $x - \frac{1}{2} = \frac{x}{3} + 7$

20. $\frac{x - 3}{4} - 9 = 11$

21. $5(x + 4) = \frac{1}{3}(x - 10)$

22. $\frac{2x + 5}{3} = 0$

23. $\frac{2}{3} - 3(x - 1) = \frac{3}{5}$

24. $2x - 3(x - 2) = \frac{1}{2}(x + 1)$

25. $\frac{2}{3} + x = 1 - \frac{x - 1}{2}$

26. $4y - 2(y - 3) = \frac{1}{4}(y + 3)$

27. $x - \frac{1}{5}x + 1 = \frac{1}{3}(x - 5)$

28. $\frac{2}{3}y + \frac{y}{2} - \frac{1}{2} = \frac{1}{3}(y - 14)$

29. $\frac{2}{3}x - \frac{1}{2}x = x + \frac{1}{6}$

30. $\frac{x}{5} - \frac{2}{3}x + \frac{1}{2} = \frac{1}{3}(x - 4)$

31. The principal and interest received on an investment totaled $7,600. If the interest was one-third the principal, what was the amount of the principal?

32. A nurse administers a dosage of a certain medication to a patient in the morning and another dose of the same medication in the evening. The evening dosage is one-half the strength of the morning medication. If the total dosage for the day is 525 milligrams, how many milligrams are administered in the morning?

33. The cost (C) of an article to a dealer is $63.00. If the margin ($M$) is to be one-fourth the selling price (S), what should the article sell for? Use the formula $C + M = S$.

34. The selling price (S) of an automobile is $8,952.30. If the dealer has set the margin (M) to be one-twentieth the cost (C), what was the cost of the vehicle? Use the formula $C + M = S$.

35. The width (w) of a rectangle is two-thirds the length (ℓ). If the perimeter (P) of the rectangle is 250 inches, find the length. Use the formula $P = 2\ell + 2w$.

36. A certain handyman estimates the cost of small jobs by assuming the amount for materials to be one-third the amount for labor. He estimated repairing Mr. Smith's porch would amount to $64.00 (labor plus materials). What did he expect to receive for his labor?

37. A sale sign states "Ladies' Dress Shoes, $\frac{1}{3}$ Off." If the sale price of a certain pair of shoes is $45.00, what was the original price?

38. By making a modification to an automobile engine, an engineer was able to reduce the fuel consumption by $\frac{1}{5}$. If the engine now uses 32 gallons of fuel, how much fuel would it have used before the modification?

39. The second side (b) of a triangle is one-half the first side (a) and the third side (c) is two-thirds the first side. If the perimeter (P) of the triangle is 39 inches, find the three sides. Use the formula $P = a + b + c$.

40. A roofing contractor orders 255 sheets of plywood to roof three houses. If the second house uses twice as many sheets as the first house and the third house uses $\frac{3}{4}$ as many sheets as the first house, how many sheets are needed for each house?

2–4 LITERAL EQUATIONS

Upon completion of this section you should be able to:

1. Identify a literal equation.
2. Apply the rules of the previous section to solve literal equations.

An equation having more than one letter is sometimes called a **literal equation.** We occasionally need to solve such an equation for one letter in terms of the others. We will use the same steps for solving an equation as we did in the previous section.

Example 1 Solve for x: $3abx + cy = 2abx + 4cy$

Solution First subtract $2abx$ from both sides.

$$3abx + cy - \boxed{2abx} = 2abx + 4cy - \boxed{2abx}$$

$$\text{or} \qquad abx + cy = 4cy$$

Subtract cy from both sides.

$$abx + cy - \boxed{cy} = 4cy - \boxed{cy}$$

$$\text{or} \qquad abx = 3cy$$

Divide both sides by ab.

$$\frac{abx}{\boxed{ab}} = \frac{3cy}{\boxed{ab}}$$

$$\text{or} \qquad x = \frac{3cy}{ab}$$

Note that we have used the step-by-step method discussed in the preceding section.

Example 2 Solve for x: $\frac{2}{3}(x + y) = 3(x + a)$

Solution First remove parentheses.

$$\frac{2}{3}x + \frac{2}{3}y = 3x + 3a$$

Multiply both sides by 3.

$$2x + 2y = 9x + 9a$$

Subtract $9x$ from both sides.

$$-7x + 2y = 9a$$

Subtract $2y$ from both sides.

$$-7x = 9a - 2y$$

Divide both sides by -7.

$$x = \frac{9a - 2y}{-7}$$

Sometimes the form of an answer may be changed. In example 2 we could multiply both numerator and denominator by -1 (this does not change the value of the answer) and obtain

$$x = \frac{9a - 2y}{-7} = \frac{(-1)(9a - 2y)}{(-1)(-7)} = \frac{-9a + 2y}{7} = \frac{2y - 9a}{7}.$$

The advantage of this last expression over the first is that there are not so many negative signs in the answer.

The most commonly used literal equations are formulas from geometry, physics, business, electronics, and so on.

Example 3 $A = \frac{1}{2}h(b + c)$ is the formula for the area A of a trapezoid, where h is the distance between the two parallel sides b and c. Solve for c.

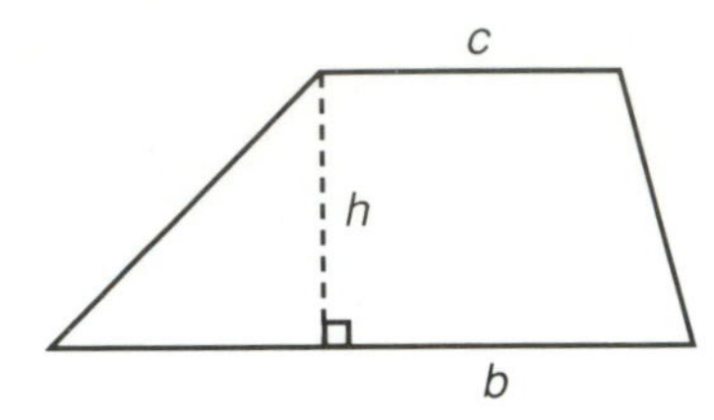

Solution First remove parentheses.

$$A = \frac{1}{2}hb + \frac{1}{2}hc$$

Multiply both sides by 2.

$$2A = hb + hc$$

Subtract hb from both sides.

$$2A - hb = hc$$

Divide both sides by h.

$$\frac{2A - hb}{h} = c$$

$$\text{or} \quad c = \frac{2A - hb}{h}$$

EXERCISE 2-4-1

Solve for x.

1. $2x + y = x + 3y$

2. $3x - 2y = x + 4y$

3. $3x + 2y = 6y + x$

4. $2x + 8y = 5x - y$

5. $2(x + a) = x - 4a$

6. $3x + a = 7(x - a)$

7. $4(x - 2y) + y = 3x - 5y$

8. $3(2x + y) = 19y + 4x$

9. $4a - 2x = 3(4x - 8a)$

10. $11x + 15y = 5(3x - y)$

11. $3ax + 2bc = 4(ax - 5bc)$

12. $2ax + bc = 2(3ax + bc)$

13. $7abx - 3y = 4abx + 5a$

14. $3(2a - 5x) = 2(a + 3x)$

15. $4x + 3a = 5(2b - x)$

16. $5(x - a) = \frac{2}{3}(x + 1)$

17. $3(2x + 5) = \frac{1}{5}(x + a)$

18. $\frac{2}{5}(x - a) = 4(x + a)$

19. $7x + a - 3b = 5x + 3a$

20. $2x + 3(x - a) = 7(x - a)$

21. $A = \frac{1}{2}bh$ is the formula for the area of a triangle. Solve for h.

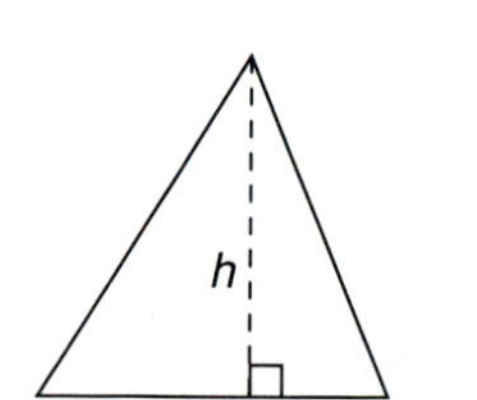

22. The formula for simple interest is $I = prt$. Solve for r.

23. A distance formula from physics is $s = \frac{1}{2}gt^2$. Solve for g.

24. $A = \frac{1}{2}h(b + c)$ is the formula for the area of a trapezoid. Solve for b.

25. $F = \frac{9}{5}C + 32$ is a formula for changing temperature from Celsius C to Fahrenheit F degrees. Solve for C.

26. A formula from physics is $s = h - vt - \frac{1}{2}at^2$. Solve for v.

27. The formula $I = \frac{nE}{nr + R}$ is used in electricity. Solve for E.

28. A formula from mathematics is $S = \frac{a}{1 - r}$. Solve for r.

29. $I = \frac{prD}{365}$ is a formula from business. Solve for p.

30. The formula $I = \frac{e + E}{nr + R}$ is used in electronics. Solve for E.

2–5 FIRST-DEGREE INEQUALITIES

Upon completion of this section you should be able to:

1. Use the inequality symbol to represent the relative positions of two numbers on the number line.
2. Graph inequalities on the number line.
3. Solve first-degree inequalities.

Given any two real numbers a and b, it is always possible to state $a = b$ or $a \neq b$. Many times we are interested only in whether or not two numbers are equal, but there are situations in which we also wish to represent the relative size of numbers that are not equal.

The symbols $<$ and $>$ are **inequality symbols** or **order relations** and are used to show the relative sizes of the values of two numbers. We usually read the symbol $<$ "less than." For instance, $a < b$ is read "a is less than b." We usually read the symbol $>$ "greater than." For instance, $a > b$ is read "a is greater than b." Notice that we have stated that we usually read $a < b$ as a is less than b. But this is only because we read from left to right. In other words, "a is less than b" is the same as saying "b is greater than a." Actually then, we have only one symbol that is written two ways for convenience of reading. One way to remember the meaning of the symbol is that the pointed end is toward the *lesser* of the two numbers.

DEFINITION
$a < b$, "a is less than b," if and only if there is a positive number c that can be added to a to give $a + c = b$.

In simpler words this definition states that a is less than b if we must add something to a to get b. Of course, the "something" must be positive.

If you think of the number line, you know that adding a positive number is equivalent to moving to the right on the number line. This gives rise to the following alternative definition, which may be easier to visualize.

DEFINITION
$a < b$ means that a is to the left of b on the real number line.

Example 1

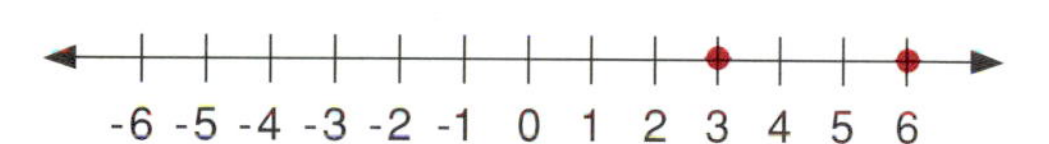

$3 < 6$ because 3 is to the left of 6.

Example 2

$-4 < 0$ because -4 is to the left of zero.

Example 3

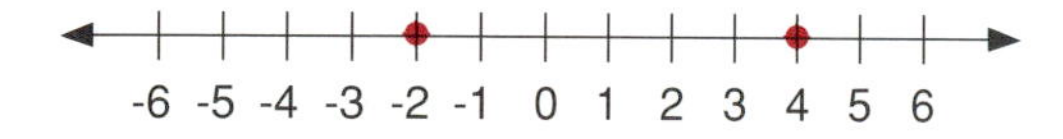

$4 > -2$ because 4 is to the right of -2
or $-2 < 4$ since -2 is to the left of 4.

Example 4

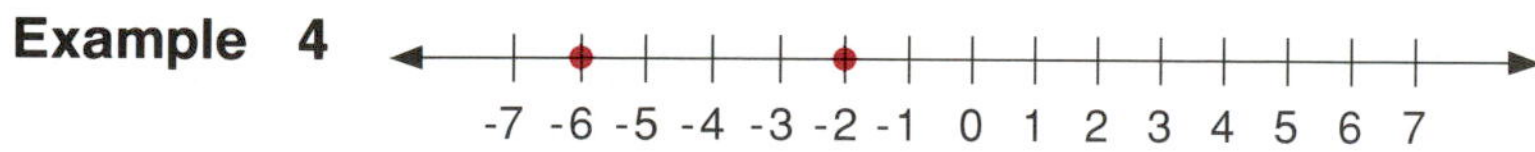

$-6 < -2$ since -6 is to the left of -2.

EXERCISE 2–5–1

Locate the following numbers on the number line and replace the question mark with $>$ or $<$.

1. $6 \ ? \ 10$

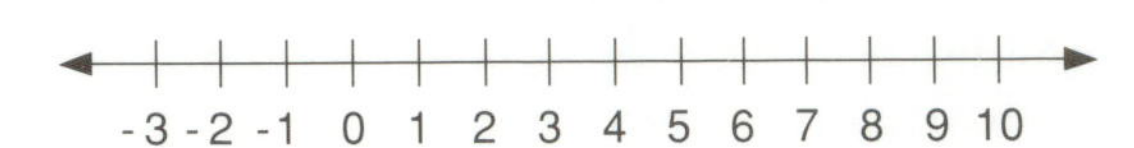

2. $-6 \ ? \ -10$

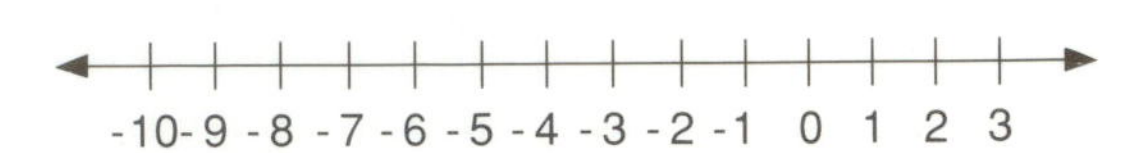

3. $-3 \ ? \ 3$

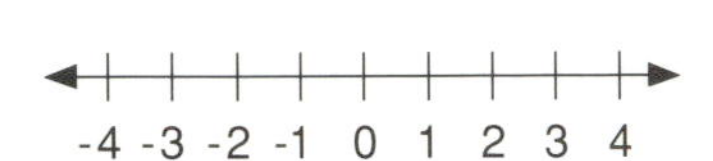

4. $-4 \ ? \ -1$

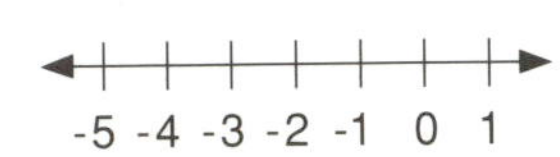

5. $4 \ ? \ 1$

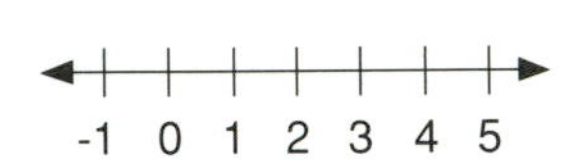

6. $1 \ ? \ 4$

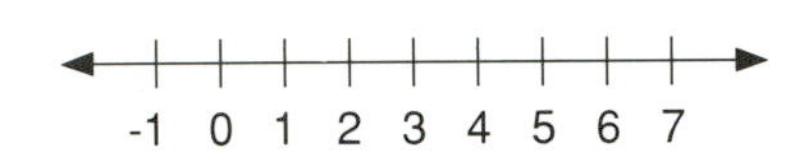

7. $-2 \ ? \ -3$

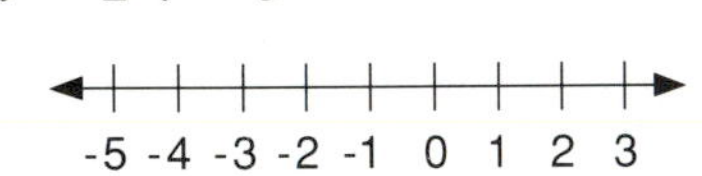

8. $-5 \ ? \ -3$

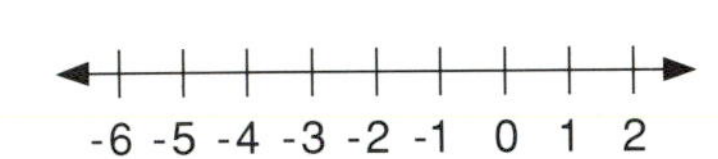

9. $0 \ ? \ 7$

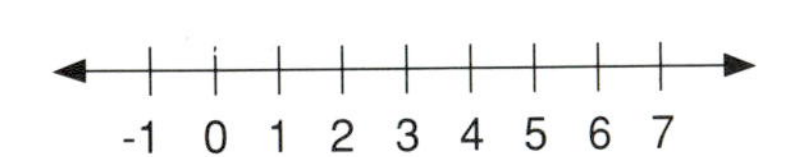

10. $0 \ ? \ -3$

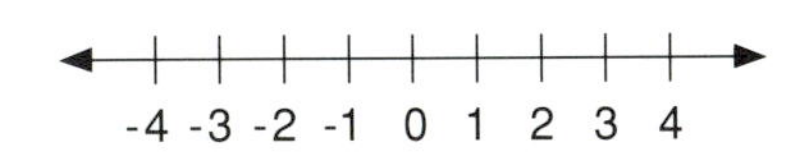

The mathematical statement $x < 3$ (read "x is less than 3") indicates that the variable x can be any number less than (or to the left of) 3. Remember, we are considering the real numbers and not just integers, so do not think of the values for x for $x < 3$ as only 2, 1, 0, -1, and so on.

As a matter of fact, to name the number x that is the largest number less than 3 is an impossible task. It can be indicated on the number line, however. To do this we need a symbol to represent the meaning of a statement such as $x < 3$.

**EFINITION**

The symbols (and) used on the number line will indicate the end point is not included in the set.

Example 5 Graph $x < 3$ on the number line.

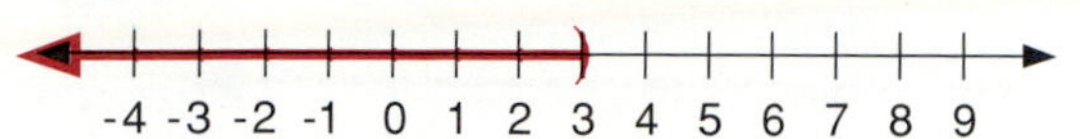

Note that the graph has an arrow indicating that the line continues without end to the left.

Example 6 Graph $x > 4$ on the number line.

Solution

Example 7 Graph $x > -5$ on the number line.

Solution

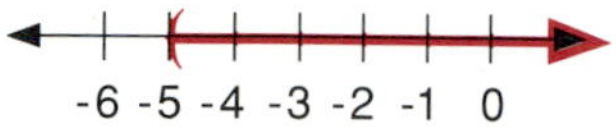

Example 8 Make a number line graph showing that $x > -1$ and $x < 5$. (Remember that "and" indicates both conditions must apply.)

Solution

The statement $x > -1$ and $x < 5$ can be condensed to read $-1 < x < 5$.

Example 9 Graph $-3 < x < 3$.

Solution

If we wish to include the endpoint in the set, we use a different symbol, $\leq$ or $\geq$. We read these symbols "less than or equal to" and "greater than or equal to."

Example 10 $x \geq 4$ indicates the number 4 *and* all real numbers to the right of 4 on the number line.

DEFINITION
The symbols [and] used on the number line will indicate the end point is included in the set.

Example 11 Graph $x \geq 1$ on the number line.

Solution

Example 12 Graph $x \leq -3$ on the number line.

Solution

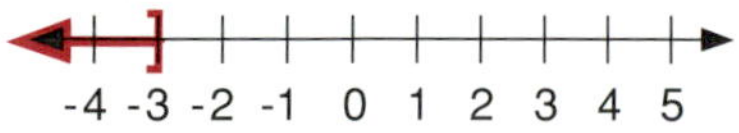

Example 13 Write an algebraic statement represented by the following graph.

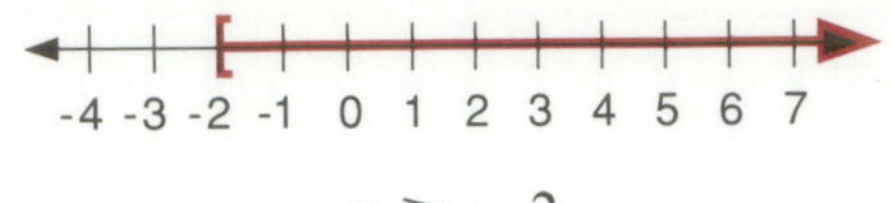

Solution

$$x \geq -2$$

Example 14 Write an algebraic statement for the following graph.

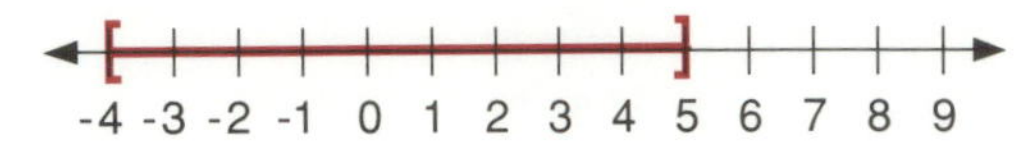

Solution

$x \geq -4$ and $x \leq 5$, which can be written $-4 \leq x \leq 5$

Example 15 Write an algebraic statement for the following graph.

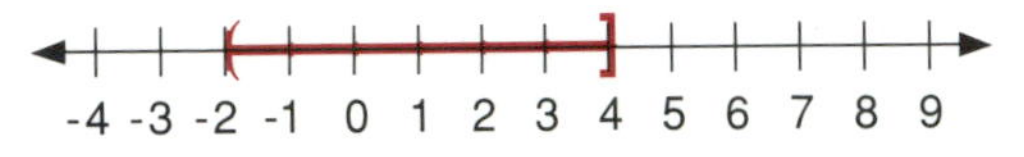

Solution

$x > -2$ and $x \leq 4$ *or* $-2 < x \leq 4$

Example 16 Graph $x > 2\frac{1}{2}$ on the number line.

Solution This example presents a small problem. How can we indicate $2\frac{1}{2}$ on the line? If we estimate the point, then another person might misread the statement. Could you possibly tell if the point represents $2\frac{1}{2}$ or maybe $2\frac{7}{16}$? Since the purpose of a graph is to clarify, always name the end point.

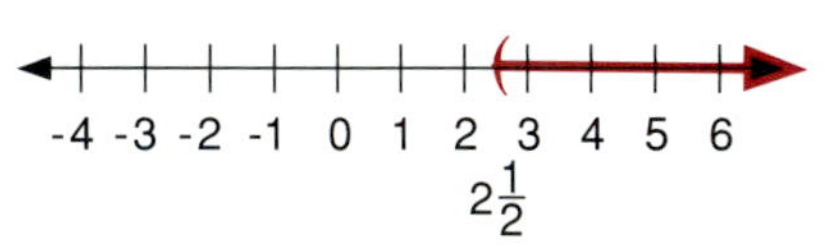

EXERCISE 2-5-2

Construct a graph on the number line.

1. $x > 7$

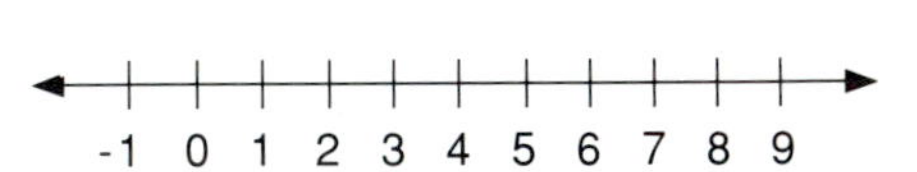

2. $x < 5$

3. $x < -1$

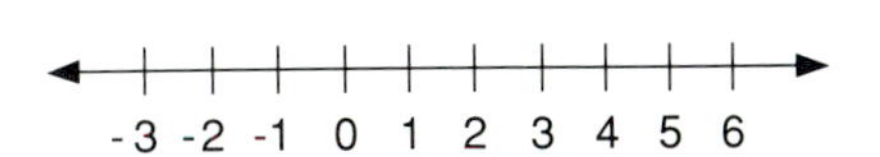

4. $x > -3$

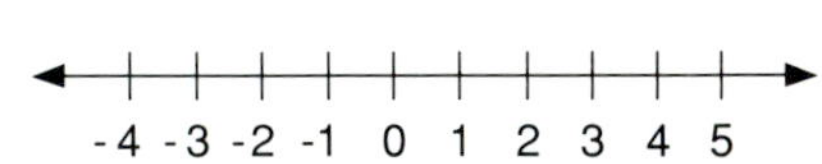

5. $x \geq 1$

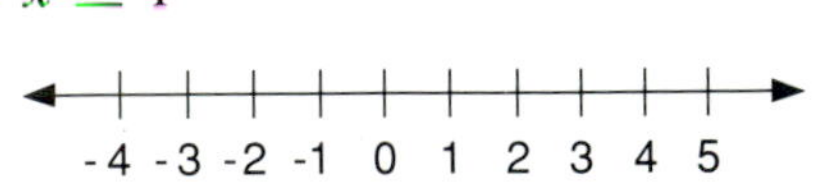

6. $x \geq -4$

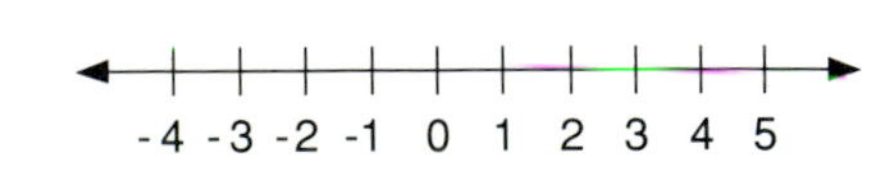

7.  $3 < x < 7$

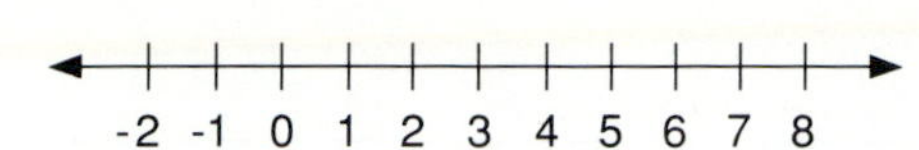

8. $-5 \leq x < 0$

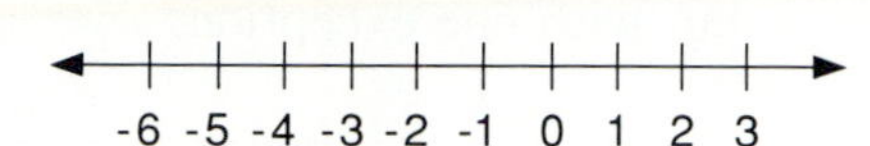

9. $x < 5.4$

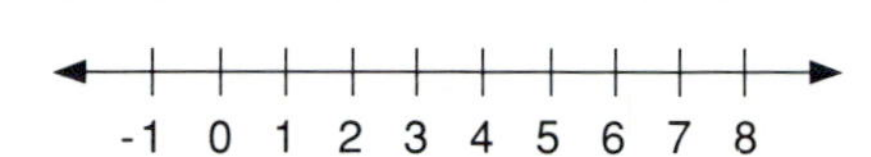

10. $6 < x < 7$

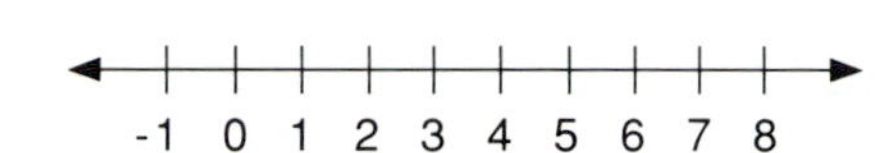

Write an algebraic statement for each graph.

11.

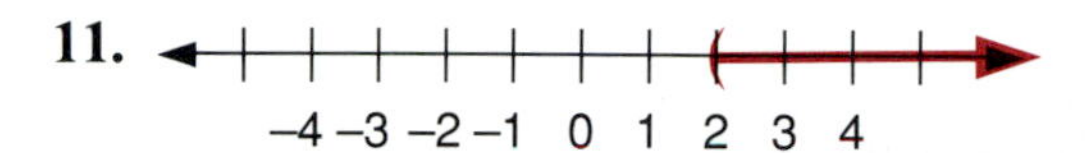

12.

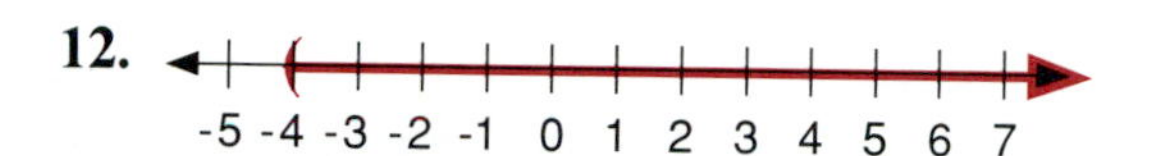

13.

–4 –3 –2 –1 0 1 2 3 4 5

14.

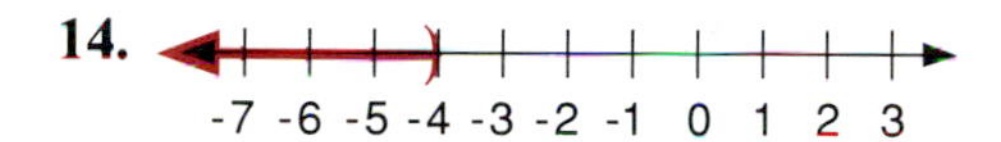

15.

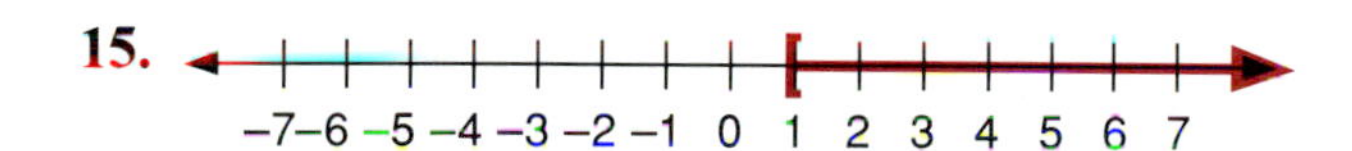

16.

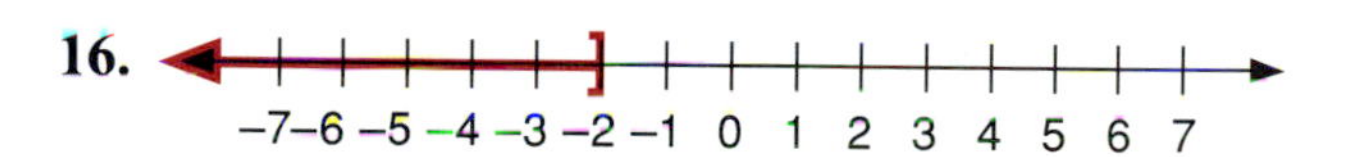

17.

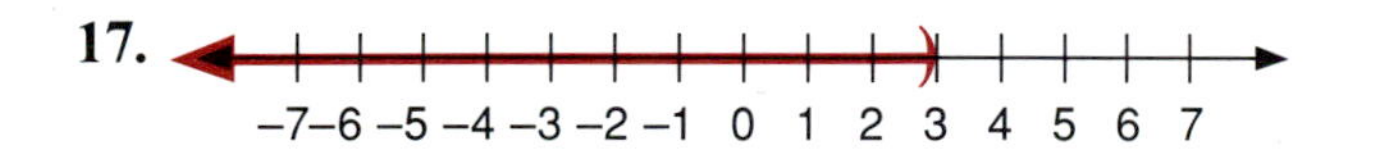

18.

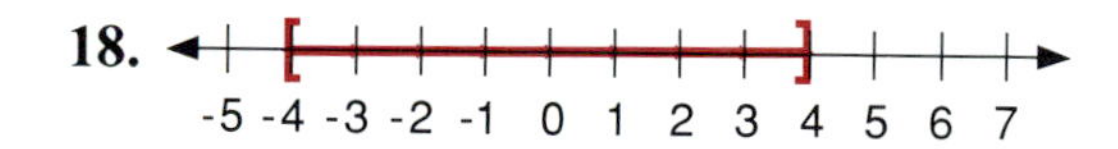

19.

-5 -4 -3 -2 -1 0 1 2 3 4 5 6

20.

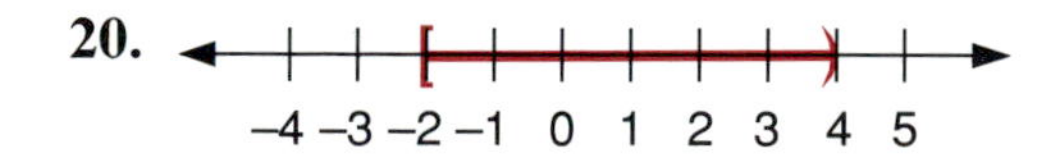

The solutions for first degree inequalities involve the same basic rules as equations—but with one exception.

> **RULE**
> If the same quantity is added to each side of an inequality, the results are unequal in the same order.

Example 17 If $5 < 8$, then $5 + 2 < 8 + 2$

Example 18 If $7 < 10$, then $7 - 3 < 10 - 3$

Example 19 If $x + 6 < 10$, then $x + 6 - 6 < 10 - 6$.

This last example indicates how we would solve the inequality $x + 6 < 10$.

Example 20 Solve the inequality and graph the solution on a number line.

Solution

$$2(x + 2) < x - 5$$
$$2x + 4 < x - 5$$
$$2x + 4 - 4 < x - 5 - 4 \quad \text{(adding } -4 \text{ to both sides)}$$
$$2x - x < -9 \quad \text{(adding } -x \text{ to both sides)}$$
$$x < -9$$

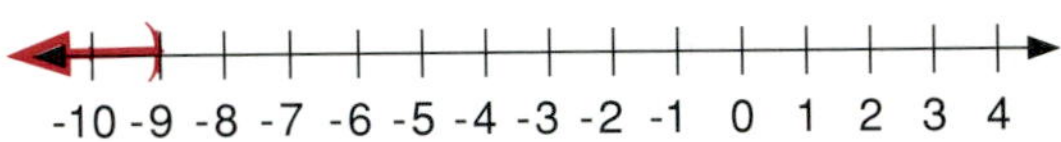

EXERCISE 2-5-3

Use the addition rule to solve the following and graph the solutions.

1. $x + 3 < 7$

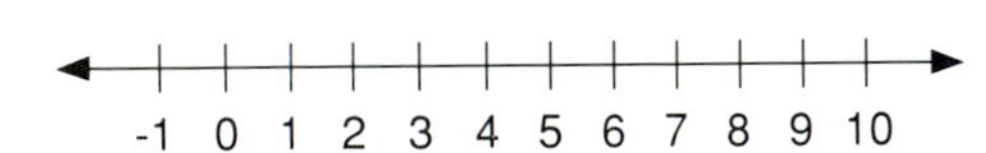

2. $x - 5 < 0$

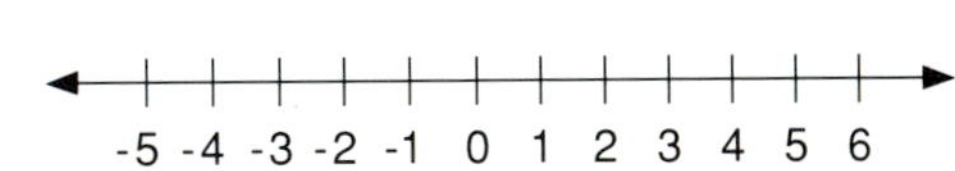

3. $3x + 4 < 2x - 1$

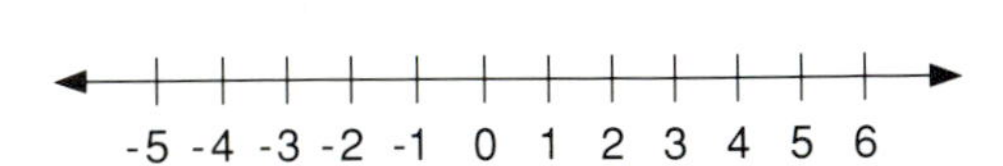

4. $5x < 4x + 1$

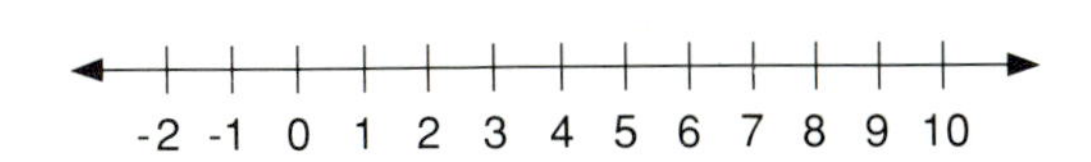

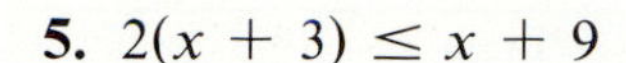

5. $2(x + 3) \leq x + 9$

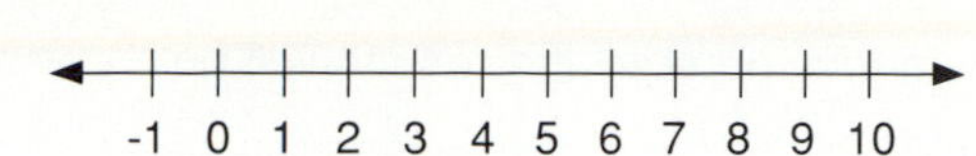

6. $-3(x - 1) \geq 2(1 - 2x)$

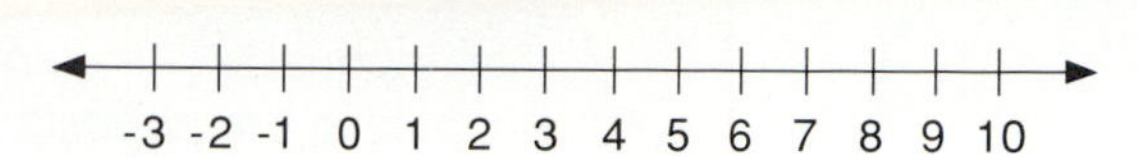

7. $5(x + 3) > 2(2x - 1)$

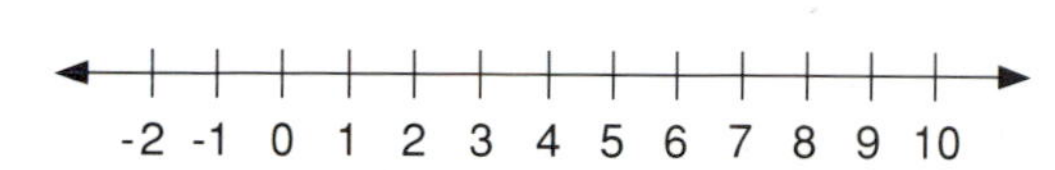

8. $2x + 3(x + 2) \leq 2(2x + 1) - 3$

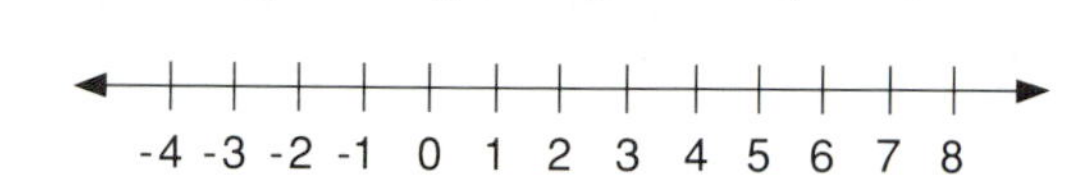

9. $3x + 5 \geq 2(x + 5) - 10$

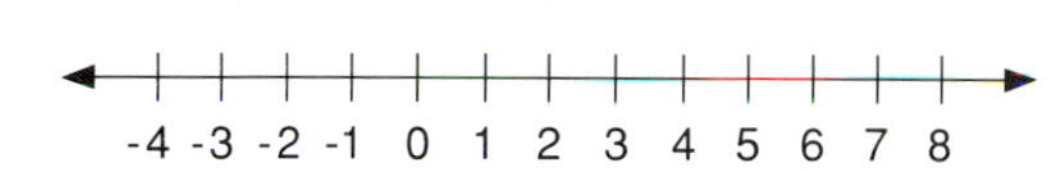

10. $3x + 7 - 2x \leq 7$

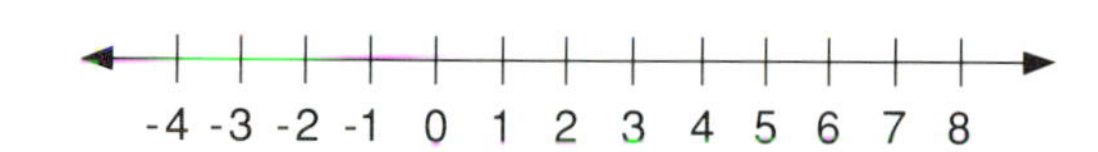

We will now use the addition rule to show an important concept concerning multiplication or division of inequalities.

Suppose $x > a$. Now add $-x$ to both sides by the addition rule.

$$x - x > a - x$$
$$0 > a - x$$

Now add $-a$ to both sides by the addition rule.

$$0 - a > a - x - a$$
$$-a > -x$$

The last statement $-a > -x$ can be rewritten $-x < -a$. We have thus established the following: "If $x > a$, then $-x < -a$." This translates into the following rule.

ULE

If an inequality is multiplied or divided by a *negative* number the results will be unequal in the *opposite* order.

Example 21 Solve for x if $-2x > 6$.

Solution To obtain x on the left side we must divide by -2.

$$\frac{-2x}{-2} < \frac{6}{-2}$$

Note the change from $>$ to $<$.

$$x < -3$$

-4 -3 -2 -1 0 1 2 3 4

Take special note of this rule. Each time you divide by a negative number, you must change the direction of the inequality symbol. This is the *only* difference between solving equations (first-degree, one variable) and solving inequalities.

Division or multiplication by a positive number will *not* change the sense of the inequality, *but* the same operations by a negative number *will* change it. This one point is a source of many errors—so be very careful!

Let us now review the step-by-step method from section 2–3 and note the differences when solving inequalities.

STEP 1 Remove all parentheses. (No change)

STEP 2 Eliminate fractions by multiplying both sides by the least common denominator of all fractions. (No change, since multiplying by a positive number)

STEP 3 Simplify by combining like terms on each side of the inequality. (No change)

STEP 4 Add or subtract quantities to get the unknown on one side, the numbers on the other. (No change)

STEP 5 Divide both sides of the inequality by the coefficient of the unknown. If the coefficient is positive, the inequality will remain the same. If the coefficient is negative, the inequality will be reversed. (This is the important difference.)

Example 22 Solve for x and graph the results: $2(x - 3) \geq 3x - 4$

Solution

$$2x - 6 \geq 3x - 4 \qquad \text{(removing parentheses)}$$

$$2x - 6 + 6 \geq 3x - 4 + 6 \qquad \text{(adding terms)}$$

$$2x \geq 3x + 2$$

$$2x - 3x \geq 3x + 2 - 3x \qquad \text{(adding terms)}$$

$$-x \geq 2$$

$$\frac{-x}{-1} \leq \frac{2}{-1} \qquad \text{(dividing by the coefficient of } x\text{)}$$

$$x \leq -2 \qquad \text{(Note the change of the symbol.)}$$

-3 -2 -1 0 1 2 3 4 5

Example 23 Solve for x and graph the results: $\frac{2}{3}x + 2 > 3(x + 1)$

Solution

$$\frac{2}{3}x + 2 > 3x + 3 \quad \text{(removing parentheses)}$$

$$2x + 6 > 9x + 9 \quad \text{(multiplying by 3)}$$

$$-7x > 3 \quad \text{(adding terms to both sides)}$$

$$\frac{-7x}{-7} < \frac{3}{-7} \quad \text{(dividing by } -7\text{)}$$

$$x < -\frac{3}{7}$$

-3 -2 -1 0 1 2 3 4 5 6

$-\frac{3}{7}$

Example 24 Solve for x and graph the results: $5(x + 3) + 2x \geq \frac{3}{4}(x - 7) + 5x - 1$

Solution

$$5x + 15 + 2x \geq \frac{3}{4}x - \frac{21}{4} + 5x - 1 \quad \text{(removing parentheses)}$$

$$20x + 60 + 8x \geq 3x - 21 + 20x - 4 \quad \text{(multiplying by 4)}$$

$$28x + 60 \geq 23x - 25 \quad \text{(combining like terms)}$$

$$5x \geq -85 \quad \text{(adding terms to both sides)}$$

$$x \geq -17 \quad \text{(dividing by the coefficient of } x\text{)}$$

Since we must graph $x \geq -17$, we can allow each unit to represent 5.

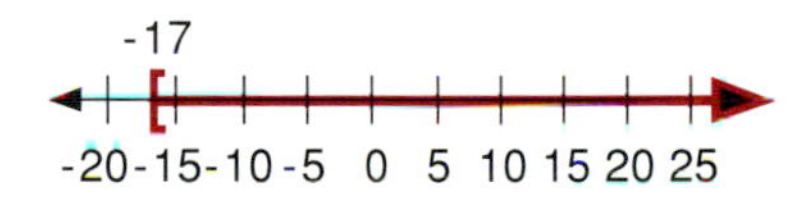

Example 25 Solve for x and graph the results: $\frac{1}{3}x + \frac{2}{3} < \frac{5}{6}x - 1$

Solution

$$2x + 4 < 5x - 6 \quad \text{(multiplying by 6)}$$

$$-3x < -10 \quad \text{(adding terms to both sides)}$$

$$\frac{-3x}{-3} > \frac{-10}{-3} \quad \text{(dividing by } -3\text{)}$$

$$x > +\frac{10}{3}$$

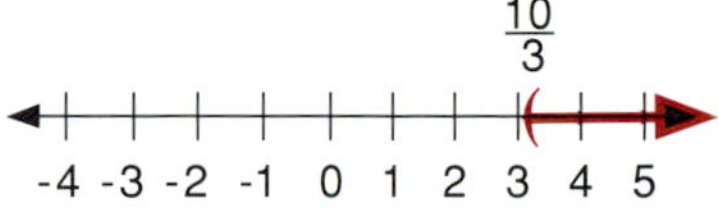

EXERCISE 2–5–4

Solve each inequality and graph the results on the number line.

1. $3x < 6$

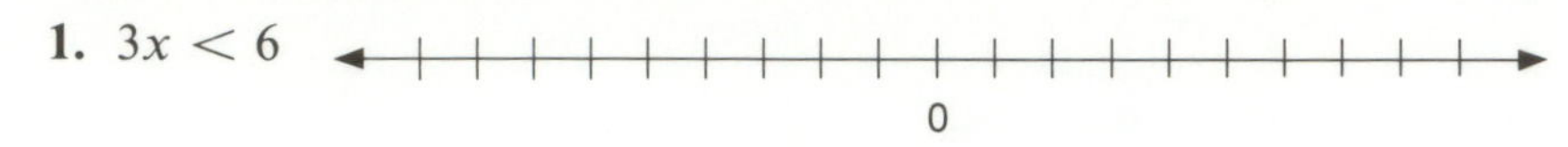

2. $3x \geq 15$

3. $2x \leq -8$

0

4. $-3x > 9$

5. $-5x < -10$

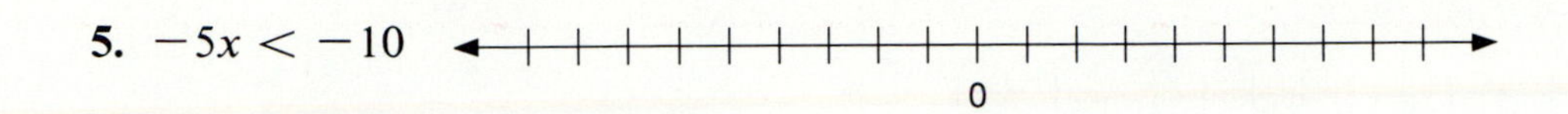

6. $-x \geq 3$

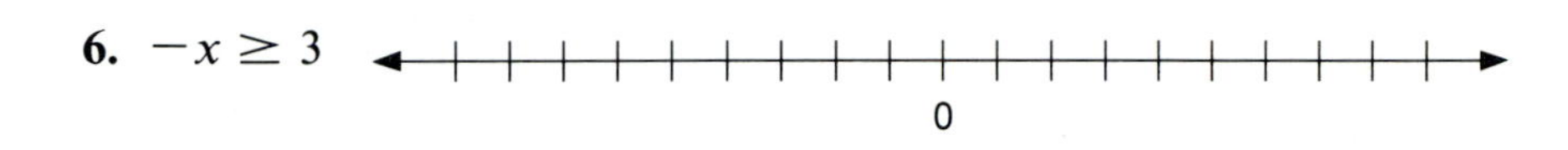

7. $\frac{1}{2}x < 2$

0

8. $\frac{1}{5}x \leq -1$

0

9. $\frac{2}{3}x > 4$

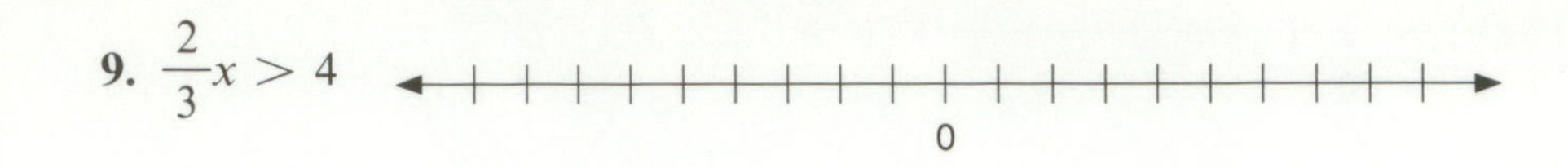

10. $-\frac{1}{2}x < 3$

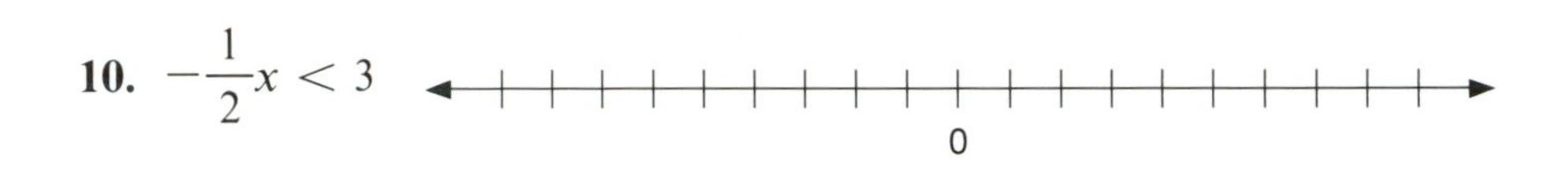

11. $-\frac{1}{3}x \geq -1$

0

12. $5x - 1 > 3(x + 1)$

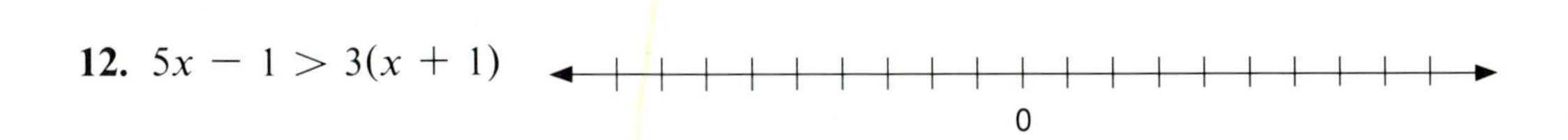

13. $2(x + 3) \leq 7(x + 2) + 2$

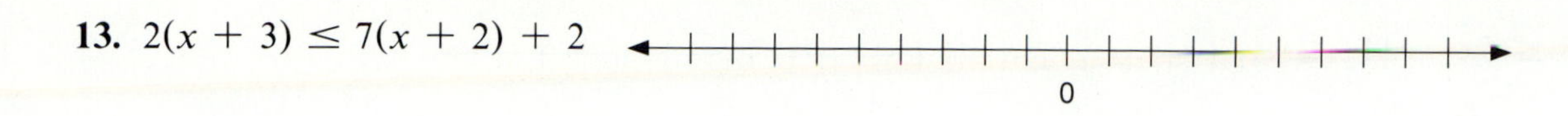

14. $\frac{2}{3}x + 1 \geq 2x - \left(\frac{x}{2} - 6\right)$

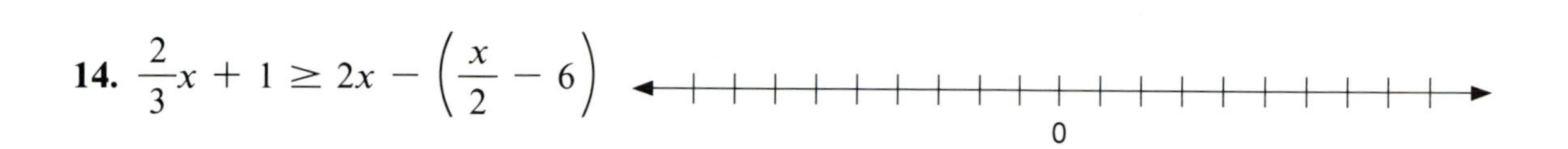

15. $6x + 2\left(\frac{1}{3}x - 10\right) > 0$

0

16. $\frac{4}{5}(x - 2) \leq \frac{1}{2}(x - 1) + \frac{1}{4}$

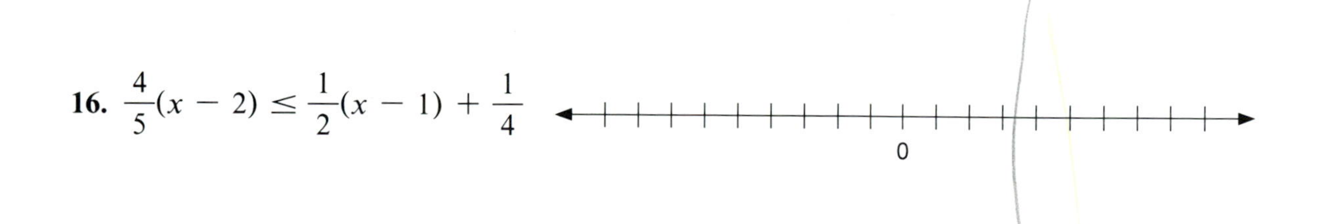

17. $3(x - 4) + 5 \leq 5x - 11$

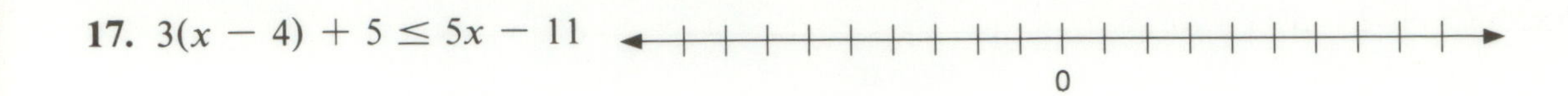

18. $\frac{2}{3}\left(x + \frac{3}{4}\right) < \frac{5}{6}$

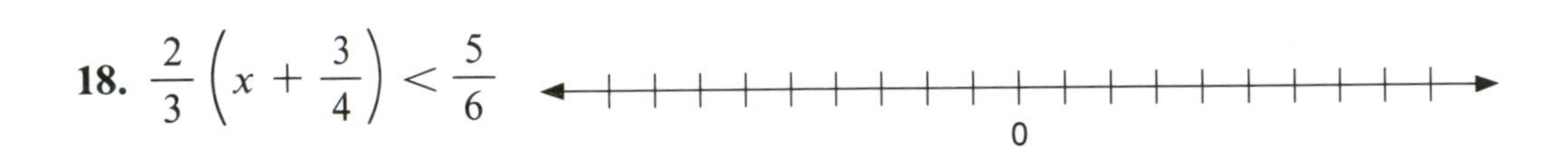

19. $3x + 7 \leq 2x + \frac{1}{3}(2x + 1)$

0

20. $2(x + 3) + 4x > 4\left(x - \frac{1}{2}\right)$

0

21. $3x + 2(x - 5) \le 7 - (x + 3)$

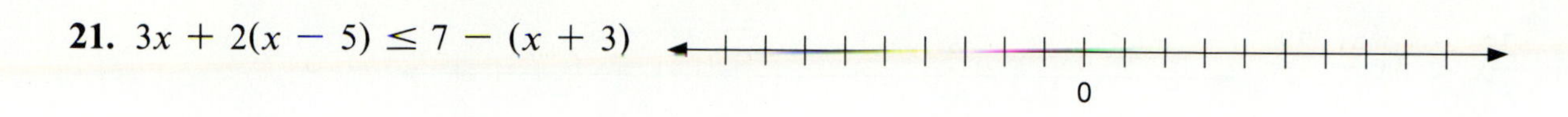

22. $\frac{5}{8}x - \frac{1}{2} < \frac{3}{5}x + 3$

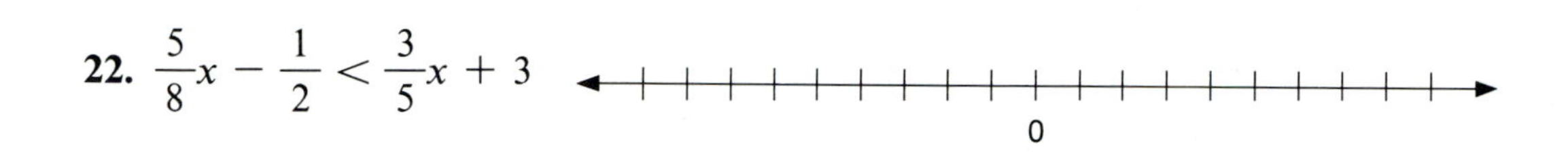

23. $x + \frac{3}{8} \ge 3x + \frac{3}{4}$

0

24. $7x + 5 > 2(x - 1) - 21$

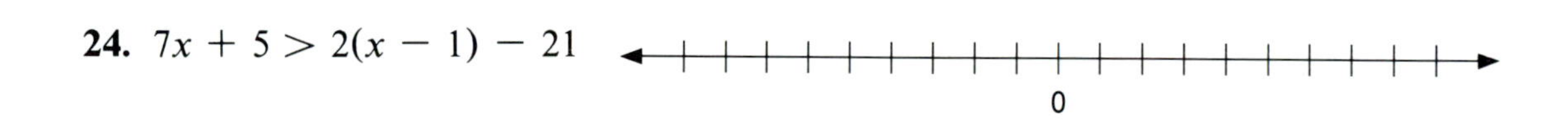

25. $x + \frac{6}{7} + \frac{2}{3}x < 3x - \frac{22}{7}$

26. $\frac{2}{3}(2x - 7) + 2 \geq \frac{1}{2}(4x - 13)$

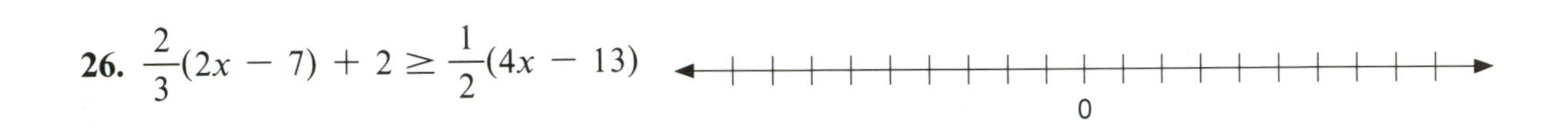

27. $3(x + 1) + 3 > 7(x - 2)$

0

28. $\frac{4}{5}x - 10 > \frac{2}{3}x + 2 - \frac{1}{5}x$

0

29. $\frac{5}{6}(x + 6) < \frac{1}{2}(10 - x)$

30. $\frac{3}{5}(4 - x) - 6 > x + 6$

2–6 ABSOLUTE VALUE

OBJECTIVES

Upon completion of this section you should be able to:

1. Solve equations involving absolute value.
2. Solve inequalities involving absolute value.

In later mathematics courses it will sometimes be necessary to think of distance without direction. This occurs when a measure of "nearness" to a point is desired.

The symbol $|x|$ is read "the **absolute value** of x." $|x|$ can be thought of as the distance of the real number x from zero on the number line. Distance is always nonnegative.

For instance,

$$|5| = 5$$

since 5 is five units from zero on the number line.

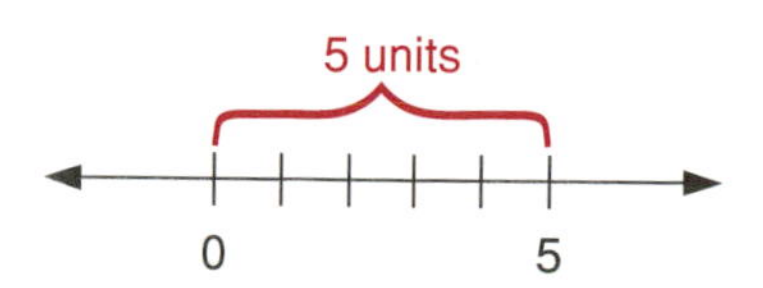

Also, $$|-7| = 7$$

since -7 is seven units from zero on the number line.

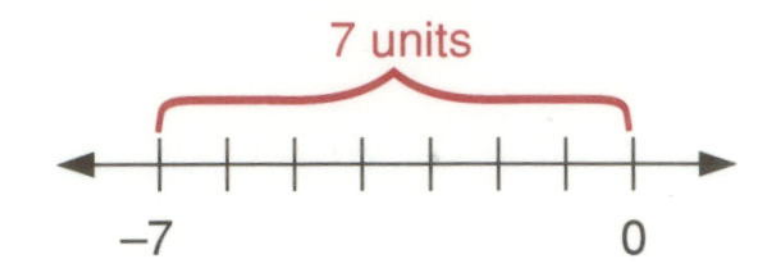

If $|x| = 4$, then $x = 4$ or $x = -4$ because both -4 and 4 are four units from zero on the number line.

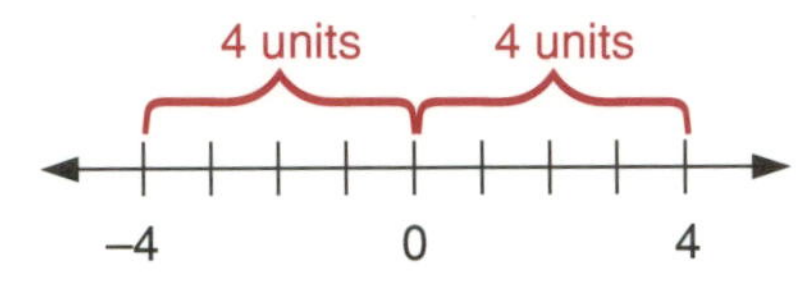

These illustrations lead us to the following definition of absolute value.

$$|x| = x \text{ if } x \geq 0$$
$$|x| = -x \text{ if } x < 0$$

Notice this definition states that the absolute value can *never* be negative. For instance, $|4| = 4$ because $4 > 0$, and $|-4| = -(-4) = 4$ because $-4 < 0$.

The preceding illustrations and definitions are the basis for solving equations involving absolute value.

Example 1 Solve for x if $|x + 3| = 7$.

Solution $|x + 3| = 7$ means that $x + 3$ is seven units from zero. Therefore $|x + 3| = 7$ is equivalent to the statement $x + 3 = 7$ or $x + 3 = -7$. Solving these two equations gives

$$x = 4 \text{ or } x = -10.$$

Substituting these values into the original equations shows us that either value is correct.

EXERCISE 2–6–1

Solve for x.

1. $|x| = 3$

2. $|x| = 0$

3. $|x + 5| = 8$

4. $|x + 7| = 2$

5. $|x - 6| = 5$

6. $|x - 3| = 0$

7. $|2x| = 6$

8. $|2x + 1| = 5$

9. $|3x - 2| = 7$

10. $|2x - 5| = 2$

Absolute value can occur in inequalities as well. For instance, the statement

$$|x| < 5$$

means the number x is within five units of zero on the number line. $|x| < 5$ is equivalent to the statement x is between -5 and 5 on the number line. Therefore the graph of $|x| < 5$ is

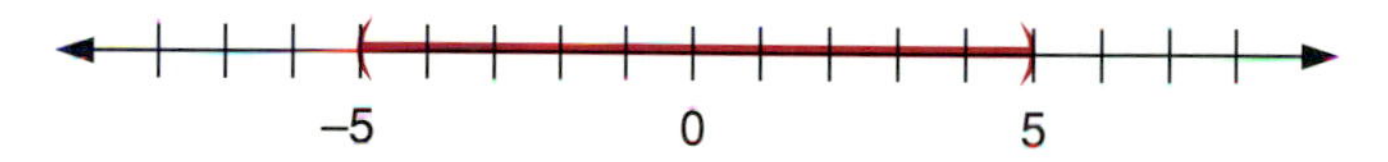

In general, $|x| < a$ is equivalent to $-a < x < a$ (read "x is greater than $-a$ and less than a").

$|x| > 5$ means that the real number x is more than five units from zero on the number line. Hence $|x| > 5$ is equivalent to $x > 5$ or $x < -5$. The graph of $|x| > 5$ is

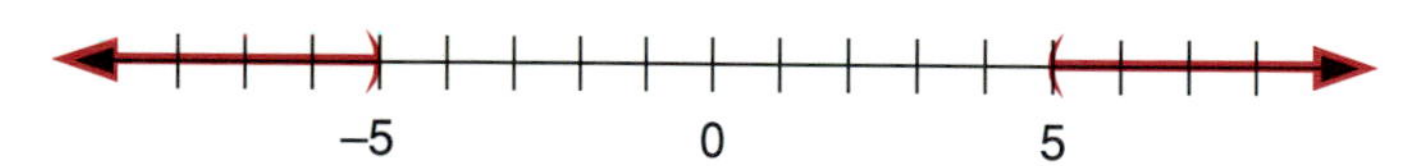

In general, $|x| > a$ is equivalent to $x > a$ or $x < -a$.

CAUTION Take careful note of the fact that an **and** statement, such as $x > -5$ and $x < 5$, can be condensed to $-5 < x < 5$. But an **or** statement, such as $x > 5$ or $x < -5$, cannot be condensed to read $5 < x < -5$ since this implies 5 is less than -5.

Example 2 Solve for x and graph $|x + 3| < 7$.

Solution From the previous discussion we see that $|x + 3| < 7$ is equivalent to the statement

$$-7 < x + 3 < 7.$$

Using the rules for solving inequalities, we subtract 3 from each part of the inequality giving

$$-10 < x < 4$$

(read "x is greater than -10 and less than 4").

The graph of this statement is

Example 3 Solve for x and graph $|x + 3| \geq 7$.

Solution From the preceding discussion we see that $|x + 3| \geq 7$ is equivalent to the statement $x + 3 \geq 7$ or $x + 3 \leq -7$. Solving these inequalities we find

$$x \geq 4 \text{ or } x \leq -10.$$

The graph of this statement is

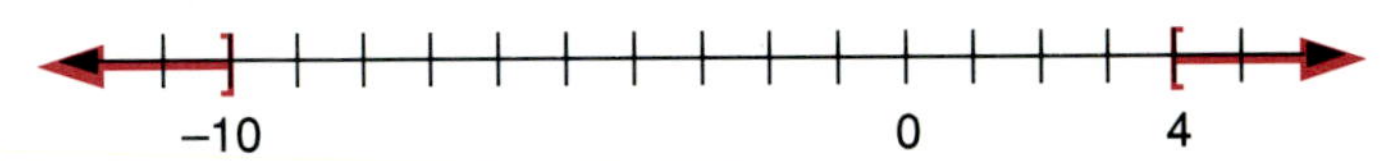

Example 4 Solve for x and graph $\left|\frac{2x - 1}{3}\right| < 5$.

Solution This statement is equivalent to

$$-5 < \frac{2x - 1}{3} < 5.$$

Multiplying each part by 3, we obtain

$$-15 < 2x - 1 < 15.$$

Adding 1 to each part gives

$$-14 < 2x < 16.$$

Finally dividing each part by 2 gives

$$-7 < x < 8.$$

The graph of this statement is

EXERCISE 2-6-2

Solve for x and graph.

1. $|x| < 3$

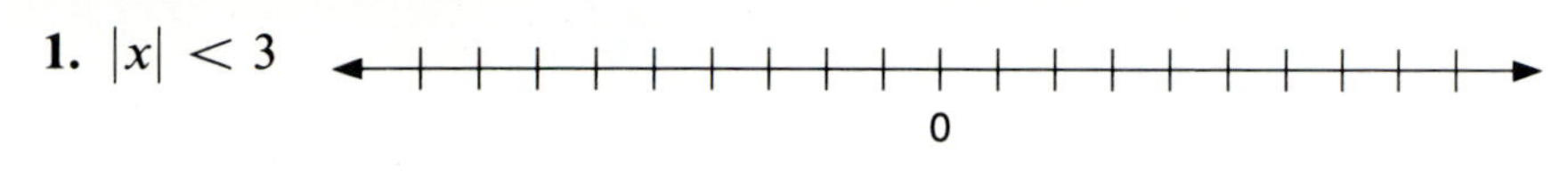

2. $|x| > 4$

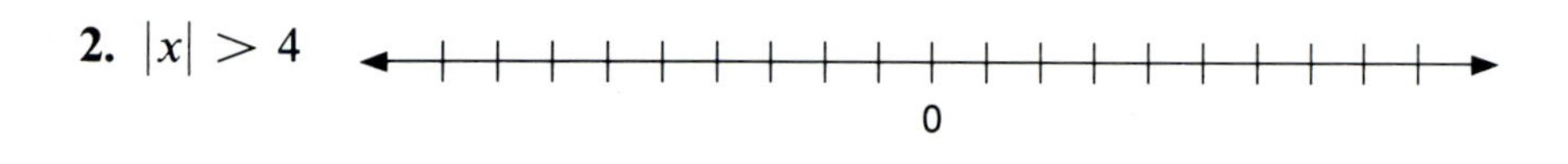

3. $|x + 5| < 8$

4. $|x + 1| > 3$

5. $|x - 2| \leq 5$

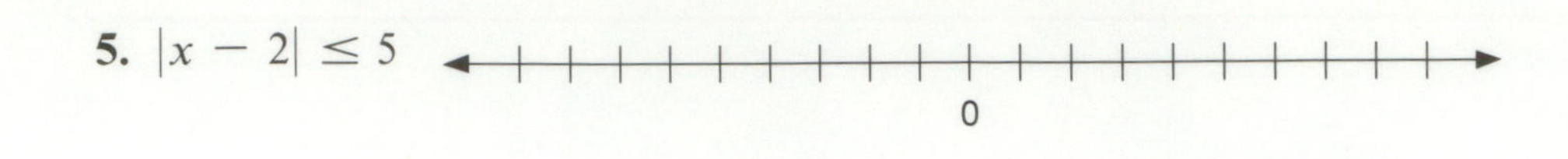

6. $|x - 4| \geq 6$

0

7. $|2x + 1| < 3$

0

8. $|3x - 2| \geq 5$

0

9. $|4x + 3| > 0$

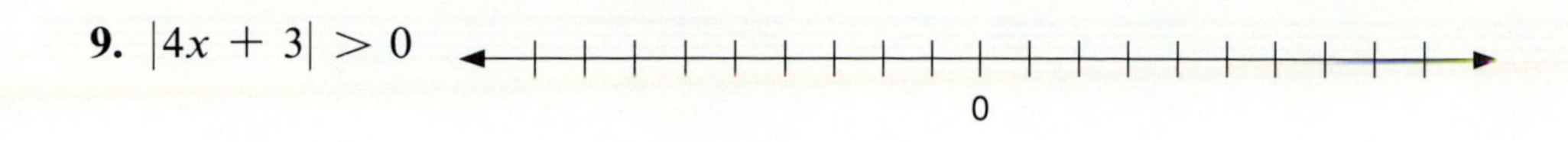

10. $|x + 5| \leq -1$

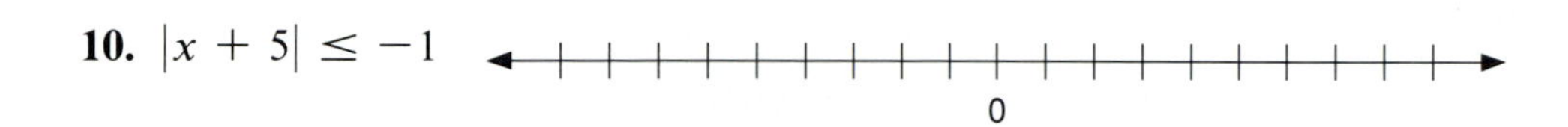

11. $\left|\dfrac{3x - 1}{5}\right| \geq 1$

0

12. $\left|\dfrac{4x + 3}{5}\right| \leq 3$

0

CHAPTER 2 SUMMARY

The number in brackets refers to the section of the chapter that discusses that concept.

Terminology

- An **equation** states that two number expressions are equal. [2–1]
- A **solution** to an equation is a value of the variable that makes the statement true. [2–1]
- Equations are **equivalent** if their solutions are identical. [2–1]
- A **ratio** is a quotient of two numbers. [2–2]
- A **proportion** is a statement that two ratios are equal. [2–2]
- A **literal equation** is an equation having more than one letter. [2–4]
- **Inequality symbols** or **order relations** are used to show the relative sizes of the values of two numbers. [2–5]
- The **absolute value** of a number is its distance from zero on the number line. [2–6]

Rules and Procedures

Solving First-Degree Equations

- If any number is added to or subtracted from both sides of an equation, the resulting equation will be equivalent to the original equation. [2–2]
- If both sides of an equation are divided by the same nonzero number, the resulting equation is equivalent to the original equation. [2–2]
- If both sides of an equation are multiplied by the same nonzero number, the resulting equation is equivalent to the original equation. [2–2]
- To solve a first-degree equation follow these steps: [2–3]

  STEP 1 Remove any parentheses.

  STEP 2 Multiply both sides by the least common denominator of all fractions appearing in the equation.

  STEP 3 Combine similar terms on each side of the equation.

  STEP 4 Add or subtract terms on both sides of the equation to get the unknown on one side and everything else on the other.

  STEP 5 Divide both sides of the equation by the coefficient of the unknown.

  STEP 6 Simplify the solution.

Inequalities

- If the same quantity is added to each side of an inequality, the results are unequal in the same order. [2–5]
- If an inequality is multiplied or divided by a positive number, the results are unequal in the same order. [2–5]
- If an inequality is multiplied or divided by a negative number, the results are unequal in the opposite order. [2–5]
- To solve an inequality follow these steps: [2–5]

  STEP 1 Remove all parentheses.

  STEP 2 Eliminate fractions by multiplying all terms by the least common denominator of all fractions.

  STEP 3 Simplify by combining like terms on each side of the inequality.

  STEP 4 Add or subtract quantities to get the unknown on one side, the numbers on the other.

  STEP 5 Divide both sides of the inequality by the coefficient of the unknown. If the coefficient is positive, the inequality will remain the same. If the coefficient is negative the inequality will be reversed.

- To graph a first-degree inequality use a number line. The critical elements of the graph of an inequality are:

  1. The starting point.
  2. Is this point included or excluded?
  3. Which direction?

CHAPTER 2 REVIEW

Which of the following pairs of equations are equivalent?

1. $x + 3 = 2x - 1$ and $x = 4$

2. $3x - 5 = 2x + 7$ and $x = 2$

3. $3(x - 2) = 2x - 5$ and $x = -3$

4. $x - 3 = 2x + 4$ and $x + 9 = 2$

5. $2x - 6 = 4$ and $3x - 1 = 14$

Solve for x.

6. $x - 6 = -11$

7. $5x + 3 = 4x - 7$

8. $27x = 3$

9. $-\frac{x}{5} = 15$

10. $\frac{2}{3}x = \frac{3}{4}$

11. $2x + 3 = 3(2x + 5)$

12. $2(x + 5) = 6 + 4(x - 3)$

13. $\frac{x}{2} + \frac{2x}{3} = 7$

14. $3x - 5(x - 2) = \frac{1}{3}(x + 2)$

15. $\frac{2}{3}x - \frac{x}{2} + \frac{1}{4} = \frac{2}{3}(x - 4)$

16. $5 - 3x = a + 7x$

17. $3(x - 2a) + 3a = 5(2x + a)$

18. $4(3x - 2) = \frac{1}{3}(x - a)$

19. $P = 2x + 2y$

20. $H = \frac{yz}{xy + w}$

Solve for x and graph.

21. $3x \leq 15$

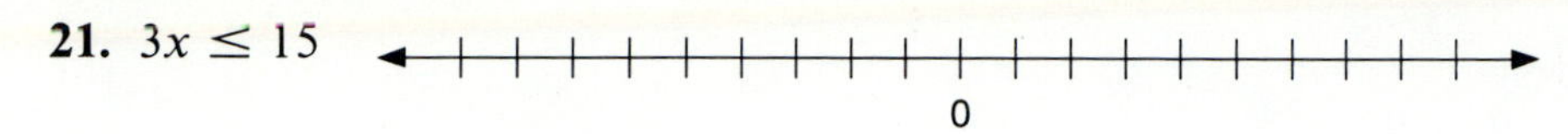

22. $-\frac{1}{2}x \leq 4$

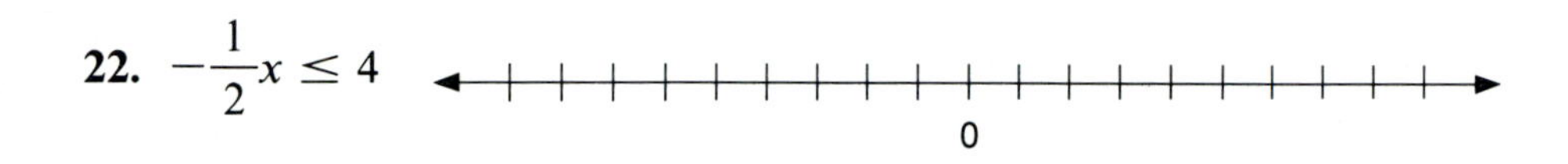

23. $3x - 5 \leq x - 11$

0

24. $2(x - 5) + 3 > 3x - 1$

0

25. $\frac{2}{5}(x + 3) - \frac{1}{2} \geq \frac{1}{3}(x - 2)$

0

Solve for x and graph the result.

26. $|x - 4| = 9$

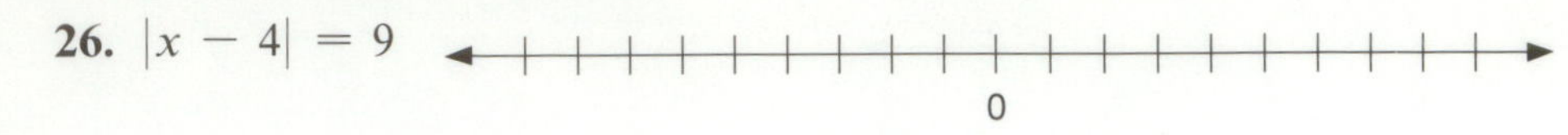

27. $|3x + 1| = 4$

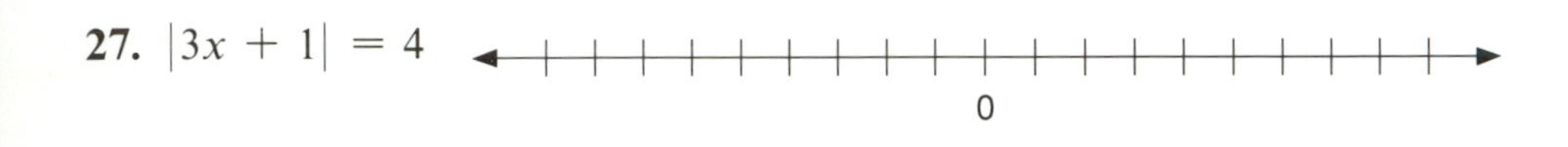

28. $|4x + 5| < 1$

0

29. $|2x - 3| \geq 5$

0

30. $\left|\frac{2x - 3}{5}\right| \leq 3$

0

31. The wages (w) made by a worker is equal to the product of the hourly rate (r) and the number of hours worked (t). The formula is given by $w = rt$. Find the hourly rate of a worker earning \$217.50 in $37\frac{1}{2}$ hours.

32. If the property tax rate is \$12.00 per \$1,000 assessed valuation, what is the tax on property assessed at \$39,500?

33. A relationship between Fahrenheit (F) and Celsius (C) temperature is given by $9C = 5F - 160$. Find the Fahrenheit temperature if the Celsius temperature is 23.5°.

34. The perimeter of a certain rectangle is $27\frac{1}{2}$ meters. Find the width of the rectangle if the width is two-thirds the length. Use the formula $P = 2\ell + 2w$.

CHAPTER

2 PRACTICE TEST

1. Which one of the following equations is equivalent to $x = 4$?
   a. $x + 4 = 0$
   b. $x + 6 = 2$
   c. $x - 2 = 6$
   d. $x - 2 = 2$

1. ______________________

2. ______________________

Solve for x.

2. $3x = -36$

3. ______________________

3. $2x + 15 = 27$

4. ______________________

4. $5x + 33 = 8$

5. ______________________

5. $10x + 7 = 8x + 25$

6. $5(2x + 4) = 7x + 5$

6. ______________________

7. $x - \frac{1}{3}x - 2 = \frac{2}{5}(x + 3)$

7. ______________________

8. $3x + 5y = 6$

8. ______________________

9. $\frac{2}{7}(x + a) = 4(2x - a)$

9. ______________________

10. $|3x + 1| = 10$

10. ______________________

Graph.

11. $x \leq 5$

−6 −5 −4 −3 −2 −1 0 1 2 3 4 5 6

12. $-1 \leq x < 4$

−6 −5 −4 −3 −2 −1 0 1 2 3 4 5 6

13. $x < -3$ or $x > 0$

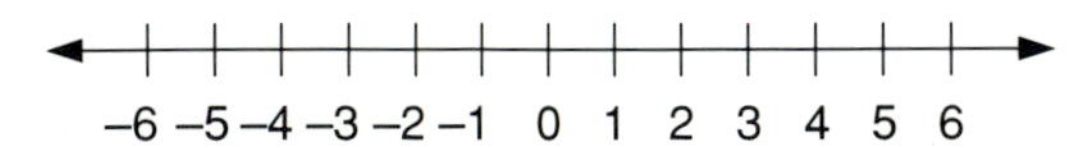

Solve for x.

14. $13 - x < 13$

14. ______________________

15. $\frac{2}{3}(x - 1) - \frac{3}{4} \geq \frac{1}{2}(x + 2)$

15. ______________________

16. $|5x - 11| < 19$

16. ______________________

17. If the ratio of passing grades to failing grades is seven to two, how many students would fail if twenty-eight passed?

17. ______________________

18. The length of a certain rectangle is $2\frac{1}{2}$ times the width. If the perimeter of the rectangle is 35, find the length and width. Use the formula $P = 2\ell + 2w$.

18. ____________________

19. ____________________

19. If $A = \frac{1}{2}h(a + b)$, solve for b.

20. Solve for x and graph the results on the number line.

$\frac{1}{2}x + 2 \geq x - \frac{1}{3}$

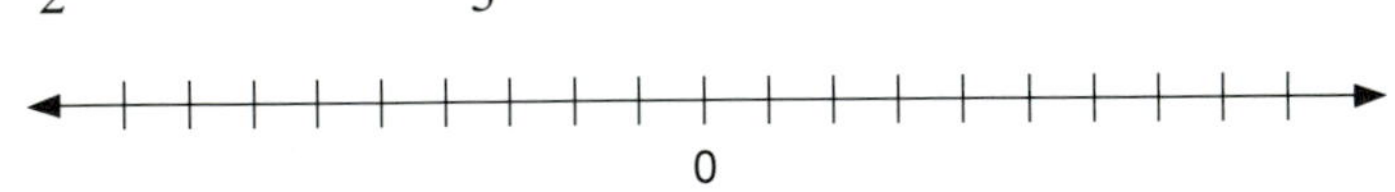

20. ____________________

SURVEY

The following questions refer to material discussed in this chapter. Work as many problems as you can and check your answers with the answer section in the back of the book. The results will direct you to the sections of the chapter in which you need to work. If you answer all questions correctly, you have a good understanding of the material in this chapter.

1. Write an algebraic expression for each.

a. Twice a number x, decreased by nine.

1. a. ____________________

b. One-third a number y, increased by one.

1. b. ____________________

c. 17.3% of a number x.

1. c. ____________________

d. The value in cents of d dimes.

1. d. ____________________

2. The second side of a triangle is twice the first, and the third side is two units more than the first. If the perimeter of the triangle is 26 units, find each side.

2. ____________________

3. A beginning tennis class has twice as many students as the advanced class. The intermediate class has three more students than the advanced class. How many students are in the advanced class if the total enrollment for the three tennis classes is 43?

3. ____________________

4. Two towns are 677 kilometers apart. A car leaves town A bound for town B at 75 kilometers per hour. Twenty minutes later a car leaves town B bound for town A at 88 kilometers per hour. In how many hours after the car leaves town B will they meet?

4. ______________________

5. How many liters of pure dye must be added to 32 liters of a 6% dye solution to obtain a 10% dye solution? (Answer correct to nearest tenth of a liter.)

5. ______________________

Solving Word Problems

In the last chapter we developed the techniques for solving equations. Equations arise as a means of solving verbal or word problems. Since word problems are a large part of algebra, it is necessary to develop techniques of solving them. In this chapter we will concentrate on ways of outlining and solving word problems.

3–1 FROM WORDS TO ALGEBRA

OBJECTIVES

Upon completion of this section you should be able to:

1. Change a word phrase to an algebraic expression.
2. Express a relationship between two or more unknowns in a given statement by using one unknown.

The primary task when attempting to solve a word problem is one of translation. The problem is written in one language and must be translated into another—the language of algebra. This translation process must be precise if we are to be successful in solving the problem.

Example 1 Statement: A number increased by seven is twelve.
Algebraic translation: $x + 7 = 12$.

Do these two statements have the same meaning? Does the algebraic equation make the same statement as the English sentence? If so, we have correctly translated from one language to another and can easily solve for the missing number.

Before we begin to outline completed sentences or solve problems we need to review the meaning of certain phrases.

Example 2 Write an algebraic expression for: A certain number decreased by four.

Solution If we let x represent "a certain number," and recognize that "decreased by" is subtraction, then our answer is $x - 4$.

Example 3 Write an algebraic expression for: Five times a certain number.

Solution If x represents "a certain number," then the expression would be $5x$.

Example 4 Write an algebraic expression for: Five more than a certain number.

Solution If x represents "a certain number," then "more than" means addition and the expression is $x + 5$.

Example 5 Write an algebraic expression for: Seven more than twice a certain number.

Solution Again let x represent the unknown number. Then twice the number would translate as $2x$ and seven more than $2x$ would be $2x + 7$.

Example 6 Write an algebraic expression for: 5% of a given number.

Solution First write 5% as the decimal .05. If x represents the given number, then the expression would be $.05x$ since "% of" implies multiplication.

Example 7 Write an algebraic expression for: The value in cents of d dimes.

Solution The value of a dime is 10 cents. Therefore to indicate the value of d dimes multiply 10 by d, obtaining $10d$.

Exercise 3-1-1

Write an algebraic expression for each.

1. A number increased by five

2. Five more than a given number

3. A given number, less nine

4. Ten greater than a given number

5. Twice a certain number

6. Twice a certain number, increased by two

7. Six times a certain number, decreased by four

8. One-fifth a certain number

9. Three less than a number

10. The product of a and 8

11. Half the product of x and 5

12. Half the sum of b and 7

13. Three times the difference of x and 9

14. 15% of a number

15. 10% of a number

16. 9.5% of a number

17. The value, in cents, of d dimes

18. The total value, in cents, of d dimes and q quarters

19. The number of months in y years

20. The number of weeks in d days

When more than one unknown number is involved in a problem, we try to outline the problem by using a letter such as x to represent one of the numbers. We then use the relationships given in the problem to express the other unknowns using the same letter.

Example 8 The Sears Tower in Chicago is eight stories taller than the Empire State Building in New York. Write algebraic expressions for the height of each building using one unknown.

Solution From the information given we do not know how many stories either building has. We will choose one of the buildings and represent the number of stories by x.

Let $x =$ number of stories in the Empire State Building.

The information states that the Sears Tower is eight stories taller than the Empire State Building. We therefore can write

$x + 8 =$ number of stories in the Sears Tower.

Example 9 The length of a rectangle is three meters more than the width. Write expressions for the length and width using one unknown.

Solution Both the length and width are unknown quantities so we must write expressions that give one of these quantities in terms of the other. First we will let x represent the width.

Let $x =$ width of the rectangle.

Now since we are told that the length is three more than the width, we can write

$x + 3 =$ length of the rectangle.

We can draw a sketch and label the sides.

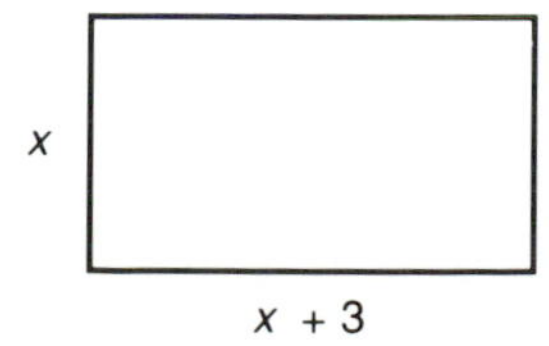

Suppose we had decided to let x represent the length of the rectangle. We would write

$$\text{let } x = \text{length of the rectangle.}$$

Now we would have to recognize that the statement "the length is three more than the width" is the same as saying "the width is three less than the length." Then $x - 3 =$ width of the rectangle. We would label the rectangle as follows.

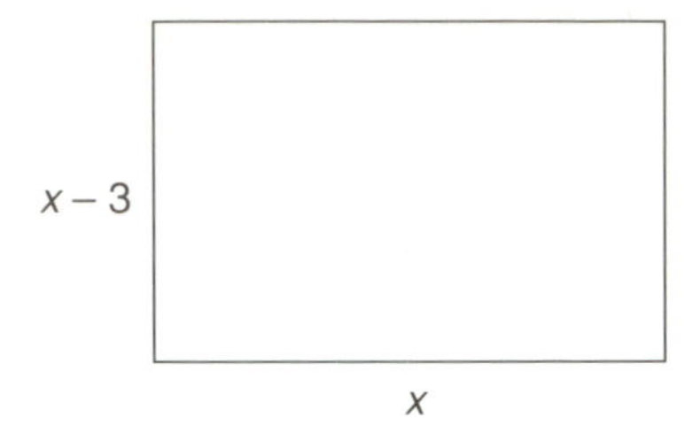

We see that there is more than one way to identify the unknowns. Therefore it is extremely important that we designate which unknown is represented by each literal expression.

Example 10 The width of a rectangle is one-fourth the length. Write expressions for the length and width using one unknown.

Solution At first glance we might think that a fraction will be necessary here. However, if we note that "the width is one-fourth the length" is the same as "the length is four times the width," we may let

$$x = \text{width of the rectangle}$$

$$\text{Then} \quad 4x = \text{length.}$$

x

$4x$

Note that if we were to let x represent the length, we would have a fraction representing the width. This certainly would not be wrong but we can often avoid the use of fractions by careful selection.

Example 11 The sum of two numbers is 20. Write expressions for both numbers using one unknown.

Solution

$$\text{Let } x = \text{first number.}$$

$$\text{Then} \quad 20 - x = \text{second number.}$$

Do these numbers add up to 20?

$$x + (20 - x) = 20$$

Example 12 Express algebraically the relationship of the unknown in this sentence: A certain number is four more than a second number and three less than a third number.

Solution To express these three numbers algebraically first decide which number will be represented by x.

For instance, if x represents the first number, we have the following:

$$\begin{aligned} x &= \text{first number} \\ x - 4 &= \text{second number} \\ x + 3 &= \text{third number.} \end{aligned}$$

Ask yourself, "Is the first number four more than the second?" and "Is the first number three less than the third?" If the answers are "yes," you have correctly outlined the sentence.

Now suppose in the same example we decide to allow x to represent the second number. We now have

$$\begin{aligned} x + 4 &= \text{first number} \\ x &= \text{second number} \\ x + 7 &= \text{third number.} \end{aligned}$$

Ask the same questions again, "Is the first number four more than the second?" and "Is the first number three less than the third?" This outline is also correct.

EXERCISE 3–1–2

Express the unknowns in terms of x:

1. A certain number is eight more than a second number. Write expressions for the numbers.

2. The Sears Tower in Chicago is 104 feet taller than the World Trade Center in New York. Write expressions for the height of each building.

3. During the second week of production of a new car, General Motors produced 612 units more than it did during the first week. Write expressions for the number of cars produced each week.

4. A 7:00 A.M. math class has eight fewer students than a 9:00 A.M. class. Write expressions for the number of students in each class.

5. The population of San Francisco is twice that of Long Beach. Write expressions for the population of each city.

6. The length of a rectangle is five meters more than the width. Write expressions for the length and width.

7. The length of a rectangle is three centimeters more than twice the width. Write expressions for the length and width.

8. The sum of two numbers is 40. Write expressions for the numbers.

9. The difference of two numbers is 18. Write expressions for the two numbers.

10. An individual makes two investments that total $10,000. Write expressions to represent the two investments.

11. This year's enrollment at a college is 7% less than last year's. Write expressions for the enrollment for last year and this year.

12. One number is four more than the second and nine less than the third. Write expressions for the three numbers.

13. One number is half the second and three times the third. Write expressions for the three numbers.

14. Write expressions for three consecutive even integers. (Examples of consecutive even integers are 2, 4, 6, 8, 10, and so on.)

15. The cost of a house today is three times its cost in 1970 and twice its cost in 1976. Write expressions for the cost in 1970, 1976, and now.

16. Jane has five dollars more than Bob and thirteen dollars less than Jim. Write expressions for the numbers of dollars each person has.

17. A Buick Regal obtains three miles per gallon more than a Chevrolet Caprice and six miles per gallon less than a Toyota Corolla. Write expressions for the mileage for the three cars.

18. A certain amount is invested in an account for one year at 14% interest. Write expressions for the original amount, the interest earned, and the total amount in the account at the end of the year.

19. A meter stick is cut into two pieces. Write expressions for the length of each piece in centimeters. (Recall that 1 m = 100 cm.)

20. A person has twice as many dimes as nickels and two more quarters than dimes. Write expressions for the number of each kind of coin the person has.

3-2 SOLVING WORD PROBLEMS

OBJECTIVES

Upon completion of this section you should be able to:

1. Correctly translate a word problem into an algebraic equation.
2. Solve the equation and find the solution to the problem.

Some problems, such as the equations in the previous chapter, are written in algebraic notation. Others involve written words that we must translate to algebraic language.

In the preceding exercises we outlined relationships between unknowns within a statement. If we are to solve a problem and find the unknown numbers, there must be within the problem a sentence that yields an equation. Thus the only difference between equations from the previous chapter and word problems is the fact that we must express the equation given in words as an equation in algebraic terms.

The first step in this process is to read and re-read the problem until we determine what is being asked for. Then we must look at the given information and outline the problem so as to arrive at an equation relating the unknown quantities.

Example 1 The Sears Tower in Chicago is eight stories taller than the Empire State Building in New York. If the total number of stories in both buildings is 212, find the number of stories in each building.

Solution Notice that we are asked to find the number of stories in each building. Looking further, we see a relationship given that leads us to the following outline.

Let $x =$ number of stories in the Empire State Building.
Then $x + 8 =$ number of stories in the Sears Tower.

In example 8 of the previous section we had enough information for this outline. This time, however, we have the added statement that the total number of stories is 212. Thus we can write the equation

$$x + (x + 8) = 212.$$

Solving this equation gives

$$\begin{aligned} 2x + 8 &= 212 \\ 2x &= 204 \\ x &= 102. \end{aligned}$$

Notice that we have found only the number of stories in the Empire State Building. The question asked us to find the number of stories in each building. Thus to find the number of stories in the Sears Tower we must substitute 102 for x in the expression $x + 8$, obtaining

$$x + 8 = (102) + 8 = 110.$$

The answers are

number of stories in the Empire State Building = 102
number of stories in the Sears Tower = 110.

Check: 110 is eight more than 102, and the sum of 110 and 102 is 212.

Example 2 The length of a rectangle is three meters more than the width. Find the length and width if the perimeter of the rectangle is 26 meters.

Solution We are asked to find the dimensions of the rectangle. From the information given we outline as follows.

Let $x =$ width.
Then $x + 3 =$ length.

x

$x + 3$

Since the perimeter is equal to the sum of twice the length and twice the width, or $P = 2\ell + 2w$, we can write the equation

$$2(x) + 2(x + 3) = 26.$$

Solving, we obtain

$$\begin{aligned} 2x + 2x + 6 &= 26 \\ 4x + 6 &= 26 \\ 4x &= 20 \\ x &= 5. \end{aligned}$$

Also, $\quad x + 3 = 8.$

Thus the width is 5 meters and the length is 8 meters.

Check: The length (8) is three more than the width (5), and the perimeter is $2(8) + 2(5) = 16 + 10 = 26$.

Example 3 The width of a rectangle is one-fourth the length. If the perimeter is 200 centimeters, find the length.

Solution Here we are asked to find the length of a rectangle and are given a relationship between the length and width. We outline as follows.

$$\begin{aligned} \text{Let } x &= \text{width.} \\ \text{Then } 4x &= \text{length.} \end{aligned}$$

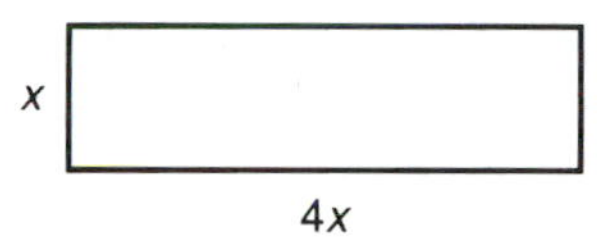

Again the perimeter is twice the width plus twice the length. Therefore

$$\begin{aligned} 2(x) + 2(4x) &= 200 \\ 2x + 8x &= 200 \\ 10x &= 200 \\ x &= 20. \end{aligned}$$

The problem asks for the length only. Thus,

$$4x = 4(20) = 80.$$

The length is 80 cm.

Check: The width (20) is one fourth the length (80). Also, the perimeter is $2(20) + 2(80) = 40 + 160 = 200$.

Example 4 The sum of two numbers is 20. Their difference is 4. Find the numbers.

Solution To find the two numbers we can let

$$x = \text{first number.}$$

Then $\quad 20 - x = \text{second number.}$

Since the difference of the two numbers is 4, we have the following equation:

$$x - (20 - x) = 4.$$

Solving, we obtain

$$\begin{aligned} x - 20 + x &= 4 \\ 2x - 20 &= 4 \\ 2x &= 24 \\ x &= 12. \end{aligned}$$

Also,

$$20 - x = 8.$$

The two numbers are 8 and 12.

Check: The sum of 8 and 12 is 20. The difference of 8 and 12 is 4.

CAUTION Do not just solve the equation for x and think you have solved the problem. Check to see what x represents and re-read the problem to see what it is asking for.

Example 5 A certain number is four more than a second number and three less than a third number. Find the numbers if their sum is 23.

Solution In this problem we are asked to find the numbers and are given a statement about their sum.

If we let x represent the first number, we have

$$\begin{aligned} x &= \text{first number} \\ x - 4 &= \text{second number} \\ x + 3 &= \text{third number}. \end{aligned}$$

The statement "their sum is 23" gives the equation

$$x + (x - 4) + (x + 3) = 23.$$

If we now solve the equation, we obtain

$$\begin{aligned} 3x - 1 &= 23 \\ 3x &= 24 \\ x &= 8. \end{aligned}$$

Leaving the answer as $x = 8$ would not be a solution to the problem. We are asked to "find the numbers." The answers must be

$$\begin{aligned} \text{first number} &= 8 \\ \text{second number} &= x - 4 = 8 - 4 = 4 \\ \text{third number} &= x + 3 = 8 + 3 = 11. \end{aligned}$$

Check: The first number (8) is four more than the second (4) and is three less than the third (11), and their sum is 23.

We have followed six basic steps in solving these examples. These steps should be observed when solving any word problem.

Step 1 Determine what is to be found.
Step 2 Use the given information to write expressions for the unknowns.
Step 3 Write an equation that relates the unknowns to each other.
Step 4 Solve the equation.
Step 5 Make sure you have answered the question.
Step 6 Check your answers to make sure they agree with the original problem.

EXERCISE 3–2–1

Utilize the steps discussed in this section to solve each problem.

1. A certain number is eight more than a second number. The sum of the two numbers is 22. Find the numbers.

2. The Sears Tower in Chicago is 104 feet taller than the World Trade Center in New York. If the sum of their heights is 2,804 feet, find the height of each building.

3. During the second week of production of a new car, General Motors produced 612 units more than it did during the first week. If the total production for the two weeks was 18,416, find the production for each week.

4. A 7:00 A.M. math class has eight fewer students than a 9:00 A.M. class. Find the number of students in each class if the total number of students in both classes is 84.

5. The population of San Francisco is twice that of Long Beach. If the total population of both cities is 1,074,000, find the population of San Francisco.

6. The length of a rectangle is five meters more than the width. Find the length and width if the perimeter is 46 meters.

7. The length of a rectangle is three centimeters more than twice the width. If the perimeter is 192 centimeters, find the length.

8. The sum of two numbers is 40. Their difference is 12. Find the numbers.

9. The difference of two numbers is 18. Their sum is 82. Find the numbers.

10. An individual makes two investments that total $10,000. If the first investment is $3,256 more than the second, find the amount of each investment.

11. A certain number is 37 more than a second number. If the sum of the two numbers is 15, find the numbers.

12. One number is four more than a second number and nine less than a third number. Find the three numbers if their sum is 50.

13. On a particular flight the cost of a coach ticket is half the cost of a first-class ticket and $30 more than an economy ticket. If the total cost of the three tickets is $406, find the cost of each.

14. A Buick Regal obtains three miles per gallon better gas mileage than a Chevrolet Caprice and six miles per gallon less than a Toyota Corolla. If the average mileage for the three cars is 25 miles per gallon, find the mileage of each.

15. One side of a triangle is twice the length of the second side and four centimeters less than that of the third side. If the perimeter of the triangle is 84 centimeters, find the length of each side.

16. Computer A is $250 more expensive than computer B. If the total cost of both computers is $11,090, find the cost of each.

17. A person has $8,000 less invested in stocks than in bonds. If the total invested in both is $86,500, find the amount invested in stocks.

18. The price of a certain model of a Texas Instruments calculator is six dollars more than a Sharp calculator and three dollars less than one made by Hewlett-Packard. If the total price of the three calculators is $46.50, find the price of each.

19. The state of Ohio has half the population of California and twice the population of Indiana. If the total population of the three states is 38,500,000, find the population of each state.

20. The atomic weight of sulfur is twice that of oxygen and eight times that of helium. If the total atomic weights of the three elements is 52, find the atomic weight of each.

3–3 NUMBER RELATION PROBLEMS

OBJECTIVES

Upon completion of this section you should be able to:

1. Identify a problem as being a number relation problem.
2. Express relationships between unknowns and determine an equation to solve a number relation problem.
3. Apply formulas to solve certain types of number relation problems.

Read the following three problems carefully.

1. Jim is two years older than Sue and four years younger than Hugh. The sum of their ages is 59. Find the age of each.
2. A length of rope is 59 meters. It is cut in three pieces such that the first is two meters longer than the second, and four meters shorter than the third. Find the length of each piece.
3. There are three numbers such that the first is two more than the second and four less than the third. If their sum is 59, find the numbers.

At first glance these may seem to be three different problems. However, closer examination shows that they are actually the same. For all three a possible outline is to let

$(x + 2)$, (x), and $(x + 6)$ represent the three unknowns.

The equation $(x + 2) + (x) + (x + 6) = 59$ will lead to the solutions.

Problems such as these are classified as **number relation** problems. If you recognize a problem in this class, then the outline for solving it must show a relation between the numbers. The equation will then be based on a statement about the sum, difference, and so forth.

The problems you solved in the last section were all number relation problems. They all involved comparing two or more numbers.

Example 1 If we wish to solve problem 3 above, we could let

$$\begin{aligned} x + 2 &= \text{first number} \\ x &= \text{second number} \\ x + 6 &= \text{third number.} \end{aligned}$$

Then $$(x + 2) + x + (x + 6) = 59.$$

Solving, we obtain

$$\begin{aligned} x &= 17 \\ x + 2 &= 19 \\ x + 6 &= 23. \end{aligned}$$

Our answers are

$$\begin{aligned} \text{first number} &= 19 \\ \text{second number} &= 17 \\ \text{third number} &= 23. \end{aligned}$$

Check: The first number (19) is two more than the second (17) and four less than the third (23). Also, the sum $19 + 17 + 23 = 59$.

Problems comparing age, height, weight, and so on, are number relation problems.

Example 2 Janet is three years older than Maria. The sum of their ages is 39. Find the age of each.

Solution Let

$$\begin{aligned} x &= \text{Maria's age} \\ x + 3 &= \text{Janet's age.} \end{aligned}$$

Then

$$x + (x + 3) = 39.$$

Solving, we obtain

$$\begin{aligned} x &= 18 \\ x + 3 &= 21. \end{aligned}$$

The answers are

$$\begin{aligned} \text{Maria's age} &= 18 \text{ years} \\ \text{Janet's age} &= 21 \text{ years.} \end{aligned}$$

Check: 21 is three more than 18 and $21 + 18 = 39$.

Example 3 Pat weighs ten pounds more than Paul and three pounds less than Pete. If the total weight of all three is 518 pounds, find the weight of each.

Solution Let

$$\begin{aligned} x + 10 &= \text{Pat's weight} \\ x &= \text{Paul's weight} \\ x + 13 &= \text{Pete's weight.} \end{aligned}$$

Then

$$(x + 10) + x + (x + 13) = 518.$$

Solving, we obtain

$$\begin{aligned} x &= 165 \\ x + 10 &= 175 \\ x + 13 &= 178. \end{aligned}$$

The answers are

$$\begin{aligned} \text{Pat's weight} &= 175 \text{ lb} \\ \text{Paul's weight} &= 165 \text{ lb} \\ \text{Pete's weight} &= 178 \text{ lb.} \end{aligned}$$

Check: Pat's weight (175 lb) is ten more than Paul's (165 lb) and three less than Pete's (178 lb). The sum is $175 + 165 + 178 = 518$.

The field of business contains number relation problems.

Example 4 A customer purchases a stereo radio for \$306.80, including a 4% sales tax. Find the price of the radio and the amount of sales tax.

Solution If x represents the price of the radio, then the sales tax is found by multiplying the price of the radio by 4%.

Thus we have

$$x = \text{price of the radio}$$
$$.04x = \text{sales tax. (Recall } 4\% = .04)$$

Then the total sale is the sum of the price of the radio and the sales tax.

$$x + .04x = 306.80$$

Solving the equation, we obtain

$$1.04x = 306.80$$
$$x = 295.$$

Then $$.04x = 11.80.$$

The answers are

$$\text{price of the radio} = \$295.00$$
$$\text{sales tax} = \$11.80.$$

Check: The total of \$295.00 and \$11.80 is \$306.80.

Some number relation problems require prior knowledge of a formula to set up the equation.

Example 5 In a certain rectangle the length is two more than the width. If the length is increased by three and the width by two, the perimeter of the new rectangle will be twice that of the original. Find the length and width of the original rectangle.

Solution Notice that to solve the problem you must know that

$$P = 2\ell + 2w$$

is the formula for the perimeter of a rectangle. We outline this as

original rectangle	*new rectangle*
$\ell = x + 2$	$\ell = (x + 2) + 3$
$w = x$	$w = x + 2$
$P_o = 2(x + 2) + 2x$	$P_n = 2[(x + 2) + 3] + 2(x + 2)$.

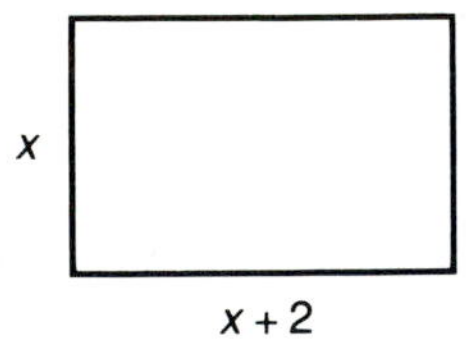

x

x + 2

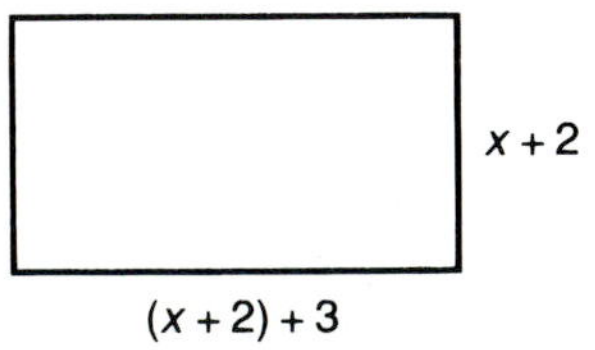

x + 2

(x + 2) + 3

Notice that we used **subscripts** o and n to distinguish the original perimeter from the new perimeter.

The equation comes from the statement that $P_n = 2P_o$.

$$\begin{aligned}
2[(x + 2) + 3] + 2(x + 2) &= 2[2(x + 2) + 2x] \\
2[x + 5] + 2(x + 2) &= 2[2x + 4 + 2x] \\
2x + 10 + 2x + 4 &= 2(4x + 4) \\
4x + 14 &= 8x + 8 \\
-4x + 14 &= 8 \\
-4x &= -6 \\
x &= \frac{-6}{-4} = \frac{3}{2}
\end{aligned}$$

Since x represents the width, we have

$$w = \frac{3}{2}.$$

Also

$$\begin{aligned}
\ell &= x + 2 \\
&= \frac{3}{2} + 2 \\
&= \frac{7}{2}.
\end{aligned}$$

Check:

$$\begin{aligned}
P_o &= 2\ell + 2w \\
&= 2\left(\frac{7}{2}\right) + 2\left(\frac{3}{2}\right) \\
&= 10
\end{aligned}
\qquad
\begin{aligned}
P_n &= 2\left(\frac{7}{2} + 3\right) + 2\left(\frac{3}{2} + 2\right) \\
&= 13 + 7 \\
&= 20
\end{aligned}$$

Thus the perimeter of the new rectangle is twice that of the original rectangle, and the answer checks.

Example 6 An individual invests \$8,000 for one year. Part is invested at 6% and the rest at 5%. If the total interest at the end of the year is \$460, how much was invested at each rate?

Solution Here we must use the interest formula

$$I = prt,$$

where I represents the interest, p the principal (amount invested), r the rate of interest, and t the time in years. We let x = amount invested at 5% and $8{,}000 - x$ = amount invested at 6%. Then the interest earned at 5% is

$$I = prt = (.05)(x)(1).$$

The interest earned at 6% is

$$I = prt = (.06)(8000 - x)(1).$$

The equation representing the total interest is

$$.05x + .06(8000 - x) = 460.$$

Solving the equation gives

$$
\begin{aligned}
.05x + 480 - .06x &= 460 \\
-.01x + 480 &= 460 \\
-.01x &= -20 \\
x &= 2{,}000 \\
8{,}000 - x &= 6{,}000
\end{aligned}
$$

Thus \$2,000 was invested at 5% and \$6,000 at 6%.

Check: 5% of \$2,000 is \$100 and 6% of \$6,000 is \$360. The two interest amounts total \$460.

Example 7 The interest due on a two-year \$300 loan amounted to \$90. What annual rate of interest was charged?

Solution Again the formula $I = prt$ is useful. Here $I = \$90$, $p = \$300$, and $t = 2$ years. We are asked to find r where r = annual rate of interest.

$$
\begin{aligned}
I &= prt \\
90 &= (300)r(2) \\
\frac{90}{600} &= \frac{600r}{600} \\
.15 &= r
\end{aligned}
$$

Thus the rate of interest was 15%.

Check: \$300 at 15% for 2 years yields \$90 interest.

EXERCISE 3–3–1

Solve.

1. A second number is five more than the first number. The sum of the two numbers is 83. Find the two numbers.

2. Jane is five years older than Louis. The sum of their ages is 41. Find the age of each.

3. One number is three more than five times another. The sum of the two numbers is 75. Find the numbers.

4. In 1979 Roger Staubach of the Dallas Cowboys completed 36 more passes than he did in 1978. If his total number of completed passes for both years was 498, how many passes did he complete in each year?

5. The length of a rectangle is 20 centimeters more than its width. Find the length and width if the perimeter is 224 centimeters.

6. The Amazon River is three times as long as the Ohio-Allegheny River. Find the length of each river if the difference in their lengths is 2,610 miles.

7. A student purchases a notebook and a calculator for $14.45. Another student purchases two notebooks and a calculator for $18.95. Find the price of a calculator and the price of a notebook.

8. A certain number is 12 more than a second number and 20 less than a third number. Find the numbers if their sum is 152.

9. Three consecutive even integers have a sum of 90. Find the integers.

10. An item sells for a certain price. If a 5% sales tax is added to the price, the total amount of the sale is $28.14. Find the original selling price of the item and the sales tax.

11. Town A has a population ten percent greater than town B. The total population of the two towns is 49,245. Find the population of each.

12. Alice is 15 centimeters taller than Jane and 3 centimeters shorter than Diane. The sum of their heights is 489 centimeters. How tall is each?

13. The selling price of car B is $1,411 more than car A and $1,922 more than car C. The total sales price of all three cars is $20,163. Find the price of each.

14. Jim weighs five kilograms less than Tom and two kilograms less than Bob. Find the weight of each person if their total weight is 223 kilograms.

15. Copy machine A produces eight more copies per minute than copy machine B. If together they produce 32 copies per minute, find the number each produces.

16. Three cyclists traveled a total distance of 223 kilometers. Cyclist A traveled five kilometers farther than cyclist B and three kilometers farther than cyclist C. Find the distance traveled by each cyclist.

17. In a given rectangle the length is twice the width. If the length is increased by seven and the width is increased by eight, the perimeter is doubled. Find the dimensions of the original rectangle.

18. An individual makes two investments that total \$10,000. One investment is at 11% and the other at 13%. If the total interest for one year is \$1,190, find the amount invested at each rate.

19. The base of a given triangle is ten. If the length of the altitude is increased so that it is four more than twice the original, and the base is decreased by two, then the area of the new triangle is twice the area of the original. Find the altitude of the original triangle. $\left(\text{Hint: Use } A = \frac{1}{2}bh.\right)$

20. A loan of \$8,500 was made for four years. The total interest paid was \$4,760. What annual rate of interest does this represent?

3–4 DISTANCE-RATE-TIME PROBLEMS

OBJECTIVES

Upon completion of this section you should be able to:

1. Identify distance-rate-time problems.
2. Apply the distance formula to solve problems in this group.

Another formula often found in word problems is

$$d = rt \text{ (distance = rate} \times \text{time)},$$

which is the **distance formula** for constant or average rate. In solving problems involving distance we generally use the substitution principle and substitute directly into the formula.

Given the rate at which an object is moving and the time that it moves at this rate, we can find the distance the object moves.

Example 1 A car travels at the constant rate of 55 miles per hour for four hours. How far does it travel?

Solution Using the formula $d = rt$, substitute $r = 55$ and $t = 4$, obtaining

$$d = (55)(4) = 220.$$

Thus the distance traveled is 220 miles.

If the distance and rate are both given, we can find the time.

Example 2 How long will it take a plane whose ground speed is 530 miles per hour to travel 2,120 miles?

Solution In this case let $d = 2{,}120$ and $r = 530$ in the formula

$$d = rt$$
$$2120 = 530t$$
$$t = 4.$$

It will therefore take 4 hours to travel the distance.

We can also solve for the rate if we are given the distance and time.

Example 3 A person walked 21 miles in $3\frac{1}{2}$ hours. What was the person's average rate?

Solution We substitute $d = 21$ and $t = 3.5$ in the formula $d = rt$, obtaining

$$21 = 3.5r$$
$$r = 6.$$

The person's average rate was 6 mph.

One type of distance problem involves two objects leaving from the same point and traveling in the same direction.

Example 4 A bank robber leaves town heading north at an average speed of 120 kilometers per hour. The sheriff leaves two hours later in a plane that travels at 200 kilometers per hour. How long will it take the sheriff to catch the robber?

Solution In this problem we are given the rate of each person and asked to find the time the sheriff will need to catch the robber. From the fact that the robber has a two hour head start we let

$$x = \text{time for the sheriff to catch the robber}$$
$$x + 2 = \text{time of the robber.}$$

This leads to the following outline using the formula $d = rt$. A table such as this is often very useful.

	r ·	t =	d
Robber	120	$2 + x$	$120(2 + x)$
Sheriff	200	x	$200x$

Notice that the distance $120(x + 2)$ and $200x$ come from the formula $d = rt$. The equation comes from the fact that the sheriff and the robber will have traveled the same distance when the sheriff catches the robber. Thus

$$\begin{aligned} 120(x + 2) &= 200x \\ 120x + 240 &= 200x \\ 80x &= 240 \\ x &= 3 \text{ hours.} \end{aligned}$$

The sheriff will catch the robber in 3 hours.

Check: The bank robber has traveled for five hours at 120km/hr for a distance of

$$5(120) = 600 \text{ km.}$$

The sheriff has traveled for three hours at 200 km/hr for a distance of

$$3(200) = 600 \text{ km.}$$

Another type of distance problem involves two objects leaving from the same point and traveling in opposite directions.

Example 5 Pamela and Sue start at the same point and walk in opposite directions. The rate at which they are moving away from one another is 11 miles per hour. At the end of three hours Pamela stops and Sue continues to walk for another hour. At the end of that time Pamela has walked twice as far as Sue. How far apart are they?

Solution First we note that we are asked to find the total distance between the two girls. The information in the problem gives us the time of each person, a total for their rates, and a comparison of their distances. If we let

$$x = \text{Sue's distance}$$
$$\text{then} \quad 2x = \text{Pamela's distance}$$

and we can use the following outline.

	r ·	t =	d
Pamela	$\frac{2}{3}x$	3	$2x$
Sue	$\frac{x}{4}$	4	x

The sum of their rates is 11 mph.

$$\frac{2}{3}x + \frac{x}{4} = 11$$

Solving, we obtain

$$x = 12$$
$$2x = 24.$$

Thus they are 36 miles apart.

Check: Pamela's rate is $\frac{2}{3}(12) = 8$ mph. Sue's rate is $\frac{12}{4} = 3$ mph. Their total rate is 11 mph. Also Pamela's distance (24) is twice Sue's distance (12).

Another approach to the same problem is to let

$$x = \text{Sue's rate.}$$
$$\text{Then} \quad 11 - x = \text{Pamela's rate.}$$

This approach leads to the following outline.

	r ·	t =	d
Sue	x	4	$4x$
Pamela	$11 - x$	3	$3(11 - x)$

Now from the fact that Pamela's distance is twice that of Sue we have the equation

$$3(11 - x) = 2(4x).$$

Solving, we obtain

$$\begin{aligned} 33 - 3x &= 8x \\ x &= 3. \end{aligned}$$

In this instance we must be very careful to note that 3 is Sue's rate and we are asked to find the total distance. So we must find d for each person and add them.

$$\begin{aligned} \text{Sue: } d &= 4x = 4(3) = 12 \text{ miles} \\ \text{Pamela: } d &= 3(11 - x) = 3(11 - 3) = 3(8) = 24 \text{ miles} \end{aligned}$$

The sum of the two distances is 36 miles.

It is important to point out that the problem can be solved correctly if the outline shows the proper relationship between the unknowns and we check to make sure the question asked has been answered.

Still another type of distance problem involves two objects that leave from two different points and travel toward each other.

Example 6 Juan and Steven started 36 miles apart and walked toward each other, meeting in three hours. If Juan walked two miles per hour faster than Steven, find the rate of each.

Solution Here we must use the statement that Juan walked two miles per hour faster than Steven to let

$$\begin{aligned} x &= \text{Steven's rate and then} \\ x + 2 &= \text{Juan's rate.} \end{aligned}$$

This leads to the following outline.

	r ·	t =	d
Juan	$x + 2$	3	$3(x + 2)$
Steven	x	3	$3x$

The total distance they have traveled is 36 miles. Thus

$$3(x + 2) + 3x = 36.$$

Solving, we obtain

$$\begin{aligned} 3x + 6 + 3x &= 36 \\ 6x &= 30 \\ x &= 5 \\ x + 2 &= 7. \end{aligned}$$

The answers are

$$\begin{aligned} \text{Juan's rate} &= 7 \text{ mph} \\ \text{Steven's rate} &= 5 \text{ mph.} \end{aligned}$$

Check: Juan's rate (7) is two miles per hour faster than Steven's (5). Also, $3(7) + 3(5) = 21 + 15 = 36$

Try making an outline for example 6 using x = Juan's distance and $36 - x$ = Steven's distance. The final result should be the same.

Within the class of problems using $d = rt$ is a subclass of problems concerned with **parallel** and **opposing forces.**

Example 7 A plane whose speed in still air is 550 miles per hour, flies against a headwind of 50 miles per hour. How long will it take to travel 1,500 miles?

Solution The speed of the plane in still air (550 mph) is reduced by the speed of the headwind (50 mph). We use the formula $d = rt$ to solve this problem.

$$\begin{aligned} d &= rt \\ 1{,}500 &= (550 - 50)t \\ 1{,}500 &= 500t \\ t &= 3 \end{aligned}$$

Thus it will take 3 hours to travel the distance.

Check: $3(550 - 50) = 3(500) = 1{,}500$

Example 8 Mike can row his boat from the hunting lodge upstream to the park in five hours. He can row back from the park to the lodge downstream in three hours. If Mike can row x kilometers per hour in still water, and if the stream is flowing at the rate of two kilometers per hour, how far is it from the lodge to the park?

Solution In working the problem we assume that the rate of the stream will increase or decrease the rate of the boat by two kilometers per hour.

If we let x = Mike's speed in still water then $x - 2$ = his upstream rate and $x + 2$ = his downstream rate. This leads to the following outline.

	r	$\cdot$ t	$=$ d
Upstream	$x - 2$	5	$5(x - 2)$
Downstream	$x + 2$	3	$3(x + 2)$

Setting the distance upstream equal to the distance downstream, we obtain

$$\begin{aligned} 5(x - 2) &= 3(x + 2) \\ 5x - 10 &= 3x + 6 \\ 2x - 10 &= 6 \\ 2x &= 16 \\ x &= 8. \end{aligned}$$

Notice that 8 is *not* the solution to the problem, but is Mike's rate of rowing in still water. However, the question asked is, "What is the distance from the lodge to the park?" To answer this use either the distance upstream or the distance downstream since they are the same. Using the upstream distance, we have

$$5(x - 2) = 5(8 - 2) = 30 \text{ km.}$$

Exercise 3–4–1

Solve.

1. A train travels at the rate of 88 miles per hour for three hours. How far has it traveled?

2. If a person walks at the rate of five miles per hour, how long will it take to walk ten miles?

3. A cyclist traveled 57 miles in three hours. What was the average speed?

4. Sheila Young, of the United States, skated 500 meters in 42.76 seconds. What was her average rate of speed in meters per second? (Give answer to nearest hundredth.)

5. Radio waves travel at the speed of light (192,000 miles per second). If it takes 1.24 seconds for a radio wave to travel from the Moon to Earth, how far is the Moon from Earth?

6. A car leaves a certain point and travels at the rate of 60 kilometers per hour. Another car leaves from the same point two hours later and travels the same route at 80 kilometers per hour. How long will it take the second car to catch the first car?

7. Two cyclists leave the same point at the same time and travel in opposite directions. One cyclist travels at the rate of 20 miles per hour and the other at 25 miles per hour. How long will it take for them to be 90 miles apart?

8. Frank and Mike live 22 miles apart. If each leaves his home at the same time and walks toward the other, Frank at a rate of six miles per hour and Mike at five miles per hour, how long will it take for them to meet?

9. A car leaves a certain point and travels east at 70 kilometers per hour. Three hours later a second car leaves the same point and travels west at 85 kilometers per hour. How long will the second car have to travel for the two cars to be 520 kilometers apart?

10. Race-car driver Al Unser is maintaining an average speed of 104 miles per hour while A. J. Foyt is two miles behind him maintaining an average speed of 108 miles per hour. How long will it take Foyt to catch Unser?

11. A train leaves a station and travels at the rate of 100 kilometers per hour. A second train leaves the same station two hours later and travels the same route at 120 kilometers per hour. How long after the second train leaves will the two trains be 60 kilometers apart?

12. A plane leaves New York bound for Houston traveling at the rate of 654 kilometers per hour. Thirty minutes later another plane leaves New York also bound for Houston traveling at the rate of 763 kilometers per hour. If both planes arrive at Houston at the same time, what is the distance between New York and Houston?

13. A man starts at a certain point and walks due east. One hour later another man starts at the same point and rides a bicycle due west, traveling nine miles per hour faster than the walker. After the cyclist has ridden for three hours they are 69 miles apart. What was the average rate of each?

14. Joan and Sally live five miles apart. They decide to meet for lunch in one-half hour at a restaurant that is located between them. They leave their homes at the same time and Joan walks at a rate two miles per hour faster than Sally. How fast does each walk if they both arrive at the restaurant in exactly the one-half hour?

15. Two towns, A and B, are 486 kilometers apart. A car leaves each town at the same time and travels toward the opposite town. The cars meet in three hours. If the car from town B traveled at eight kilometers per hour faster than the car from town A, what was the speed of each car?

16. A stream is flowing at the rate of four kilometers per hour. A motorboat that can travel at thirty-two kilometers per hour in still water runs downstream. How long will it take to travel 99 kilometers?

17. A plane travels a certain distance with a 30 kilometer per hour tailwind in three hours and returns against the wind in three hours, thirty minutes. Find the speed of the plane in still air.

18. A plane flew for $3\frac{1}{2}$ hours with a 30 miles per hour tailwind. It took five hours for the return trip. Find the speed of the plane in still air.

19. A motorboat travels upstream in five hours and makes the return trip downstream in three hours. If the rate of the stream's current is ten kilometers per hour, find the rate of the boat in still water.

20. A motorboat traveling at the rate of thirty kilometers per hour in still water takes four hours to go upstream and three hours to return. Find the rate of the current.

3-5 MIXTURE PROBLEMS

**BJECTIVES**

Upon completion of this section you should be able to:

1. Identify mixture problems.
2. Construct an outline that allows for easy comparison of items and the formation of an equation.
3. Find the solution to mixture problems.

The final type of problem we will discuss in this chapter is the mixture problem. Mixture problems come in various settings. They may involve mixing candy, coffee, coins, liquids, and so on. Regardless of the items being mixed together, we must again be sure that our outline and equation properly relate the facts given in the problem.

Example 1 A merchant has one brand of coffee that sells for \$3.40 per pound and a second brand that sells for \$4.20 per pound. The merchant wishes to make a blend of the two coffees. How many pounds of each price coffee must be used to make 20 pounds of a mixture that would sell for \$3.60 per pound?

Solution If we use the fact that the merchant wishes to obtain 20 pounds of the mixture and let $x =$ the number of pounds of one brand, then $20 - x =$ the number of pounds of the second brand. This leads to the following outline.

Type	Pounds ·	Price =	Total Value
First brand	x	3.40	$3.40x$
Second brand	$20 - x$	4.20	$4.20(20 - x)$
Mixture	20	3.60	$3.60(20)$

Looking at the "total value" column of the table, notice that the sum of the total values of the two brands must equal the total value of the mixture. This fact gives us the equation

$$3.40x + 4.20(20 - x) = 3.60(20).$$

Solving, we obtain

$$\begin{aligned} 3.40x + 84 - 4.20x &= 72 \\ -0.8x + 84 &= 72 \\ -0.8x &= -12 \\ x &= 15 \\ 20 - x &= 5. \end{aligned}$$

Thus the merchant must use 15 pounds of the first brand and 5 pounds of the second brand.

Check: $3.40(15) + 4.20(5) = 72.00$ and $72.00 \div 20 = 3.60$

Example 2 A total of 100 coins in nickels and quarters has a value of \$14.40. How many of each type of coin are there?

Solution We are given that there are 100 coins, so if we let $x =$ the number of nickels, then $100 - x =$ the number of quarters. We must also know that nickels are valued at five cents each and quarters are valued at twenty-five cents each and we must express these values as decimals. This leads to our outline for the problem. Again, a table format is helpful.

Type	Number	· Value	= Total Value
Nickels	x	.05	$.05x$
Quarters	$100 - x$	.25	$.25(100 - x)$
Mixture	100		14.40

Again we see that the sum of the total value of nickels and the total value of quarters is equal to the total value of the mixture (\$14.40). We therefore write the equation

$$.05x + .25(100 - x) = 14.40.$$

Solving, we obtain

$$\begin{aligned} .05x + 25 - .25x &= 14.40 \\ -.20x + 25 &= 14.40 \\ -.20x &= -10.6 \\ x &= 53 \\ 100 - x &= 47 \end{aligned}$$

Thus there are 53 nickels and 47 quarters.

Check: $.05(53) + .25(47) = 2.65 + 11.75 = 14.40$

Example 3 A merchant has three types of hard candy. The first sells for \$1.10 per kilogram, the second for \$1.50 per kilogram, and the third for \$1.60 per kilogram. How many kilograms of each candy would be used to make 100 kilograms that would sell for \$1.36 per kilogram, if twice as much of the second type must be used as the third?

Solution To outline this problem we must note the relationship between the second and third types of candy as well as the total number of kilograms of candy. If we let $x =$ the number of kilograms of the third type of candy, then $2x =$ the number of kilograms of the second type. The fact that we have a total of 100 kilograms of candy leaves the first type to contain $[100 - (x + 2x)]$ or $100 - 3x$ kilograms of candy. This leads to the following outline.

Type	Kg $\cdot$	Price per Kg $=$	Total Value
First	$100 - 3x$	1.10	$1.10(100 - 3x)$
Second	$2x$	1.50	$1.50(2x)$
Third	x	1.60	$1.60(x)$
Mixture	100	1.36	136.00

Notice that the equation must come from the value column. Any equation must have like quantities on each side. You can *never* equate pounds with value, number of coins with the dollar value of the coins, and so forth.

The equation is

$$1.10(100 - 3x) + 1.50(2x) + 1.60(x) = 136.$$

Solving, this gives

$$\begin{aligned} 110 - 3.3x + 3x + 1.60x &= 136 \\ 1.3x + 110 &= 136 \\ 1.3x &= 26 \\ x &= 20. \end{aligned}$$

Also $2x = 40$

and $100 - 3x = 40.$

Thus the merchant would use

40 kg of the first type
40 kg of the second type
20 kg of the third type.

Check:

40 kg at \$1.10	= \$ 44
40 kg at \$1.50	= \$ 60
20 kg at \$1.60	= \$ 32
100 kg at \$1.36	= \$136

Example 4 How much pure alcohol must be added to nine quarts of water to obtain a mixture of 40% alcohol?

Solution Here we are adding pure alcohol to 9 quarts of water. If we let x = number of quarts of pure alcohol to be added, then $x + 9$ = the total number of quarts in the resulting mixture.

We set up the table.

	Quarts ·	% Alcohol =	Amount of Alcohol
Alcohol	x	100%	$1x$
Mixture	$x + 9$	40%	$.40(x + 9)$

Notice that the amount of alcohol in the mixture will be equal to the amount added. We can therefore write the equation

$$x = .40(x + 9).$$

Solving, we obtain

$$\begin{aligned} x &= .04x + 3.6 \\ .60x &= 3.6 \\ x &= 6. \end{aligned}$$

Therefore six quarts of alcohol must be added.

Check: If we add 6 quarts to 9 quarts there will be a total of 15 quarts in the mixture, of which 6 quarts, or 40%, is alcohol.

Example 5 The capacity of a car's radiator is nine liters. The mixture of antifreeze and water is 30% antifreeze. How much of the mixture in the radiator must be drawn off and replaced with pure antifreeze to raise the percentage of the mixture to 65% antifreeze?

Solution In outlining this problem again keep in mind that the equation must equate like quantities. We can choose to make an equation equating antifreeze with antifreeze, or water with water, but *not* water with antifreeze. If we choose to equate antifreeze with antifreeze, we proceed as follows. If x represents the number of liters of the mixture to be drawn off, we obtain the following outline.

	Liters ·	% Antifreeze =	Liters of Antifreeze
Originally	9	30%	$.30(9)$
Drawn off	x	30%	$.30(x)$
Added back	x	100%	$1.00(x)$
Final	9	65%	$.65(9)$

Note that the original amount of antifreeze, less the amount of antifreeze drained off, plus the amount of antifreeze added back will equal the final amount of antifreeze. Thus

$$\begin{aligned} .30(9) - .30(x) + 1.00(x) &= .65(9) \\ 2.7 + .7x &= 5.85 \\ .7x &= 3.15 \\ x &= 4.5 \text{ liters.} \end{aligned}$$

Check: $.30(9) - .30(4.5) + 1.00(4.5) = 2.7 - 1.35 + 4.5 = 5.85$

EXERCISE 3–5–1

Solve.

1. A candy dealer wishes to mix caramels that sell for $3.50 per pound with creams that sell for $2.75 per pound to obtain 30 pounds of a mixture that will sell for $3.00 per pound. How many pounds of each must be used?

2. If peanuts sell for $2.50 per pound and cashews sell for $4.50 per pound, how many pounds of each must be used to produce 20 pounds of mixed nuts to sell for $3.25 per pound?

3. A total of 75 coins in nickels and quarters has a value of $15.15. How many coins of each type are there?

4. A total of 35 coins in dimes and quarters has a value of $5.60. How many of each type of coin are there?

5. A fruit market wishes to prepare 35 bushels of a mixture of oranges and grapefruits to sell for $6.50 per bushel. If oranges sell for $6.00 per bushel and grapefruits sell for $7.75 per bushel, how many bushels of each must be used?

6. A candy store owner sells pecans for $6.00 per kilogram and cashews for $7.50 per kilogram. A 50-kilogram mixture of the two kinds of nuts is to be made that will sell for $6.60 per kilogram. How much of each kind should be used?

7. A collection of coins has three times as many nickels as quarters. If the total value of the two types is \$8.40, find the number of each type.

8. How many liters of pure alcohol must be added to ten liters of water to obtain a mixture of 60% alcohol?

9. A college sold tickets for a play. The tickets sold at \$1.00 per child and \$2.50 per adult. If 210 tickets were sold for a total of \$450, how many of each type of ticket were sold?

10. A merchant mixes three brands of coffee that sell for \$3.20, \$3.80, and \$4.50 per pound to produce 50 pounds of a blend that will sell for \$3.69 per pound. If twice as much of the \$3.20 brand is used as the \$4.50 brand, how much of each brand is used?

11. A collection of 53 coins contains nickels, dimes, and quarters. The total value is \$7.50. If there are seven more dimes than quarters, find the number of each type of coin.

12. Mary has \$4.25 in nickels, dimes, and quarters. She has twice as many dimes as quarters. If the total number of coins is 37, find the number of each kind of coin.

13. A dairy wishes to produce 225 gallons of a "low-fat" milk, which is 2% butterfat, by adding skim milk (no butterfat) to milk that is 3.6% butterfat. How much of each type of milk must the dairy use?

14. How much water should be added to five liters of a 25% salt solution to obtain a 15% salt solution?

15. An airline sold 176 tickets on a flight. A first-class ticket costs $250, a coach fare $200, and a super saver ticket $120. If they sold eight more super saver tickets than first-class tickets and the total revenue for the fares was $33,840, how many tickets of each class did they sell?

16. A nursery wishes to prepare 100 bags of a "weed and feed" mixture by mixing fertilizer and weed killer. If the mixture is to sell for $7.80 per bag, how many bags of fertilizer that sells for $7.00 per bag and weed killer which sells for $9.00 per bag must be used?

17. How many liters of pure alcohol must be added to ten liters of 50% alcohol to obtain a mixture of 60% alcohol?

18. A nurse has twelve milliliters of a 16% solution and wishes to dilute it to a 6% solution. How much distilled water must be added to obtain the desired solution?

19. A 10% salt solution is added to a 25% salt solution to obtain ten liters of a 15% salt solution. How much of each is to be used?

20. A tank contains five liters of a 30% acid solution. How much must be drained off and replaced with a 50% acid solution to obtain five liters of 38% acid solution?

CHAPTER 3 SUMMARY

The number in brackets refers to the section of the chapter that discusses that concept.

Terminology

- An **outline** expresses all unknowns in a problem in terms of one unknown. [3–1]
- **Number relation problems** involve comparing two or more numbers. [3–3]
- **Distance-rate-time problems** involve the distance formula for a constant rate given by $d = rt$. [3–4]
- **Mixture problems** involve combining quantities of various concentrations. [3–5]

Rules and Procedures

Solving Word Problems

- Translate from words to algebraic expressions to properly outline a word problem. [3–1]
- Outline a problem completely and accurately before attempting to solve it. [3–2]
- To solve a word problem follow these steps: [3–2]
  - *STEP 1* Determine what is to be found.
  - *STEP 2* Use the given information to write expressions for the unknowns.
  - *STEP 3* Write an equation that relates the unknowns to each other.
  - *STEP 4* Solve the equation.
  - *STEP 5* Make sure you have answered the question.
  - *STEP 6* Check your answers to make sure they agree with the original problem.

CHAPTER

3 REVIEW

This chapter review contains problems representing all of the topics discussed in this chapter. You will find it easier if you classify the problem before proceeding to outline and solve it.

Write an algebraic expression for problems 1–5.

1. Twice a number, decreased by seven

2. Half a number, increased by three

3. The sum of a given number and 5

4. The value in cents of x quarters

5. The number of meters in y centimeters

6. A certain number is seven more than three times another number. Write expressions for the numbers.

7. The sum of two numbers is 84. Write expressions for the numbers.

8. Write an expression for the interest when a certain amount is invested for a year at 16%.

9. The sum of three numbers is 181. The first number is twice the third. Write expressions for the three numbers.

10. Write an expression for the total amount received at the end of one year if x dollars is invested at 17.3% interest.

11. If x represents one number and $x + 5$ represents another number, write a mathematical statement that indicates their sum is 23.

12. If one number is represented by x and a second number is represented by $3x - 1$, write a mathematical statement that indicates their sum is 19.

13. Write an equation to show that a number represented by $3x$ less a number represented by $x - 7$ is equal to 13.

14. Write an equation to show that the sum of two numbers, represented by x and $x - 6$, less a third number represented by $3x$, is equal to -11.

15. Express algebraically the fact that a certain number added to 9% of the number is 1.26.

16. One item costs $1.25 more than another item. Write an equation to show that the sum of the costs of the two items is $5.00.

17. A certain number, less a second number that is five more than three times the first number, is equal to -7. Find the numbers.

18. A certain number less 35% of that number is 312. Find the number.

19. How many kilograms of water must be evaporated from 90 kilograms of salt water that contains 5% salt to obtain a solution that is 7% salt? (Answer correct to nearest tenth of a kilogram.)

20. A certain amount of money is invested for one year at 8% interest. The total principal and interest at the end of the year is $1,620. Find the original amount invested.

21. The length of one side of a triangle is twice that of the second side and four less than that of the third side. If the perimeter is 64, find the three sides.

22. A meter stick is cut into two pieces so that one piece is six centimeters longer than the other. Find the lengths of the two pieces.

23. A man has twice as many dimes as nickels, and two more quarters than dimes. The total value of the coins is $3.50. How many of each kind of coin does he have?

24. Roger is seven years older than Dave. The sum of their ages is 63. Find their ages.

25. A grocer has three grades of hamburger: Grade A at $3.80 per kilogram, Grade B at $3.50 per kilogram, and Grade C at $3.30 per kilogram. Using equal amounts of Grades A and C, the grocer wishes to mix the three grades to obtain 50 kilograms of hamburger that will sell for $3.53 per kilogram. How many kilograms of each grade must be used?

26. Ellen is four years older than Kathy, and Barbara is two years younger than twice Kathy's age. How old is each of the women if the sum of their ages is 58?

27. A boat travels upstream and back in 12 hours. If the speed of the boat in still water is 25 kilometers per hour and the current in the river is 10 kilometers per hour, how far upstream did the boat travel?

28. At the start of the semester there were twice as many students enrolled in psychology as in algebra, and three fewer students in history than in psychology. After the first week of classes, the algebra class had doubled in enrollment, four students had dropped psychology, and five had added history. If the total enrollment after the first week of class was 148, how many students were enrolled in each class at the start of the semester?

29. Two towns, A and B, are 435 kilometers apart. A car leaves town B and travels toward town A at a speed of 65 kilometers per hour. At the same time a car leaves town A and travels toward town B. At what speed must this car travel to meet the other car in three hours?

30. A candy dealer has three types of candy selling at \$4.50 a box, \$6.95 a box, and \$8.00 a box. The dealer has four more \$6.95 boxes than \$8.00 boxes, and the \$4.50 boxes total two more than twice the number of \$8.00 boxes. The dealer wishes to mix the candy in the same total number of boxes and sell the new boxes for \$6.00 each. How many boxes of each candy does the dealer have? (Assume his total revenue remains the same.)

31. A 40-meter length of rope is cut into three pieces. The second piece is three times as long as the first, and the third piece is five meters shorter than twice the length of the second. Find the length of each piece.

32. A plane with a still air speed of 680 kilometers per hour makes an outgoing trip against a headwind in five hours and the return trip in three hours and thirty minutes. Find the speed of the headwind.

33. A girl has 107 coins consisting of nickels, dimes, and quarters. She has seven less than twice as many quarters as nickels. If the total value of the coins is \$17.15, how many of each type of coin does she have?

34. The length of a rectangle is four more than twice the width. If the width is tripled and the length is doubled, the perimeter of the new rectangle is twelve more than twice the perimeter of the original. Find the dimensions of the original rectangle.

35. A tank contains 20 liters of a 10% acid solution. How much pure acid should be added to the solution to obtain a 15% acid solution? (Answer correct to the nearest tenth of a liter.)

36. The sum of three numbers is 123. The second number is five less than twice the first. The third number is seven more than the sum of the first two numbers. Find the numbers.

37. A boat leaves a dock and travels at the rate of 36 kilometers per hour. Twenty minutes later a second boat leaves the same dock and travels the same route at 50 kilometers per hour. How long will it take the second boat to be five kilometers behind the first boat?

38. A full two-liter bottle contains a 20% alcohol solution. How much of the solution must be poured out and replaced with pure alcohol to obtain two liters of a 30% alcohol solution?

CHAPTER 3

PRACTICE TEST

1. Write an algebraic expression for each.

a. Three times a number x, increased by six

1. a. ____________

b. Half a number y, decreased by seven

1. b. ____________

c. 9.4% of a number x

1. c. ____________

d. The number of weeks in d days

1. d. ____________

2. Sally has three more dimes than nickels and twice as many quarters as dimes. If she has a total of $5.05, find how many she has of each coin.

2. ____________

3. A piece of rope that is 63 meters long is cut into three pieces such that the second piece is twice as long as the first and the third piece is three meters longer than the second. Find the length of each piece.

3. ____________

4. A man can row at the rate of 12 kilometers per hour in still water. He rows upstream for two hours and returns in one hour and twelve minutes. Find the rate of the current.

4. ______________________

5. How many liters of water must be added to 10 liters of a 12% alcohol solution to obtain an 8% alcohol solution?

5. ______________________

CHAPTER

SURVEY

The following questions refer to material discussed in this chapter. Work as many problems as you can and check your answers with the answer section in the back of the book. The results will direct you to the sections of the chapter in which you need to work. If you answer all questions correctly, you have a good understanding of the material contained in this chapter.

Multiply.

1. $(2x - 1)(x + 5)$ **1.** ____________

2. $(x - 2)(x^2 + 3x - 5)$ **2.** ____________

Factor completely.

3. $35x^2 - 28x$ **3.** ____________

4. $6x^3y^3 - 8x^3y^2 + 10x^2y^3$ **4.** ____________

5. $a(x - y) - (x - y)$ **5.** ____________

6. $ax - 3b + 3a - bx$ **6.** ____________

7. $x^2 - 13x + 40$ **7.** ____________

8. $3x^2 + 17x + 20$ **8.** ____________

9. $4x^2 - 25$

9. ______________________

10. $x^2 + 20x + 100$

10. ______________________

11. $6x^2 - 27x - 15$

11. ______________________

12. $16x^3 - 48x^2 + 36x$

12. ______________________

13. Supply the missing term so that the expression $4x^2 +$ ____ $+ 9$ will be a perfect square trinomial.

13. ______________________

14. Find the remainder when $2x^3 + 13x^2 + 11x - 14$ is divided by $(x + 5)$.

14. ______________________

15. Is $(4x - 1)$ a factor of $4x^3 - 9x^2 + 14x - 3$? Show your work.

15. ______________________

Products and Factoring

The process of factoring is essential to the simplification of many algebraic expressions and is a useful tool in solving higher degree equations. In fact, the process of factoring is so important that very little of algebra beyond this point can be accomplished without understanding it.

In earlier chapters the distinction between *terms* and *factors* has been stressed. You should remember that terms are added or subtracted and factors are multiplied. Three important definitions follow.

DEFINITION

Terms occur in an indicated sum or difference. **Factors** occur in an indicated product.

DEFINITION

An expression is in **factored form** only if the *entire* expression is an indicated product.

	In Factored Form	Not in Factored Form
Example 1	$2x(x + y)$	$2x + 3y + z$
Example 2	$(x + y)(3x - 2y)$	$2(x + y) + z$
Example 3	$(x + 4)(x^2 + 3x - 1)$	$(x + y)(2x - y) + 5$

Note in these examples that we must always regard the entire expression. Factors can be made up of terms and terms can contain factors, but *factored form* must conform to the definition.

DEFINITION

Factoring is a process of changing an expression from a sum or difference of terms to a product of factors.

Note that in this definition it is implied that the value of the expression is not changed—only its form.

4–1 FINDING COMMON FACTORS OF A POLYNOMIAL

OBJECTIVES

Upon completion of this section you should be able to:

1. Determine which factors are common to all terms in a polynomial expression.
2. Write the expression in factored form.

In chapter 1 we multiplied an expression such as $5(2x + 1)$ to obtain $10x + 5$. In general, factoring will "undo" multiplication. Each term of $10x + 5$ has 5 as a factor, and $10x + 5 = 5(2x + 1)$.

To factor a polynomial expression by finding common factors, proceed as in the following example.

Example 1 Factor $3x^2 + 6xy + 9xy^2$.

Solution First list the factors of each term.

$3x^2$ has factors $1, 3, x, x^2$, $\textcircled{3x}$, and $3x^2$
$6xy$ has factors $1, 2, 3, 6, x, 2x$, $\textcircled{3x}$, $6x, y$, and so on.
$9xy^2$ has factors $1, 3, 9, x$, $\textcircled{3x}$, $9x, xy, xy^2$, and so on.

Next look for factors that are common to all terms, and search out the greatest of these. This is the **greatest common factor.** In this case, the greatest common factor is $3x$.

Proceed by placing $3x$ before a set of parentheses.

$$3x(\qquad\qquad)$$

The terms within the parentheses are found by *dividing* each term of the original expression by $3x$.

$$3x^2 + 6xy + 9xy^2 = 3x(x + 2y + 3y^2)$$

The original expression is now changed to factored form. To check the factoring keep in mind that factoring changes the *form* but not the value of an expression. If the answer is correct, it must be true that $3x(x + 2y + 3y^2) = 3x^2 + 6xy + 9xy^2$. Multiply to see that this is true. A second check is also necessary for factoring—we must be sure that the expression has been **completely factored.** In other words, "Did we find all common factors? Can we factor further?"

In the example $3x^2 + 6xy + 9xy^2$, if we had only used the factor 3, the answer would be

$$3(x^2 + 2xy + 3xy^2).$$

Multiplying to check, we find the answer is actually equal to the original expression. However, the factor x is still present in all terms. Hence the expression is not completely factored.

RULE

For factoring to be correct the solution must meet two criteria.

1. It must be possible to multiply the factored expression and get the original expression.
2. The expression must be *completely* factored.

Example 2 Factor $12x^3 + 6x^2 + 18x$.

Solution At this point it should not be necessary to list the factors of each term. You should be able to mentally determine the greatest common factor. A good procedure to follow is to think of the elements individually. In other words, don't attempt to obtain all common factors at once but get first the number, then each letter involved. For instance, 6 is a factor of 12, 6, and 18, and x is a factor of each term. Hence $12x^3 + 6x^2 + 18x = 6x(2x^2 + x + 3)$. Multiplying, we get the original and can see that the terms within the parentheses have no other common factor, so we know the solution is correct.

Example 3 Factor $a^2b^2c + 2ab^2c^2 - 3ab^3c$.

Solution We note that a, b^2, and c are common factors.

Hence $$a^2b^2c + 2ab^2c^2 - 3ab^3c = ab^2c(a + 2c - 3b).$$

Check: $ab^2c(a + 2c - 3b) = a^2b^2c + 2ab^2c^2 - 3ab^3c$

Example 4 Factor $2x - 4y + 2$.

Solution The only common factor is 2.

$$2x - 4y + 2 = 2(x - 2y + 1)$$

Example 5 Factor $5x^3y + 10x^2y^2 + 5xy^2$.

Solution $$5x^3y + 10x^2y^2 + 5xy^2 = 5xy(x^2 + 2xy + y)$$

DEFINITION

If an expression cannot be factored it is said to be **prime.**

Example 6 Factor $3x^2y + 5x + 2y^2$.

Solution Since there is no common factor (except 1), this expression is *prime.*

EXERCISE 4–1–1

Factor each of the following by finding all common factors.

1. $12a + 10b$ **2.** $12x - 20y$ **3.** $a^2 + 5a$

4. $x^2 - xy$

5. $2a^2 + 6a$

6. $15x^2 - 35x$

7. $x^2 - x$

8. $6a^2 + 3a$

9. $8x^2y + 12xy$

10. $x^3y - x^2y^2$

11. $10a^2b - 2ab^2 + 6ab$

12. $15x^4y - 20x^2y^2 + 5x^2y$

13. $3x + 4y - 5z$

14. $6x + 21y - 27z$

15. $14a^2b - 35ab - 63a$

16. $12x^3y^2 + 20x^2y + 28x^2y^2$

17. $11a^3b^4 + 44ab^3 - 33a^2b^4$

18. $3a - 27b + 9c$

19. $4a^2 - 29b + 10ab$

20. $27x^3y^2 + 18x^2y^2 - 36x^2y^3$

4-2 THE PRODUCT OF POLYNOMIALS

OBJECTIVES

Upon completion of this section you should be able to:

1. Multiply two binomials using the shortcut method.
2. Multiply a binomial and trinomial.
3. Multiply two polynomials.

In this section we wish to learn how to find the product of any two polynomials. We will first turn our attention to the product of two binomials.

As defined earlier, a binomial is a polynomial having two terms. We will use the distributive property $a(b + c) = ab + ac$ to establish a useful pattern for the product of two binomials.

Given the problem

$$\text{“expand } (2x + 3)(x + 4)\text{”}$$

we proceed as follows. Using the distributive law as a formula, let

$$a = (2x + 3)$$
$$b = x$$
$$c = 4.$$

Then $a(b + c) = ab + ac$ will be

$$(2x + 3)(x + 4) = (2x + 3)x + (2x + 3)4.$$

Using the distributive law again on each of these terms, we obtain

$$2x^2 + 3x + 8x + 12.$$

Now combining similar terms yields

$$(2x + 3)(x + 4) = 2x^2 + 11x + 12.$$

By inspecting this final answer in relationship to the problem, we can establish a pattern that becomes a very important shortcut.

1. $(2x + 3)(x + 4)$ The product of the first terms in each parentheses is $2x^2$, which is the first term of the answer.
2. $(2x + 3)(x + 4)$ The product of the outer terms of each parentheses is $8x$.
3. $(2x + 3)(x + 4)$ The product of the inner terms in each parentheses is $3x$.
4. $(2x + 3)(x + 4)$ The product of the last terms in each parentheses is 12, which is the last term of the answer.

When the outer and inner products are similar, they combine to give the middle term of the answer. In this case, $8x + 3x = 11x$.

This method of multiplying two binomials is sometimes called the FOIL method. FOIL stands for the First, Outer, Inner, and Last terms in each parentheses.

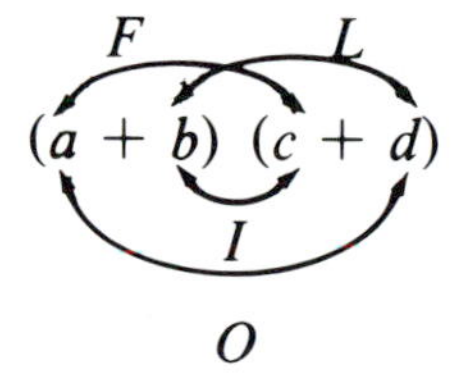

Example 1 Expand $(3x + y)(2x - y)$.

Solution $(3x + y)(2x - y) = 6x^2 - 3xy + 2xy - y^2 = 6x^2 - xy - y^2$

(F, O, I, L)

If the product of the two outside terms is not similar to the product of the two inside terms, our answer would contain four terms instead of three.

Example 2 Expand $(2a + b)(3a + c)$.

Solution $(2a + b)(3a + c) = 6a^2 + 2ac + 3ab + bc$

(F, O, I, L)

This pattern is extremely important since it is the basis for factoring. It is important that you be able to use this pattern rapidly and correctly to develop a satisfactory factoring ability.

EXERCISE 4-2-1

Expand.

1. $(a + 3)(a + 2)$

2. $(a + 1)(a + 5)$

3. $(x + 5)(x + 2)$

4. $(x - 1)(x - 4)$

5. $(a - 3)(a - 2)$

6. $(a - 4)(a - 2)$

7. $(a - 3)(a - 5)$

8. $(x + 2)(x - 3)$

9. $(x - 1)(x + 5)$

10. $(x + 7)(x - 2)$

11. $(a - 5)(a + 9)$

12. $(2x + 1)(x + 2)$

13. $(3x + 2)(2x + 1)$

14. $(a + 2)(3a - 5)$

15. $(2a - 1)(3a + 4)$

16. $(5x + 2)(x - 4)$

17. $(3x - 5)(2x - 3)$

18. $(2x + y)(3x + 5y)$

19. $(2a - b)(5a + 3)$

20. $(3x + 1)(x + 7y)$

21. $(a - 5)(4a + c)$

22. $(x + 2)(x + 2)$

23. $(x + 5)(x + 5)$

24. $(x + 3)^2$

25. $(x - 7)^2$

26. $(x - 1)^2$

27. $(x + 3)(x - 3)$

28. $(x - 5)(x + 5)$

29. $(2x + 7)(2x - 7)$

30. $(a + b)(a - b)$

We should mention here that the distributive property can be used to find the product of any two polynomials. This is illustrated in the following examples.

Example 3 Expand $(x + 2)(x^2 - 3x + 4)$.

Solution We apply the distributive property by multiplying each term of the trinomial $(x^2 - 3x + 4)$ by $(x + 2)$, obtaining

$$x^2(x + 2) - 3x(x + 2) + 4(x + 2)$$

or

$$x^3 + 2x^2 - 3x^2 - 6x + 4x + 8.$$

Combining similar terms, we obtain

$$x^3 - x^2 - 2x + 8$$

as the final product.

Example 4 Expand $(x - 2)(x^2 + 2x + 4)$.

Solution Proceeding as in example 3, we obtain

$$\begin{aligned}(x - 2)(x^2 + 2x + 4) &= x^2(x - 2) + 2x(x - 2) + 4(x - 2)\\ &= x^3 - 2x^2 + 2x^2 - 4x + 4x - 8\\ &= x^3 - 8.\end{aligned}$$

EXERCISE 4–2–2

Expand.

1. $(x + 5)(x^2 + 3x + 1)$

2. $(2x - 1)(x^2 - 4x + 3)$

3. $(5a + b)(3a^2 + ab - 4b^2)$

4. $(x + 3)(x^2 - 3x + 9)$

5. $(x - 6)(2x^2 - x + 1)$

6. $(x + 1)(x^3 + 4x^2 - x + 3)$

7. $(x - 5)(x^2 + 5x + 25)$

8. $(2a - b)(4a^2 + 2ab + b^2)$

9. $(x^2 + xy + 1)(x - xy + 5)$

10. $(x + y + 4)(x - y - 2)$

4-3 FACTORING BY GROUPING

OBJECTIVES

Upon completion of this section you should be able to:

1. Factor expressions when the common factor involves more than one term.
2. Factor by grouping.

An extension of the ideas presented in section 4–1 applies to a method of factoring called **grouping.**

First we must note that a common factor does not need to be a single term. For instance, in the expression $2y(x + 3) + 5(x + 3)$ we have two terms. They are $2y(x + 3)$ and $5(x + 3)$. In each of these terms we have a factor $(x + 3)$ that is made up of terms. This factor $(x + 3)$ is a common factor.

Example 1 Factor $2y(x + 3) + 5(x + 3)$.

Solution Since $(x + 3)$ is a common factor, we have

$$2y(x + 3) + 5(x + 3) = (x + 3)(2y + 5).$$

Multiplying as in the previous section, we find

$$(x + 3)(2y + 5) = 2y(x + 3) + 5(x + 3).$$

Example 2 Factor $3x^2(x + y) + 5x(x + y)$.

Solution We see that x and $(x + y)$ are common factors.

$$3x^2(x + y) + 5x(x + y) = x(x + y)(3x + 5)$$

Example 3 Factor $5x(a + b) + (a + b)$.

Solution

$$5x(a + b) + (a + b) = (a + b)(5x + 1)$$

EXERCISE 4-3-1

Factor.

1. $2x(x + 4) + 3(x + 4)$
2. $4a(a + b) + 3(a + b)$
3. $a(a - 1) + 5(a - 1)$
4. $2x(x - 5) + 7(x - 5)$
5. $6x(y + 4) - 7(y + 4)$
6. $3a(b - c) - 4(b - c)$
7. $3x(x - 2) + (x - 2)$
8. $10x(y + 1) - (y + 1)$
9. $7a(a - 4) - 15(a - 4)$
10. $14a(b + 1) + 5(b + 1)$
11. $x(x + 3) - (x + 3)$
12. $a(a + b) + b(a + b)$
13. $3x^2(x + 1) + 2x(x + 1)$
14. $a^2(a - 1) + 3a(a - 1)$
15. $4a(a + b) + a^2(a + b)$
16. $5a(a - b) - (a - b)$
17. $x(x - 1) - (x - 1)$
18. $6a^2(a + 1) + 9(a + 1)$
19. $8a^2(a + 4) + 2a(a + 4)$
20. $15x^3(x - 3) - 12x^2(x - 3)$

Sometimes when there are four or more terms we must insert an intermediate step or two to factor.

Example 4 Factor $3ax + 6y + a^2x + 2ay$.

Solution First note that not all four terms in the expression have a common factor, but that some of them do. For instance, we can factor 3 from the first two terms, giving $3(ax + 2y)$. If we factor a from the remaining two terms, we get $a(ax + 2y)$. The expression is now $3(ax + 2y) + a(ax + 2y)$, and we have a common factor of $(ax + 2y)$ and can factor as $(ax + 2y)(3 + a)$. Multiplying $(ax + 2y)(3 + a)$, we get the original expression $3ax + 6y + a^2x + 2ay$ and see that the factoring is correct.

$$3ax + 6y + a^2x + 2ay = (ax + 2y)(3 + a)$$

This is an example of *factoring by grouping* since we "grouped" the terms two at a time.

Example 5 Factor $ax - ay + 2x - 2y$.

Solution

$$ax - ay + 2x - 2y = a(x - y) + 2(x - y) = (x - y)(a + 2)$$

Example 6 Factor $2ax + 3a + 4x + 6$.

Solution

$$2ax + 3a + 4x + 6 = a(2x + 3) + 2(2x + 3) = (2x + 3)(a + 2)$$

Sometimes the terms must first be rearranged before factoring by grouping can be accomplished.

Example 7 Factor $3ax + 2y + 3ay + 2x$.

Solution The first two terms have no common factor, but the first and third terms do, so we will rearrange the terms to place the third term after the first. Always look ahead to see the order in which the terms could be arranged.

$$\begin{aligned} 3ax + 2y + 3ay + 2x &= 3ax + 3ay + 2x + 2y \\ &= 3a(x + y) + 2(x + y) \\ &= (x + y)(3a + 2) \end{aligned}$$

In all cases it is important to be sure that the factors within parentheses are exactly alike. This may require factoring a negative number or letter.

Example 8 Factor $ax - ay - 2x + 2y$.

Solution Note that when we factor a from the first two terms we get $a(x - y)$. Looking at the last two terms, we see that factoring $+2$ would give $2(-x + y)$ but factoring -2 gives $-2(x - y)$. We want the terms within parentheses to be $(x - y)$, so we proceed in this manner.

$$ax - ay - 2x + 2y = a(x - y) - 2(x - y) = (x - y)(a - 2)$$

EXERCISE 4-3-2

Factor.

1. $ax + ay + 3x + 3y$

2. $ax + ay + 4x + 4y$

3. $ab + ac - 2b - 2c$

4. $3a - 3b + a^2 - ab$

5. $2ax + ay + 6x + 3y$

6. $2ax + 3a + 4x + 6$

7. $ax - ay + 2x - 2y$

8. $ax + 2ay - 3x - 6y$

9. $5x + 10y + ax + 2ay$

10. $2ax + 2ay - 5x - 5y$

11. $3ax - 2y - 3ay + 2x$

12. $ax + y + ay + x$

13. $2ax - y + 2ay - x$

14. $xy + 3y + 2x + 6$

15. $6ax - y + 2ay - 3x$

16. $cx - 4y + cy - 4x$

17. $3x - 2y - 6 + xy$

18. $xy - 8 - 2x + 4y$

19. $5y - 2x - 10 + xy$

20. $3y - x - 3 + xy$

4–4 FACTORING TRINOMIALS

OBJECTIVES

Upon completion of this section you should be able to:

1. Factor a trinomial having a first term coefficient of 1.
2. Find the factors of any factorable trinomial.

A large number of future problems will involve factoring trinomials as products of two binomials. In section 4–2 we established a pattern for multiplying two binomials. Using this technique, we are now ready to factor trinomials.

Example 1 Factor $x^2 + 11x + 24$.

Solution Since this is a trinomial and has no common factor we will use the multiplication pattern to factor.

First write parentheses under the problem.

$$\begin{array}{c} x^2 + 11x + 24 \\ (\quad)(\quad) \end{array}$$

We now wish to fill in the terms so that the pattern will give the original trinomial when we multiply. The first term is easy since we know that $(x)(x) = x^2$.

$$\overset{x^2}{(x \quad)(x \quad)}$$

We must now find numbers that multiply to give 24 and *at the same time* add to give the middle term. Notice that in each of the following we will have the correct first and last term.

$$\begin{aligned} (x + 1)(x + 24) &= x^2 + 25x + 24 \\ (x + 2)(x + 12) &= x^2 + 14x + 24 \\ (x + 4)(x + 6) &= x^2 + 10x + 24 \\ (x + 3)(x + 8) &= x^2 + 11x + 24 \end{aligned}$$

Only the last product has a middle term of $11x$, and the correct solution is

$$x^2 + 11x + 24 = (x + 3)(x + 8).$$

This method of factoring is called **trial and error**—for obvious reasons.

Example 2 Factor $x^2 - 11x + 24$.

Solution Here the problem is only slightly different. We must find numbers that multiply to give 24 and *at the same time* add to give -11. You should always keep the pattern in mind. The last term is obtained strictly by multiplying, but the middle term comes finally from a sum. Knowing that the product of two negative numbers is positive, but that the sum of two negative numbers is negative, we obtain

$$x^2 - 11x + 24 = (x - 8)(x - 3).$$

Example 3 Factor $x^2 - 5x - 24$.

Solution We are here faced with a negative number for the third term, and this makes the task slightly more difficult. Since -24 can only be the product of a positive number and a negative number, and since the middle term must come from the sum of these numbers, we must think in terms of a difference. We must find numbers whose product is 24 and that differ by 5. Furthermore, the larger number must be negative, because when we add a positive and negative number the answer will have the sign of the larger. Keeping all of this in mind, we obtain

$$x^2 - 5x - 24 = (x - 8)(x + 3).$$

The following points will help as you factor trinomials:

1. When the sign of the third term is positive, both signs in the factors must be alike—and they must be like the sign of the middle term.
2. When the sign of the last term is negative, the signs in the factors must be unlike—and the sign of the larger must be like the sign of the middle term.

EXERCISE 4-4-1

Factor.

1. $x^2 + 41x + 40$

2. $x^2 + 25x + 24$

3. $x^2 + 29x - 30$

4. $x^2 + 9x - 22$

5. $x^2 + 2x - 48$

6. $x^2 + 4x - 21$

7. $x^2 + 7x - 18$

8. $x^2 + 11x + 18$

9. $x^2 - 8x + 12$

10. $x^2 + x - 12$

11. $x^2 + 34x - 35$

12. $x^2 + 13x + 36$

13. $x^2 + 9x - 36$

14. $x^2 - 12x + 35$

15. $x^2 - 11x + 24$

16. $x^2 + 22x + 21$

17. $x^2 + 13x - 48$

18. $x^2 - 13x + 40$

19. $x^2 + 6x - 40$

20. $x^2 - 13x + 30$

21. $x^2 + 15x + 50$

22. $x^2 + 55x + 54$

23. $x^2 + 49x - 50$

24. $x^2 + 10x - 56$

25. $x^2 + 22x + 72$

26. $x^2 + 21x - 72$

27. $x^2 - 5x - 24$

28. $x^2 + 18x - 40$

29. $x^2 + 2x - 15$

30. $x^2 + 13x + 22$

31. $x^2 + 14x + 48$

32. $x^2 + 20x - 96$

33. $x^2 + 31x + 84$

34. $x^2 - 28x + 96$

35. $x^2 + 83x - 84$

36. $x^2 + x - 30$

37. $x^2 - 3x - 40$

38. $x^2 + 16x + 15$

39. $x^2 + 21x - 22$

40. $x^2 - 20x + 96$

41. $x^2 + 8x - 84$

42. $x^2 + 131x + 130$

43. $x^2 - 12x - 28$

44. $x^2 - 39x + 140$

45. $x^2 + 31x + 130$

46. $x^2 - 8x + 15$

47. $x^2 + 3x - 130$

48. $x^2 + 16x + 48$

49. $x^2 - 50x + 96$

50. $x^2 + 23x - 140$

In the previous exercise the coefficient of all the first terms was 1. When the coefficient of the first term is not 1, the problem of factoring is much more complicated because the number of possibilities is greatly increased.

Example 4 Factor $6x^2 + 17x + 12$.

Solution Factors of $6x^2$ are x, $2x$, $3x$, and $6x$. Factors of 12 are 1, 2, 3, 4, 6, and 12.

Notice that there are twelve ways to obtain the first and last terms, but only one has $17x$ as a middle term.

$(6x + 1)(x + 12)$	Middle term $73x$
$(6x + 2)(x + 6)$	Middle term $38x$
$(6x + 3)(x + 4)$	Middle term $27x$
$(6x + 12)(x + 1)$	Middle term $18x$
$(6x + 6)(x + 2)$	Middle term $18x$
$(6x + 4)(x + 3)$	Middle term $22x$
$(2x + 2)(3x + 6)$	Middle term $18x$
$(2x + 6)(3x + 2)$	Middle term $22x$
$(2x + 1)(3x + 12)$	Middle term $27x$
$(2x + 12)(3x + 1)$	Middle term $38x$
$(2x + 4)(3x + 3)$	Middle term $18x$
$(2x + 3)(3x + 4)$	Middle term $17x$

There is only one way to obtain all three terms:

$$(2x + 3)(3x + 4) = 6x^2 + 17x + 12.$$

In this example one out of twelve possibilities is correct. Thus *trial and error* can be very time-consuming.

Even though the method used is one of guessing, it should be "educated guessing" in which we apply all of our knowledge about numbers and exercise a great deal of mental arithmetic. In example 4 we would immediately dismiss many of the combinations. Since we are searching for $17x$ as a middle term we would not attempt those possibilities that multiply 6 by 6, or 3 by 12, or 6 by 12, and so on, as those products will be larger than 17. Also, since 17 is odd, we know it is the sum of an even number and an odd number. All of these things help reduce the number of possibilities to try.

Example 5 Factor $6x^2 - 23x + 15$.

Solution First we should analyze the problem.

1. The last term is positive, so two like signs.
2. The middle term is negative, so both signs will be negative.
3. The factors of $6x^2$ are x, $2x$, $3x$, $6x$. The factors of 15 are 1, 3, 5, 15.
4. Eliminate as too large the product of 15 with $2x$, $3x$, or $6x$. Try some reasonable combinations.

$(2x - 5)(3x - 3)$	Middle term $(-21x)$ Incorrect
$(2x - 3)(3x - 5)$	Middle term $(-19x)$ Incorrect
$(6x - 3)(x - 5)$	Middle term $(-33x)$ Incorrect
$(6x - 5)(x - 3)$	Middle term $(-23x)$ CORRECT

so $$6x^2 - 23x + 15 = (6x - 5)(x - 3).$$

Example 6 Factor $4x^2 - 5x - 6$.

Solution Analyze.

1. The last term is negative, so unlike signs.
2. We must find products that differ by 5 with the larger number negative.
3. We eliminate a product of $4x$ and 6 as probably too large.
4. Try some combinations.

$(2x - 3)(2x + 2)$ Incorrect

$(4x - 2)(x + 3)$ Incorrect

$(4x - 3)(x + 2)$ Here the middle term is $+5x$, which is the right number but the wrong sign. Be careful not to accept this as the solution, but switch signs so the larger product agrees in sign with the middle term.

$(4x + 3)(x - 2)$ gives the correct sign for the middle term.

$$4x^2 - 5x - 6 = (4x + 3)(x - 2)$$

EXERCISE 4-4-2

Factor.

1. $2x^2 + 5x + 2$ **2.** $2x^2 + 11x + 5$ **3.** $2x^2 + 5x + 3$ **4.** $3x^2 + 8x + 4$

5. $2x^2 - 7x + 3$ **6.** $2x^2 + 7x + 3$ **7.** $6x^2 + 7x + 2$ **8.** $10x^2 + 19x + 6$

9. $2x^2 + 3x - 20$ **10.** $6x^2 - 11x - 10$ **11.** $3x^2 - x - 10$ **12.** $15x^2 + 23x + 4$

13. $8x^2 - 2x - 3$ **14.** $8x^2 + 22x + 5$ **15.** $6x^2 + 31x + 5$ **16.** $3x^2 - 13x - 10$

17. $5x^2 + 8x + 3$ **18.** $12x^2 - 11x + 2$ **19.** $16x^2 - 6x - 1$ **20.** $9x^2 + 9x + 2$

21. $4x^2 - x - 18$ **22.** $3x^2 - 4x - 7$ **23.** $4x^2 + 4x + 1$ **24.** $3x^2 + x - 14$

25. $6x^2 - 49x + 30$ **26.** $10x^2 + 21x + 9$ **27.** $9x^2 - 12x + 4$ **28.** $10x^2 + 21x - 10$

29. $6x^2 - x - 1$ **30.** $8x^2 - 14x - 15$ **31.** $6x^2 + 13x + 6$ **32.** $4x^2 - 11x + 6$

33. $15x^2 - 19x - 10$ **34.** $6x^2 - 5x - 6$ **35.** $12x^2 - 16x + 5$ **36.** $16x^2 + 24x + 9$

37. $6x^2 + x - 15$ **38.** $8x^2 - 2x - 15$ **39.** $15x^2 + 2x - 24$ **40.** $32x^2 + 4x - 3$

4-5 SPECIAL CASES IN FACTORING

OBJECTIVES

Upon completion of this section you should be able to:

1. Identify and factor the difference of two perfect squares.
2. Identify and factor a perfect square trinomial.
3. Identify and factor the sum or difference of two cubes.

In this section we wish to examine some special cases of factoring that occur often in problems. If these special cases are recognized, the factoring is then greatly simplified.

The first special case we will discuss is the **difference of two perfect squares.**

Recall that in multiplying two binomials by the pattern, the middle term comes from the sum of two products.

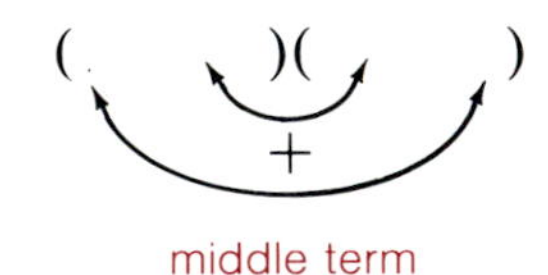

From our experience with numbers we know that the sum of two numbers is zero only if the two numbers are negatives of each other.

Example 1 $(3x - 4)(3x + 4) = 9x^2 - 16$

Example 2 $(2x + 1)(2x - 1) = 4x^2 - 1$

Example 3 $(5x - 6)(5x + 6) = 25x^2 - 36$

In each example the middle term is zero. Note that if two binomials multiply to give a binomial (middle term missing), they must be in the form of $(a - b)(a + b)$.

RULE
$(a - b)(a + b) = a^2 - b^2$

Reading this rule from right to left tells us that if we have a problem to factor and if it is in the form of $a^2 - b^2$, the factors will be $(a - b)(a + b)$.

Example 4 Factor $25x^2 - 16$.

Solution Here both terms are perfect squares and they are separated by a negative sign.

The form of this problem is $(5x)^2 - (4)^2$ (the form of $a^2 - b^2$). So

$$25x^2 - 16 = (5x + 4)(5x - 4).$$

Special cases do make factoring easier, but be certain to recognize that a special case is just that—very special. In this case *both* terms must be perfect squares *and* the sign must be negative, hence "the difference of two perfect squares."

Example 5 $x^2 - 7$ Not the special case. Why?

Example 6 $4x^2 + 9$ Not the special case. Why?

You must also be careful to recognize perfect squares. Remember that perfect square numbers are numbers that have square roots that are integers. Also, perfect square exponents are even.

Example 7 $4x^6 - 9 = (2x^3 - 3)(2x^3 + 3)$

Example 8 $x^4 - 25 = (x^2 - 5)(x^2 + 5)$

Example 9 $(a + b)^2 - c^2 = [(a + b) - c][(a + b) + c]$

Example 10 $(x + y)^2 - (p + k)^2 = [(x + y) - (p + k)][(x + y) + (p + k)]$

CAUTION Students often overlook the fact that 1 *is* a perfect square. Thus an expression such as $x^2 - 1$ is the difference of two perfect squares and can be factored by this method.

EXERCISE 4-5-1

Factor.

1. $x^2 - 4$
2. $x^2 - 1$
3. $a^2 - 81$
4. $x^2 - 100$
5. $y^2 - 121$
6. $4x^2 - 9$
7. $16x^2 - 1$
8. $4x^2 - y^2$

9. $4x^2 - 25y^2$ **10.** $16a^2 - 121b^2$ **11.** $(x + y)^2 - 64$ **12.** $x^4 - 9$

13. $x^6 - 36$ **14.** $25a^4 - 49$ **15.** $(x + y)^4 - z^2$

Another special case in factoring is the **perfect square trinomial.** Observe that squaring a binomial gives rise to this case.

$$(a + b)^2 = (a + b)(a + b) = a^2 + 2ab + b^2$$

We recognize this case by noting the special features in the example. Three things are evident.

1. The first term is a perfect square.
2. The third term is a perfect square.
3. The middle term is twice the product of the square root of the first and third terms.

perfect squares

$$\boxed{a^2} + \boxed{2ab} + \boxed{b^2}$$

twice the product of the square root of the first and third terms

Example 11 Factor $25x^2 + 20x + 4$.

Solution
1. $25x^2$ is a perfect square since $(5x)^2 = (5x)(5x) = 25x^2$.
2. 4 is a perfect square since $2^2 = (2)(2) = 4$.
3. $20x$ is twice the product of the square roots of $25x^2$ and 4. $20x = 2(5x)(2)$

ULE

To factor a perfect square trinomial form a binomial with the square root of the first term, the square root of the last term and the sign of the middle term, and indicate the square of this binomial.

$$\boxed{25x^2} + 20x + \boxed{4}$$

square root square root

$(5x + 2)^2$

Thus $$25x^2 + 20x + 4 = (5x + 2)^2.$$

Example 12 $25x^2 - 20x + 4 = (5x - 2)^2$

Example 13 $9x^2 - 6x + 1 = (3x - 1)^2$

Example 14 $x^2 + 8x + 16 = (x + 4)^2$

Example 15 $4x^2 + 15x + 9$ Not the special case of a perfect square trinomial. The middle term would need to be $12x$ to have a perfect square trinomial.

EXERCISE 4–5–2

Square the following binomials.

1. $(x + y)^2$

2. $(x + 4)^2$

3. $(x + 6)^2$

4. $(a + 1)^2$

5. $(a - 2)^2$

6. $(x - 5)^2$

7. $(2x + 7)^2$

8. $(2x - 1)^2$

9. $(3a + 1)^2$

10. $(4x - 3)^2$

Supply the missing middle terms so the following expressions are perfect square trinomials.

11. $x^2 +$ ______ $+ 4$

12. $x^2 +$ ______ $+ 9$

13. $4x^2 +$ ______ $+ 25$

14. $9x^2 -$ ______ $+ 4$

15. $x^2 +$ ______ $+ 1$

16. $25a^2 +$ ______ $+ 1$

17. $4x^2 -$ ______ $+ y^2$

18. $16x^2 +$ ______ $+ 4y^2$

19. $x^2 -$ ______ $+ 49y^2$

20. $36x^4 +$ ______ $+ 25y^2$

Factor.

21. $x^2 + 6x + 9$

22. $x^2 + 10x + 25$

23. $x^2 - 14x + 49$

24. $a^2 - 2a + 1$

25. $4x^2 + 12x + 9$

26. $4a^2 - 36ab + 81b^2$

27. $9x^2 + 30x + 25$

28. $49a^2 + 14a + 1$

29. $36x^2 - 60xy + 25y^2$

30. $25x^2 + 90x + 81$

Another special case of factoring is the **sum or difference of two cubes.** Notice the products obtained in the following two examples.

Example 16 Multiply $(a + b)(a^2 - ab + b^2)$.

Solution Using the distributive law as in section 4–2, we have

$$\begin{aligned}(a + b)(a^2 - ab + b^2) &= a^2(a + b) - ab(a + b) + b^2(a + b)\\ &= a^3 + a^2b - a^2b - ab^2 + ab^2 + b^3\\ &= a^3 + b^3.\end{aligned}$$

Example 17 Multiply $(a - b)(a^2 + ab + b^2)$.

Solution Again using the distributive law, we have

$$\begin{aligned}(a - b)(a^2 + ab + b^2) &= a^2(a - b) + ab(a - b) + b^2(a - b)\\ &= a^3 - a^2b + a^2b - ab^2 + ab^2 - b^3\\ &= a^3 - b^3.\end{aligned}$$

These examples give rise to the following rule for factoring the sum or difference of two cubes.

RULE

$$a^3 + b^3 = (a + b)(a^2 - ab + b^2)$$
$$a^3 - b^3 = (a - b)(a^2 + ab + b^2)$$

Example 18 Factor $x^3 + 27$.

Solution We must first recognize that this is the sum of two cubes.

$$x^3 + 27 = x^3 + (3)^3$$

Following the pattern given in the rule, we have

$$\begin{aligned} x^3 + (3)^3 &= (x + 3)[x^2 - 3x + (3)^2] \\ &= (x + 3)(x^2 - 3x + 9). \end{aligned}$$

Example 19 Factor $8x^3 - 125$.

Solution Here we note that

$$\begin{aligned} 8x^3 - 125 &= (2x)^3 - (5)^3 \\ &= (2x - 5)[(2x)^2 + (2x)(5) + (5)^2] \\ &= (2x - 5)(4x^2 + 10x + 25). \end{aligned}$$

EXERCISE 4-5-3

Factor.

1. $x^3 + y^3$

2. $x^3 - y^3$

3. $a^3 - 8$

4. $x^3 + 8$

5. $x^3 - 1$

6. $a^3 + 1$

7. $27a^3 + 1$

8. $8x^3 - y^3$

9. $27a^3 + 8b^3$

10. $a^3 - 125$

11. $64x^3 - 27y^3$

12. $x^3 - y^6$

4–6 COMPLETE FACTORIZATION

Upon completion of this section you should be able to factor a trinomial using the following two steps:

1. Look for common factors.
2. Factor the remaining trinomial by applying the methods of this chapter.

We have now studied all of the usual methods of factoring found in elementary algebra. However, you must be aware that a single problem can require more than one of these methods. Remember that there are two checks for correct factoring.

1. Will the factors multiply to give the original problem?
2. Are all factors prime?

Example 1 Factor $2x^2 + 10x + 12$.

Solution We notice that 2 is a common factor.

$$2x^2 + 10x + 12 = 2(x^2 + 5x + 6)$$

Now if we leave the problem at this point, the first check works but how about the second?

We see that the factor $(x^2 + 5x + 6)$ can now be factored into

$$(x + 3)(x + 2).$$

The complete factorization then is

$$2x^2 + 10x + 12 = 2(x + 3)(x + 2).$$

Suppose in this example we had looked at the problem $2x^2 + 10x + 12$ and simply used the trial and error method. We could get

$$2x^2 + 10x + 12 = (2x + 4)(x + 3).$$

Once again the first check works because the product gives the original. However, the factor $(2x + 4)$ has a common factor of 2.

$$2x + 4 = 2(x + 2)$$

Thus $2x^2 + 10x + 12 = 2(x + 2)(x + 3)$.

Nothing short of this complete factorization is correct.

Example 2 Factor $12x^2 - 27$.

Solution We note a common factor of 3.

$$12x^2 - 27 = 3(4x^2 - 9)$$

At this point we recognize $4x^2 - 9$ as the difference of two perfect squares.

$$3(4x^2 - 9) = 3(2x - 3)(2x + 3)$$
$$\text{So } 12x^2 - 27 = 3(2x - 3)(2x + 3).$$

A good procedure to follow in factoring is to always remove the greatest common factor first and then factor what remains, if possible.

EXERCISE 4-6-1

Factor completely. If the expression is prime, so state.

1. $4x^2 + 14x + 6$

2. $6x^2 + 5x + 1$

3. $3x^2 + 6x - 144$

4. $3x^2 - 8x - 3$

5. $3x^2 - 12$

6. $4x^2 + 24x + 36$

7. $x^2 + x + 1$

8. $3x^3 + 10x^2 - 8x$

9. $42x^2 - 7x - 7$

10. $5x^2 + 5x - 60$

11. $x^3 - 8x^2 + 12x$

12. $20x^2 - 5$

13. $ax^2 - 3 + 3x^2 - a$

14. $14x^2 + 11x + 2$

15. $8x^2 - 8x + 2$

16. $3x^3 - 39x^2 + 120x$

17. $x^2 + 5x - 14$

18. $6x^2 - 3x + 2$

19. $4x^2 - 4x - 3$

20. $a^2x + a^2 - 9x - 9$

21. $27x^3 - 75x$

22. $2x^2 + 4x + 6$

23. $x^2 + x - 30$

24. $3x^2 + 93x + 252$

25. $7x^2 + 3x - 2$

26. $4x^4 - 8x$

27. $5x^3 - 5$

28. $2x^4 + 16x$

4–7 SHORTCUTS TO TRIAL AND ERROR FACTORING

OBJECTIVES

Upon completion of this section you should be able to:

1. Find the key number of a trinomial.
2. Use the key number to factor a trinomial.

In this section we wish to discuss some shortcuts to trial and error factoring. These shortcuts may not always be practical for large numbers, but they will increase speed and accuracy for those who master them.

The first step in these shortcuts is finding the **key number.** After you have found the key number it can be used in more than one way.

DEFINITION

In a trinomial to be factored the **key number** is the product of the coefficients of the first and third terms.

Example 1 In $6x^2 - 22x + 12$ the key number is $(6)(12) = 72$.

Example 2 In $5x^2 + 13x - 6$ the key number is $(5)(-6) = -30$.

EXERCISE 4–7–1

Find the key number in each trinomial.

1. $2x^2 + 7x + 3$

2. $3x^2 - x - 10$

3. $2x^2 + 3x - 20$

4. $10x^2 + 19x + 6$

5. $16x^2 - 6x - 1$

6. $8x^2 + 22x + 5$

7. $3x^2 - 4x - 7$

8. $6x^2 - 49x + 30$

9. $4x^2 - 11x + 6$

10. $15x^2 - 19x - 10$

The first use of the key number is shown in example 3.

Example 3 Use the key number to factor $4x^2 + 3x - 10$.

Solution

Step 1 Find the key number. In this example $(4)(-10) = (-40)$.

Step 2 Find factors of the key number (-40) that will add to give the coefficient of the middle term $(+3)$. In this case $(+8)(-5) = -40$ and $(+8) + (-5) = +3$.

Step 3 These factors $(+8)$ and (-5) will be the outside and inside products in the multiplication pattern.

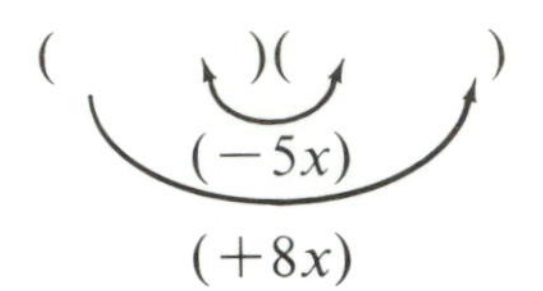

Step 4 Using only the outside product, find factors of the first and third terms that will multiply to give the product. In this example we must find factors of $4x^2$ and -10 that will multiply to give $+8x$. These are $4x$ from $4x^2$ and $(+2)$ from (-10). Place these factors in the first and last positions in the pattern.

$$(4x \quad)(\quad +2)$$
$$8x$$

Step 5 Forget the key number at this point and look back at the original problem. Since the first and last positions are correctly filled, it is now only necessary to fill the other two positions.

$$4x^2 + 3x - 10$$
$$4x^2$$
$$(4x \quad)(\quad +2)$$

We know the product of the two first terms must give $4x^2$ and $4x$ is already in place. There is no choice other than x.

$$4x^2$$
$$(4x \quad)(x + 2)$$

We know that the product of the two second terms must be (-10) and $(+2)$ is already in place. We have no choice other than (-5).

$$-10$$
$$(4x - 5)(x + 2)$$

We now have the factors of $4x^2 + 3x - 10$ as $(4x - 5)(x + 2)$.

Example 4 Factor $3x^2 - 10x - 8$.

Solution Key number $= (3)(-8) = -24$.

Factors of -24 that add to give (-10) are $(-12)(+2)$.

Using $(-12x)$ as the product of the outside terms, we find $(3x)(-4) = -12x$.

$$(3x \qquad)(\qquad -4)$$

Looking back to the original problem $3x^2 - 10x - 8$, we find the other terms to be x and $+2$.

Therefore $3x^2 - 10x - 8 = (3x + 2)(x - 4)$.

EXERCISE 4-7-2

Use the method discussed to factor.

1. $2x^2 + 7x + 6$

2. $2x^2 + 13x + 20$

3. $3x^2 + 14x + 8$

4. $3x^2 + 4x - 4$

5. $2x^2 - 9x - 5$

6. $5x^2 + 19x - 4$

7. $3x^2 - 17x + 10$

8. $4x^2 - 11x + 6$

9. $6x^2 + 7x + 2$

10. $6x^2 - x - 2$

11. $4x^2 - 33x - 27$

12. $8x^2 + 18x - 5$

13. $6x^2 - 41x + 30$

14. $8x^2 + 10x - 7$

15. $18x^2 - 3x - 10$

A second use for the key number as a shortcut involves factoring by grouping. It works as in example 5.

Example 5 Factor $4x^2 + 3x - 10$.

Solution **Step 1** Find the key number $(4)(-10) = -40$.

Step 2 Find factors of (-40) that will add to give the coefficient of the middle term $(+3)$. They are $(+8)$ and (-5) since $(+8) + (-5) = +3$.

Step 3 Rewrite the original problem by breaking the middle term into the two parts found in step 2. $8x - 5x = 3x$, so we may write

$$4x^2 + 3x - 10 = 4x^2 + 8x - 5x - 10.$$

Step 4 Factor this problem from step 3 by the grouping method studied in section 4–3.

$$\begin{aligned} 4x^2 + 8x - 5x - 10 &= 4x(x + 2) - 5(x + 2) \\ &= (x + 2)(4x - 5) \end{aligned}$$

Hence $$4x^2 + 3x - 10 = (x + 2)(4x - 5).$$

Example 6 Factor $3x^2 - 10x - 8$.

Solution Key number $= -24$.

Factors of key number that add to give -10 are -12 and 2. Hence

$$\begin{aligned} 3x^2 - 10x - 8 &= 3x^2 - 12x + 2x - 8 \\ &= 3x(x - 4) + 2(x - 4) \\ &= (x - 4)(3x + 2). \end{aligned}$$

The *key number* method is superior to *trial and error* for two reasons.

1. You are able to make a more direct search for factors rather than using haphazard guessing.
2. If no factors of the key number will combine to obtain the coefficient of the middle term, the trinomial cannot be factored.

EXERCISE 4–7–3

Use the method discussed to factor.

1. $x^2 + 9x + 20$ **2.** $x^2 - x - 6$ **3.** $x^2 - 8x + 12$ **4.** $2x^2 + 7x + 3$

5. $3x^2 + 5x + 2$ **6.** $2x^2 - 5x - 12$ **7.** $5x^2 + 9x - 2$ **8.** $2x^2 - 11x + 15$

9. $3x^2 - 14x + 8$ **10.** $3x^2 - 28x + 32$ **11.** $4x^2 + 9x - 9$ **12.** $x^2 + 8x + 16$

13. $4x^2 + 8x + 3$ **14.** $6x^2 - x - 12$ **15.** $16x^2 - 8x - 3$

4–8 LONG DIVISION OF POLYNOMIALS

**BJECTIVES**

Upon completion of this section you should be able to:

1. Divide a polynomial by a binomial.
2. Determine if a binomial is an exact divisor of a polynomial.

In our discussion of the relation between the factored form and the expanded form of a polynomial it is clear that a factor is also an *exact divisor*. For example, 7 is a factor of 35 because dividing 35 by 7 will not leave a remainder. 7 is *not* a factor of 45 because dividing 45 by 7 will leave a remainder of 3.

It is sometimes useful to be able to determine whether a particular polynomial is an exact divisor of another polynomial.

The technique explained here is called *long division*. It is useful in the solving of higher degree equations and in factoring expressions of higher degree. We will concentrate on the technique itself and leave discussion of its use for future topics.

Example 1 Is $(x + 3)$ an exact divisor (factor) of $2x^3 + 11x^2 + 8x - 21$?

Solution **Step 1** Be sure that both polynomials are in descending powers of the variable and supply a zero for any missing terms. Then arrange as follows:

$$x + 3\overline{\smash{)}2x^3 + 11x^2 + 8x - 21}$$

Step 2 To obtain the first term of the quotient divide the first term of the dividend by the first term of the divisor. In this case, $2x^3 \div x = 2x^2$. We record this as follows:

$$\begin{array}{r l} & 2x^2 \\ x + 3 & \overline{\smash{)}2x^3 + 11x^2 + 8x - 21} \end{array}$$

Step 3 Multiply the *entire* divisor by the term obtained in step 2. Subtract the result from the dividend as follows:

$$\begin{array}{r|l}
 & 2x^2 \\ \hline
x + 3 & 2x^3 + 11x^2 + 8x - 21 \\
 & \underline{2x^3 + 6x^2} \\
 & \phantom{2x^3 + {}} 5x^2 + 8x - 21
\end{array}$$

Step 4 Divide the first term of the remainder by the first term of the divisor to obtain the next term of the quotient. Then multiply the entire divisor by the resulting term. Finally subtract again as follows:

$$\begin{array}{r|l}
 & 2x^2 + 5x \\ \hline
x + 3 & 2x^3 + 11x^2 + 8x - 21 \\
 & \underline{2x^3 + 6x^2} \\
 & \phantom{2x^3 + {}} 5x^2 + 8x - 21 \\
 & \phantom{2x^3 + {}} \underline{5x^2 + 15x} \\
 & \phantom{2x^3 + 5x^2 + {}} -7x - 21
\end{array}$$

This process is repeated until either the remainder is zero or the power of the first term of the remainder is less than the power of the first term of the divisor.

$$\begin{array}{r|l}
 & 2x^2 + 5x - 7 \\ \hline
x + 3 & 2x^3 + 11x^2 + 8x - 21 \\
 & \underline{2x^3 + 6x^2} \\
 & \phantom{2x^3 + {}} 5x^2 + 8x - 21 \\
 & \phantom{2x^3 + {}} \underline{5x^2 + 15x} \\
 & \phantom{2x^3 + 5x^2 + {}} -7x - 21 \\
 & \phantom{2x^3 + 5x^2 + {}} \underline{-7x - 21} \\
 & \phantom{2x^3 + 5x^2 + -7x - 2{}} 0
\end{array}$$

Since the remainder is zero, we know that $(x + 3)$ is an exact divisor (factor) of $2x^3 + 11x^2 + 8x - 21$. In fact, we know that

$$(x + 3)(2x^2 + 5x - 7) = 2x^3 + 11x^2 + 8x - 21.$$

We can use this information to factor the expression $2x^3 + 11x^2 + 8x - 21$ as

$$(x + 3)(2x^2 + 5x - 7)$$

and finally as

$$(x + 3)(2x + 7)(x - 1).$$

Example 2 Is $5x + 1$ a factor of $5x^3 - 24x^2 + 7$?

Solution Notice that the x term is missing in the dividend. This means that the coefficient of that term is zero and we must enter it as a term as follows:

$$\begin{array}{r}
x^2 - 5x + 1 \\
5x + 1 \overline{)5x^3 - 24x^2 + 0x + 7} \\
\underline{5x^3 + x^2 } \\
-25x^2 + 0x + 7 \\
\underline{-25x^2 - 5x } \\
5x + 7 \\
\underline{5x + 1} \\
6
\end{array}$$

Since the remainder is not zero, we know that $5x + 1$ is *not* a factor of $5x^3 - 24x^2 + 7$.

EXERCISE 4-8-1

Find the quotient and remainder. Then answer the question.

1. Is $x + 7$ a factor of $x^2 + 5x - 14$?

2. Is $x + 2$ a factor of $x^3 - x^2 - 2x + 2$?

3. Is $x - 3$ an exact divisor of $2x^3 - 2x^2 - 5x + 6$?

4. Is $x - 5$ an exact divisor of $x^3 + 2x^2 - 30x + 2$?

5. Is $x + 4$ an exact divisor of $x^3 + 2x^2 - 32$?

6. Is $x + 3$ an exact divisor of $x^4 - x^2 - 90$?

7. Is $x - 10$ a factor of $x^3 + x^2 + 10x + 100$?

8. Is $3x + 2$ a factor of $3x^3 + 11x^2 - 9x - 10$?

9. Is $2x + 1$ a factor of $4x^3 + 8x^2 - x - 2$?

10. If $2x + 4$ is a factor of $6x^3 - 18x + 12$, find the other factor. Can this second factor be refactored?

CHAPTER 4 SUMMARY

The number in brackets refers to the section of the chapter that discusses that concept.

Terminology

- An expression is in **factored form** only if the entire expression is an indicated product. [4–1]
- **Factoring** is a process of changing an expression from a sum or difference of terms to a product of factors. [4–1]
- The **greatest common factor** is the largest factor common to all terms. [4–1]
- A **prime** expression cannot be factored. [4–1]
- An expression is **completely factored** when no further factoring is possible. [4–1]
- The **FOIL** method can be used to multiply two binomials. [4–2]
- The possibility of **factoring by grouping** exists when an expression contains four or more terms. [4–3]
- The **key number** is the product of the coefficients of the first and third terms of a trinomial. [4–7]
- An **exact divisor** of a polynomial is a factor of the polynomial. [4–8]

Rules and Procedures

Factoring Criteria

- For factoring to be correct the solution must meet two criteria:
  1. It must be possible to multiply the factored expression and obtain the original expression.
  2. The expression must be completely factored. [4–1]

Common Factors

- To remove common factors find the greatest common factor (GCF) and divide each term by it. The factored form is the indicated product of the GCF and the sum or difference of terms found by this division. [4–1]

Multiplying Polynomials

- To multiply two binomials use the FOIL method. [4–2]

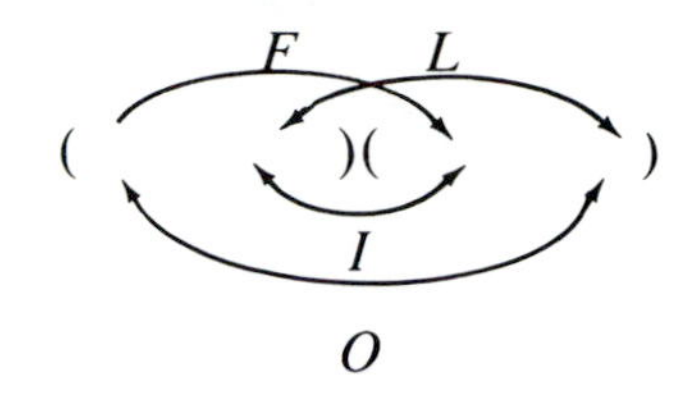

- The distributive property can be used to find the product of any two polynomials. [4–2]

Factoring a General Trinomial

- To factor a trinomial by trial and error use the FOIL pattern for multiplication to find factors that will give the original trinomial. [4–4]
- To factor a trinomial using the key number multiply the first and third term coefficients to obtain the key number. Then:

1. Find the factors of the key number whose sum is the coefficient of the middle term. These factors are the necessary products of the outside and inside terms that give the necessary factors of the trinomial.
2. Find the factors of the key number whose sum is the coefficient of the middle term. Rewrite the original trinomial by making the middle term the sum of the factors of the key number and then factor the four-term expression by grouping. [4–7]

Special Cases in Factoring

- To factor the difference of two perfect squares use

$$a^2 - b^2 = (a - b)(a + b). \qquad [4\text{–}5]$$

- To factor a perfect square trinomial use

$$a^2 + 2ab + b^2 = (a + b)^2$$
or
$$a^2 - 2ab + b^2 = (a - b)^2. \qquad [4\text{–}5]$$

- To factor the sum or difference of two cubes use

$$a^3 + b^3 = (a + b)(a^2 - ab + b^2)$$
or
$$a^3 - b^3 = (a - b)(a^2 + ab + b^2). \qquad [4\text{–}5]$$

Exact Divisors

- If a polynomial is divided by a binomial and the remainder is zero, the binomial is an exact divisor of the polynomial. [4–8]

CHAPTER 4 REVIEW

Expand.

1. $(2x + 3)(x - 2)$

2. $(3a - 2)^2$

3. $(3x + 8)(3x - 8)$

4. $(x + 3)(2x^2 - x + 4)$

5. $(x + y + 3)(x - y - 1)$

Completely factor.

6. $9x + 6y$

7. $a^2 - a$

8. $6x^3y + 10x^2y^2$

9. $6a^2b + 9ab^2 - 6ab$

10. $x(a + b) - 2(a + b)$

11. $25x^2 - 49$

12. $x^2 + 8x + 16$

13. $9x^2 - 30x + 25$

14. $5x^2 - 45$

15. $24x^3 - 3$

16. $x^2 + 5x + 6$

17. $x^2 + 5x - 14$

18. $x^2 - 4x + 3$

19. $x^2 + 2x - 15$

20. $x^2 + x - 72$

21. $2x^2 + 5x + 2$

22. $2x^2 - x - 3$

23. $3x^2 + 13x - 10$

24. $6x^2 - 5x - 4$

25. $6x^2 - 7x + 2$

26. $4x^2 + 14x + 6$

27. $18x^2 + 3x - 6$

28. $6x^3 + 8x^2 - 8x$

29. $3x^2 + 9x + 15$

30. $5x^2 - 2x + 3$

31. $cx + dx + cy + dy$

32. $ax - 3y - 3x + ay$

33. $ax + 20 - 5a - 4x$

34. $a^2x - 8 + 2a^2 - 4x$

35. $2a^2x - 3 - 2x + 3a^2$

Find the quotient and remainder. Then answer the question.

36. Is $x - 1$ a factor of $x^3 + 2x^2 - x - 2$?

37. Is $x + 5$ an exact divisor of $x^3 - 2x^2 - 20x + 75$?

38. Is $x - 3$ an exact divisor of $2x^3 - 4x^2 + 5x - 3$?

39. Is $3x - 1$ a factor of $3x^3 + 2x^2 - 4x + 1$?

40. Is $2x + 3$ a factor of $4x^3 + 4x^2 - x - 1$?

CHAPTER

PRACTICE TEST

Expand.

1. $(3x - 1)(2x + 3)$

1. ______________________

2. $(2x - 1)(x^2 - 2x + 3)$

2. ______________________

Factor completely.

3. $22a^2 + 33a$

3. ______________________

4. $10x^3y - 6x^2y^2 + 2xy^3$

4. ______________________

5. $2x(a - 1) - 3(a - 1)$

5. ______________________

6. $x^2 - 81$

6. ______________________

7. $x^2 - 22x + 121$

7. ______________________

8. $ax + bx + 3a + 3b$

8. ______________________

9. $x^2 - 14x + 49$

9. ______________________

10. $4x^2 - 19x + 12$

10. ______________________

11. $2x^2 + 13x + 15$

11. ______________________

12. $x^2 + 11x + 18$

12. ______________________

13. $25x^2 - 144$

13. ______________________

14. $x^2 - 6x - 27$

14. ____________________

15. $ax^2 - 27 + 3x^2 - 9a$

15. ____________________

16. $6x^2 - 10x - 4$

16. ____________________

17. $32x^3 - 72x$

17. ____________________

18. $x^2 + 2x - 63$

18. ____________________

19. $6x^3 + 3x^2 - 18x$

19. ____________________

20. $27x^3 + 36x^2 + 12x$

20. ____________________

21. $6x^2 - x - 15$

21. ______________________

22. $24x^2 + 52x + 20$

22. ______________________

Find the quotient and remainder. Then answer the question.

23. Is $x - 5$ an exact divisor of $3x^3 - 13x^2 - 8x + 10$?

23. ______________________

24. Is $2x - 1$ a factor of $4x^3 + 4x^2 - 7x - 2$?

24. ______________________

CHAPTERS 1-4

CUMULATIVE TEST

1. Combine: $+7 - (-3) + (-8)$

1. ______________________

2. Divide: $\dfrac{-32x^7}{-4x}$

2. ______________________

3. Multiply: $5x^2y(3xy^3 - 2x^2)$

3. ______________________

4. Simplify: $10x^2 - [7x - 2x(5x + 2)]$

4. ______________________

5. Evaluate $6x^3 - 2x^2y + 3y^3$ when $x = -2$, $y = 3$.

5. ______________________

Solve for x.

6. $5(x + 3) - 3 = 2(x - 3)$

6. ______________________

7. $\dfrac{x}{2} - 2 = \dfrac{x}{3}$

7. ______________________

8. $3(x - a) = \dfrac{1}{2}(x + 3a)$

8. ______________________

Solve for x and graph the solution on the number line.

9. $-2x > +8$

9. ______

10. $|x - 4| \leq 5$

10. ______

11. ______

Factor completely.

11. $14a^2b^3 - 7a^4b^2 + 21a^3b^3$

12. ______

12. $18x^2 - 8$

13. ______

13. $4x^2 - 20x + 25$

14. $14x^2 + 21x - 14$

14. ______

15. Is $x - 3$ a factor of $x^3 + 2x^2 - 17x + 6$? Show the division.

15. ______

16. Write an algebraic expression for the number of hours in m minutes.

16. ______

17. The sum of two numbers is 40. Their difference is 12. Find the numbers.

17. ______________________

18. A certain amount of money was invested at an annual rate of interest of 9.5%. Twice that amount was invested at 8.6%. If the total interest for the year on the two investments was \$934.50, how much was invested at each rate?

18. ______________________

19. A motorboat travels up a river against a 5 mph current for 2 hours and returns in $1\frac{1}{2}$ hours. Find the rate of the motorboat in still water.

19. ______________________

20. How many liters of pure alcohol must be added to fifteen liters of water to obtain a solution of 40% alcohol?

20. ______________________

CHAPTER

SURVEY

The following questions refer to material discussed in this chapter. Work as many problems as you can and check your answers with the answer section in the back of the book. The results will direct you to the sections of the chapter in which you need to work. If you answer all questions correctly, you have a good understanding of the material contained in this chapter.

1. Simplify: $\dfrac{x^2 - 9}{2x^3 - x^2 - 15x}$ 1. ______

2. Multiply: 2. ______
$$\frac{x^2 - 8x + 16}{2x^2 + 11x + 5} \cdot \frac{4x^2 - 1}{x^2 - x - 12} \cdot \frac{x^2 + 8x + 15}{2x^2 - 9x + 4}$$

3. Divide: $\dfrac{3x^2 - 4x - 4}{x^2 + 3x - 10} \div \dfrac{3x^2 + 11x + 6}{x^2 - x - 12}$ 3. ______

4. Find the LCD for the following fractions: 4. ______
$$\frac{1}{x^2 + 10x + 24}, \quad \frac{x - 1}{3x^2 + 10x - 8}$$

5. Find the missing numerator: 5. ______
$$\frac{x - 3}{x + 3} = \frac{?}{2x^2 - x - 21}$$

6. Add: $\dfrac{5}{x^2 + 2x - 8} + \dfrac{x + 3}{x^2 - 3x + 2}$ 6. ______

7. Subtract: $\dfrac{x + 5}{x^2 + 2x - 3} - \dfrac{1}{x^2 + 4x + 3}$ 7. ______

8. Simplify: $\dfrac{\dfrac{1}{x^2 - 2x - 15} - \dfrac{1}{x + 3}}{1 - \dfrac{1}{x - 5}}$ 8. ______

9. Solve: $\dfrac{x}{x + 2} + \dfrac{3}{x - 2} = 1 - \dfrac{1}{x^2 - 4}$ 9. ______

10. Person A can complete a job in 3 hours and person B can complete the same job in 6 hours. How long will it take both of them working together to complete the job? 10. ______

Algebraic Fractions

Fractions occur in algebra both in expressions and equations. Your knowledge of the arithmetic of fractions will form a foundation for work in algebraic fractions. The methods used in reducing, multiplying, dividing, adding, and subtracting fractions and simplifying complex fractions are identical with the methods from arithmetic. Equations with algebraic fractions are also covered in this chapter.

To work the problems in this chapter you must know the processes of factoring from the previous chapter. Mastery of one topic is almost always necessary for understanding another topic in algebra.

5–1 SIMPLIFYING FRACTIONS

OBJECTIVES

Upon completion of this section you should be able to:

1. Factor the numerator and denominator of a fraction.
2. Simplify algebraic fractions.

The **fundamental principle of fractions** states that the numerator and denominator of a fraction can be multiplied by the same nonzero number without changing the value of the fraction. In symbols this principle is stated as

$$\frac{a}{b} = \frac{ax}{bx}, \ x \neq 0.$$

The fundamental principle is used to simplify (or reduce) fractions as well as to change one denominator to some other denominator. If we reverse the equality of the above statement

$$\frac{ax}{bx} = \frac{a}{b},$$

we can restate it in words as "common factors of the numerator and denominator of a fraction may be *cancelled* without changing the value of the fraction."

A fraction is in its **simplified form** if the numerator and denominator have no prime factors in common.

**ULE**

To simplify (or reduce) a fraction factor both the numerator and denominator into prime factors, then cancel all common factors.

Example 1 Simplify: $\dfrac{12}{18}$

Solution In studying algebraic fractions we will frequently use examples from arithmetic to illustrate a principle.

$$\frac{12}{18} = \frac{(2)(2)(3)}{(2)(3)(3)}$$

Cancelling like factors, we obtain

$$\frac{(\cancel{2})(2)(\cancel{3})}{(\cancel{2})(3)(\cancel{3})} = \frac{2}{3}.$$

Example 2 Simplify: $\dfrac{5x + 20}{25}$

Solution $\dfrac{5x + 20}{25} = \dfrac{5(x + 4)}{(5)(5)}$

Cancelling like factors, we obtain

$$\frac{\cancel{5}(x + 4)}{(\cancel{5})(5)} = \frac{x + 4}{5}.$$

Example 3 Simplify: $\dfrac{a^2 - b^2}{a^2 + 3ab + 2b^2}$

Solution $\dfrac{a^2 - b^2}{a^2 + 3ab + 2b^2} = \dfrac{(a - b)(a + b)}{(a + b)(a + 2b)}$

Cancelling like factors, we obtain

$$\frac{(a - b)\cancel{(a + b)}}{\cancel{(a + b)}(a + 2b)} = \frac{a - b}{a + 2b}.$$

CAUTION Note carefully that *only factors* may be cancelled. Terms can *never* be cancelled. For instance,

$$\frac{2 + 4}{2} = \frac{6}{2} = \frac{\cancel{2} \cdot 3}{\cancel{2}} = 3.$$

However, if we cancelled *terms* we would get

$$3 = \frac{6}{2} = \frac{\not{2} + 4}{\not{2}} = 1 + 4 = 5$$

which is *NOT* correct.

$$3 \neq 5!$$

Example 4 Simplify: $\frac{x + 4}{3x^2 + 5x - 28}$

Solution $\frac{x + 4}{3x^2 + 5x - 28} = \frac{(1)\cancel{(x + 4)}}{\cancel{(x + 4)}(3x - 7)} = \frac{1}{(3x - 7)}$

Example 5 Simplify: $\frac{x^3 - 2x^2 - 15x}{x^3 - 3x^2 - 10x}$

Solution

$$\begin{aligned}\frac{x^3 - 2x^2 - 15x}{x^3 - 3x^2 - 10x} &= \frac{x(x^2 - 2x - 15)}{x(x^2 - 3x - 10)} \\ &= \frac{\cancel{x}\cancel{(x - 5)}(x + 3)}{\cancel{x}\cancel{(x - 5)}(x + 2)} \\ &= \frac{x + 3}{x + 2}\end{aligned}$$

If there are no common factors present, the original problem is already in simplified form.

Example 6 $\frac{x^2 + x - 6}{x^2 + 3x + 2} = \frac{(x + 3)(x - 2)}{(x + 2)(x + 1)}$

Solution There are no common factors, hence the fraction is in simplest form.

EXERCISE 5-1-1

Simplify when possible.

1. $\frac{x^2 + 2x}{x^2 + 3x + 2}$ **2.** $\frac{x}{x^2 + 3x}$ **3.** $\frac{x^2 - 1}{x^2 + x - 2}$ **4.** $\frac{x + 3}{x^2 - 9}$

5. $\frac{x^2 + 2x + 1}{x^2 - 4x - 5}$ **6.** $\frac{x^2 - 4}{x^2 + 4x + 4}$ **7.** $\frac{x^2 - 4x + 3}{x^2 - 6x + 9}$ **8.** $\frac{x^2 + 4x + 4}{x^2 - 4}$

9. $\dfrac{x^2 - x - 12}{x^2 + 3x}$

10. $\dfrac{x^2 - x - 2}{x^2 + x - 2}$

11. $\dfrac{x^2 + 5x}{x^2 + 8x + 15}$

12. $\dfrac{x^2 - 2x - 8}{x + 2}$

13. $\dfrac{a^2 + 12a + 35}{a^2 + 4a - 21}$

14. $\dfrac{x^3 - 2x^2 - 3x}{x^3 + 3x^2 + 2x}$

15. $\dfrac{x^2 - 9}{x^2 - x - 12}$

16. $\dfrac{x^2 + 3x}{x^3 + 9x^2 + 18x}$

17. $\dfrac{9x^2 - 16}{3x^2 - 16x + 16}$

18. $\dfrac{2x^2 - 5x - 3}{2x^2 + 11x + 5}$

19. $\dfrac{3a^2 + 17a + 10}{2a^2 + 7a - 15}$

20. $\dfrac{6x^2 - x - 2}{6x^2 + 5x - 6}$

5–2 MULTIPLICATION AND DIVISION OF ALGEBRAIC FRACTIONS

Upon completion of this section you should be able to:

1. Multiply two or more fractions using the three-step rule.
2. Change a division problem to a related multiplication problem and use the three-step rule to simplify.

Now that you have learned to simplify fractions, we wish to present the basic operations that can be performed on them. You will find that the process of simplifying is closely related to the operations of multiplication and division presented in this section.

The product of two fractions is equal to the product of their numerators divided by the product of their denominators. In symbols this definition is written

$$\frac{a}{b} \cdot \frac{c}{d} = \frac{ac}{bd}.$$

For instance,

$$\frac{2}{7} \cdot \frac{7}{12} = \frac{(2)(7)}{(7)(12)} = \frac{14}{84}.$$

The final answer would now need to be simplified. This means that 14 and 84 must be factored and the common factors cancelled.

$$\frac{14}{84} = \frac{(7)(2)}{(7)(2)(2)(3)} = \frac{1}{6}$$

(Note that 1 is always a factor of any expression, although it usually is not written. Thus if all other factors are cancelled, the factor 1 is still present.)

This example should make it clear that applying the definition directly will lead to a longer process for finding the answer. It is a duplication of effort if we first multiply and then proceed to simplify. A more direct process is to factor and cancel first, then multiply the remaining factors.

RULE

To multiply algebraic fractions:

Step 1 Factor all numerators and denominators into prime factors.
Step 2 Cancel any factor that is common to a numerator and a denominator.
Step 3 Multiply the remaining factors in the numerator and place this product over the product of the remaining factors in the denominator.

Example 1 Multiply: $\frac{2x^2 - 5x - 12}{x^2 - 3x - 4} \cdot \frac{x^2 - 1}{2x^2 + 7x + 6}$

Solution **Step 1** $= \frac{(2x + 3)(x - 4)}{(x - 4)(x + 1)} \cdot \frac{(x + 1)(x - 1)}{(2x + 3)(x + 2)}$

Step 2 $= \frac{(2x + 3)(x - 4)}{(x - 4)(x + 1)} \cdot \frac{(x + 1)(x - 1)}{(2x + 3)(x + 2)}$

Step 3 $= \frac{x - 1}{x + 2}$

In step 1 the factoring may be any of the types studied in the previous chapter.

Example 2 Multiply: $\frac{x^2 - 6x + 8}{x^2 + 5x - 14} \cdot \frac{x^2 - x - 6}{8 + 2x - x^2}$

Solution **Step 1** $= \frac{(x - 4)(x - 2)}{(x - 2)(x + 7)} \cdot \frac{(x + 2)(x - 3)}{(4 - x)(2 + x)}$

Note that the factor $(x - 4)$ is the negative of $(4 - x)$. If we factor a (-1) from $(x - 4)$, it will then be in the form of $(-1)(4 - x)$ and the factors $(4 - x)$ will cancel.

Step 2 $= \frac{(-1)(4 - x)(x - 2)}{(x - 2)(x + 7)} \cdot \frac{(x + 2)(x - 3)}{(4 - x)(2 + x)}$

Step 3 $= -\frac{(x - 3)}{(x + 7)}$

This answer could have different forms. For instance,

$$-\frac{x - 3}{x + 7} = \frac{-x + 3}{x + 7} = \frac{3 - x}{x + 7} = \frac{x - 3}{-x - 7}.$$

Any one of these forms is considered correct.

This process can be extended to the product of any number of fractions. Steps 1, 2, and 3 will still apply.

Example 3 Multiply: $\frac{x^2 + 4x + 4}{x^2 - x - 6} \cdot \frac{2x - 6}{4x - 3} \cdot \frac{2x^2 + x - 6}{2x^2 + 8x + 8}$

Solution **Step 1** $= \frac{(x + 2)(x + 2)}{(x + 2)(x - 3)} \cdot \frac{2(x - 3)}{4x - 3} \cdot \frac{(2x - 3)(x + 2)}{2(x + 2)(x + 2)}$

Step 2 $= \frac{\cancel{(x + 2)}\cancel{(x + 2)}}{\cancel{(x + 2)}\cancel{(x - 3)}} \cdot \frac{\cancel{2}\cancel{(x - 3)}}{4x - 3} \cdot \frac{(2x - 3)\cancel{(x + 2)}}{\cancel{2}\cancel{(x + 2)}\cancel{(x + 2)}}$

Step 3 $= \frac{2x - 3}{4x - 3}$

Division is defined as multiplication by the reciprocal. The reciprocal of the fraction is the invert of the fraction. Thus division by a fraction is performed by inverting the divisor and then multiplying. In symbols this is stated

$$\frac{a}{b} \div \frac{c}{d} = \frac{a}{b} \cdot \frac{d}{c}.$$

The divisor follows the division sign and it is always the divisor that is inverted.

Example 4 Divide: $\frac{2x^2 + 7x + 5}{3x^2 - x - 14} \div \boxed{\frac{2x^2 - x - 15}{x^2 - x - 6}}$ ← Divisor

Solution $= \frac{2x^2 + 7x + 5}{3x^2 - x - 14} \cdot \frac{x^2 - x - 6}{2x^2 - x - 15}$

Now proceed as in multiplication.

$$= \frac{\cancel{(2x + 5)}(x + 1)}{(3x - 7)\cancel{(x + 2)}} \cdot \frac{\cancel{(x + 2)}\cancel{(x - 3)}}{\cancel{(x - 3)}\cancel{(2x + 5)}}$$

$$= \frac{x + 1}{3x - 7}$$

Example 5 Divide: $\frac{6x^2 - 5x - 4}{3x^2 - 10x + 8} \div (2x^2 + 5x + 2)$

Solution In this case we must recognize that the divisor, $2x^2 + 5x + 2$, can be written $\frac{2x^2 + 5x + 2}{1}$. When we invert and multiply, we have

$$= \frac{6x^2 - 5x - 4}{3x^2 - 10x + 8} \cdot \frac{1}{2x^2 + 5x + 2}$$

$$= \frac{\cancel{(2x + 1)}\cancel{(3x - 4)}}{\cancel{(3x - 4)}(x - 2)} \cdot \frac{1}{\cancel{(2x + 1)}(x + 2)}$$

$$= \frac{1}{(x - 2)(x + 2)} \text{ or } \frac{1}{x^2 - 4}.$$

EXERCISE 5-2-1

Perform the indicated operation.

1. $\dfrac{2x + 2}{x + 3} \cdot \dfrac{x + 3}{2x - 6}$

2. $\dfrac{3x + 15}{x^2 + 6x + 5} \cdot \dfrac{x^2 + 2x + 1}{x^2 - 1}$

3. $\dfrac{x^2 - 1}{5x - 5} \cdot \dfrac{5}{x^2 + 5x + 4}$

4. $\dfrac{x^2 + 4x + 3}{x - 5} \cdot \dfrac{x^2 - 3x - 10}{x + 3}$

5. $\dfrac{x^2 + 3x - 10}{x^2 + x - 6} \div \dfrac{x + 7}{x + 3}$

6. $\dfrac{x^2 + 3x - 4}{x^2 - 1} \div \dfrac{x^2 + 6x + 8}{x + 1}$

7. $\dfrac{x^2 - 25}{x^2 - 3x - 10} \div \dfrac{x^2 - x - 30}{x^2 + 4x + 4}$

8. $\dfrac{x^2 + 8x + 15}{x^2 - 4x - 21} \div \dfrac{x + 5}{x^2 - 3x - 28}$

9. $\dfrac{x + 3}{x^2 + 3x - 10} \cdot \dfrac{x^2 - 3x + 2}{x^2 + 4x + 3}$

10. $\dfrac{x^2 + 5x - 14}{x - 3} \cdot \dfrac{6 - 2x}{x^2 + 12x + 35}$

11. $\dfrac{2x^2 + 11x + 12}{4x^2 + 16x + 15} \div \dfrac{2x^2 + 3x - 20}{4x^2 - 25}$

12. $\dfrac{3x^2 + 11x + 10}{x^2 - 4} \div \dfrac{2x^2 + 7x - 4}{x^2 + 2x - 8}$

13. $\dfrac{1 - 2x}{3x^2 - 2x - 8} \cdot \dfrac{3x + 4}{2x^2 + x - 1}$

14. $\dfrac{x^2 + 6x + 5}{2x^2 - 2x - 12} \cdot \dfrac{4x^2 - 36}{x^2 + 8x + 15}$

15. $\dfrac{x^2 - 2x - 63}{x^2 - 81} \div \dfrac{2x^2 - 5x - 12}{2x + 3}$

16. $\dfrac{3x^2 + 9x - 54}{2x^2 - 2x - 12} \div \dfrac{3x^2 + 21x + 18}{4x^2 - 12x - 40}$

17. $\dfrac{2x^2 - 9x - 5}{20 + x - x^2} \cdot \dfrac{3x^2 + 14x + 8}{2x^2 - 11x - 6}$

18. $\dfrac{9x^2 + 9x - 4}{6x^2 - 11x + 3} \cdot \dfrac{15 - 10x}{3x + 4}$

19. $\dfrac{6 + x - x^2}{2x^2 + 3x - 2} \div \dfrac{x^2 + 4x - 21}{3x^2 + 25x + 28}$

20. $\dfrac{4x^2 + 12x + 9}{3x^2 + 2x - 1} \div \dfrac{4x^2 - 9}{3x^2 + 14x - 5}$

21. $\dfrac{3x^2 + 10x + 3}{x^2 + 10x + 21} \cdot \dfrac{2x^2 + 9x - 35}{6x^2 - 13x - 5}$

22. $\dfrac{x + 3}{x^2 - 4x - 5} \cdot \dfrac{x^2 - 2x - 3}{x^2 + 5x + 6} \cdot \dfrac{x^2 + x - 30}{x^2 + x - 12}$

23. $\dfrac{x^2 - 7x - 18}{2x^2 + 9x + 10} \div (2x + 5)$

24. $\dfrac{14x^2 + 23x + 3}{2x^2 + x - 3} \div \dfrac{7x^2 + 15x + 2}{2x^2 - 3x + 1}$

25. $\dfrac{2x^2 + 9x - 5}{x^2 + 10x + 21} \cdot \dfrac{3x^2 + 11x + 6}{2x^2 - 13x + 6} \cdot \dfrac{x^2 + 3x - 28}{3x^2 - 10x - 8}$

26. $\dfrac{2x^2 - 7x - 15}{5x^2 - 24x - 5} \cdot \dfrac{20x^2 + 14x + 2}{2x^2 + 11x + 12} \cdot \dfrac{3x^2 + x - 2}{10x^2 + 35x + 15}$

27. $\dfrac{3x^2 - 17x + 10}{9x^2 - 12x + 4} \div \dfrac{20 + x - x^2}{6x^2 - 7x + 2}$

28. $\dfrac{25x^2 - 1}{10x^2 + 17x + 3} \div (1 - 5x)$

29. $\dfrac{3x^2 + 17x + 10}{x^2 + 10x + 25} \cdot \dfrac{x^2 - 25}{3x^2 + 11x + 6} \cdot \dfrac{5x^2 + 16x + 3}{5x^2 - 24x - 5}$

30. $\dfrac{3x^2 + 22x - 16}{2x^2 + 19x + 24} \div \dfrac{3x^2 + 13x - 10}{2x^2 + 13x + 15}$

5-3 FINDING THE LEAST COMMON DENOMINATOR

Upon completion of this section you should be able to:

1. Combine algebraic fractions having the same denominator.
2. Find the LCD (least common denominator) of several algebraic fractions.
3. Change an algebraic fraction to an equal fraction having a different denominator.

We have stated that in all of mathematics only like quantities can be combined (added or subtracted). In this section we will find the least common denominator of two or more fractions. This will enable us to change unlike fractions to like fractions so that they may be combined.

Like fractions are those that have the same denominators. Only like fractions can be combined. When two fractions have the same denominator their sum or difference is the sum or difference of their numerators over their common denominator. In symbols this is expressed as

$$\frac{a}{b} + \frac{c}{b} = \frac{a + c}{b} \quad \text{or} \quad \frac{a}{b} - \frac{c}{b} = \frac{a - c}{b}.$$

Example 1 $\dfrac{1}{5} + \dfrac{3}{5} = \dfrac{1 + 3}{5} = \dfrac{4}{5}$

Example 2 $\frac{2}{c} + \frac{3}{c} + \frac{x}{c} = \frac{2 + 3 + x}{c} = \frac{5 + x}{c}$

Example 3 $\frac{2}{3} - \frac{x}{3} = \frac{2 - x}{3}$

Example 4 $\frac{7}{8} - \frac{x + 5}{8} = \frac{7 - (x + 5)}{8}$

$$= \frac{7 - x - 5}{8}$$

$$= \frac{2 - x}{8}$$

EXERCISE 5-3-1

Add the fractions.

1. $\frac{1}{7} + \frac{4}{7}$
2. $\frac{2}{13} + \frac{8}{13}$
3. $\frac{3}{x} + \frac{5}{x}$
4. $\frac{7}{a} - \frac{2}{a}$
5. $\frac{1}{y} - \frac{a}{y}$
6. $\frac{b}{x} + \frac{1}{x}$
7. $\frac{3}{x} - \frac{4}{x} + \frac{a}{x}$
8. $\frac{2}{y} - \frac{c}{y} + \frac{4}{y}$
9. $\frac{a}{x} + \frac{b}{x} + \frac{c}{x}$
10. $\frac{a}{c} + \frac{3}{c} + \frac{a}{c}$
11. $\frac{6}{x} - \frac{x + 2}{x}$
12. $\frac{5}{a} - \frac{a + 4}{a}$

When fractions do not have the same denominators, it is necessary to change them to a common denominator before they can be added. The fundamental principle of fractions—numerator and denominator may be multiplied by the same number—allows us to change to a new denominator. The new denominator will always contain the original denominator as a factor. For instance, the fraction $\frac{2}{3}$ can be changed to other forms by multiplying numerator and denominator by the same number.

$$\frac{2}{3} = \frac{\boxed{5} \cdot 2}{\boxed{5} \cdot 3} = \frac{10}{15}$$

$$\frac{2}{3} = \frac{\boxed{7} \cdot 2}{\boxed{7} \cdot 3} = \frac{14}{21}$$

The fraction $\frac{2}{3}$ could not be changed to a simple fraction having a denominator of 10, since 3 is not a factor of 10. (A simple fraction is a fraction whose numerator and denominator are not fractional in form.)

A common denominator for $\frac{2}{3}$ and $\frac{5}{8}$ would be a number that contained 3 and 8 as factors. Some possible common denominators would be 24, 48, 72, and so on, all of which are divisible by both 3 and 8.

We are interested in the **least common denominator** (**LCD**) for two fractions. Using the least common denominator will make our addition easier. For the fractions $\frac{2}{3}$ and $\frac{5}{8}$, 24 is the least common denominator.

Example 5 Find the LCD for $\frac{x+5}{x-2}$ and $\frac{3x+4}{x+6}$.

Solution Since $(x - 2)$ will not factor and $(x + 6)$ will not factor, the least common denominator is the product of $(x - 2)$ and $(x + 6)$, or $(x - 2)(x + 6)$. The factored form is usually preferred, but our answer could be written as $x^2 + 4x - 12$.

Example 6 Find the LCD for $\frac{2x-3}{6x^2-5x-4}$ and $\frac{x+1}{3x^2-10x+8}$.

Solution Of course, the product

$$(6x^2 - 5x - 4)(3x^2 - 10x + 8)$$

is a common denominator, but it may not be the *least* common denominator. If the two denominators contain like factors, then the least common denominator will *not* be their product. Thus we must first factor the denominators.

$$\frac{2x-3}{6x^2-5x-4} = \frac{2x-3}{(3x-4)(2x+1)}$$

$$\frac{x+1}{3x^2-10x+8} = \frac{x+1}{(3x-4)(x-2)}$$

We are now looking for the least expression that contains both denominators as factors. Thus we know that the first denominator must be contained in the common denominator. In other words, the common denominator must contain at least the factors $(3x - 4)$ and $(2x + 1)$.

Now we look at the next denominator. Its factors must also be contained in the common denominator. This means that we need to include any factors of the second denominator that are not already present. Since $(3x - 4)$ is already present, we need only to include $(x - 2)$.

Thus our least common denominator is $(3x - 4)(2x + 1)(x - 2)$.

Notice in each of the preceding examples that the factors of each denominator are contained in the common denominator. Also note that no other factors are present and therefore this makes it the least common denominator.

RULE

To find the LCD (least common denominator) of several fractions:

Step 1 Factor each denominator into prime factors.
Step 2 Write the first denominator as the proposed common denominator.
Step 3 By inspection, determine which factors of the second denominator are not already in the proposed denominator and include them.
Step 4 Repeat step 3 for each fraction.

The resulting expression is the *least* common denominator.

EXERCISE 5-3-2

Find the LCD for the fractions in each of the following problems.

1. $\frac{1}{x}, \quad \frac{1}{y}, \quad \frac{3}{z}$

2. $\frac{1}{a}, \quad \frac{2}{a^2b}, \quad \frac{1}{b^2}$

3. $\frac{1}{x}, \quad \frac{1}{x + 2}$

4. $\frac{1}{x + 1}, \quad \frac{1}{x + 3}$

5. $\dfrac{a}{a-3}, \quad \dfrac{1}{a^2-9}$

6. $\dfrac{1}{x+3}, \quad \dfrac{1}{x^2-x-12}$

7. $\dfrac{1}{a-2}, \quad \dfrac{2a}{a^2-3a-10}$

8. $\dfrac{x-1}{x^2+5x+6}, \quad \dfrac{x}{x^2-x-12}$

9. $\dfrac{3x+1}{2x^2+11x+12}, \quad \dfrac{x-6}{2x^2+5x-12}$

10. $\dfrac{2x-1}{2x^2-x-15}, \quad \dfrac{x+4}{4x^2-25}$

11. $\dfrac{x-7}{x^2-3x-18}, \quad \dfrac{2x+3}{x^2+8x+12}$

12. $\dfrac{x+1}{x^2-x-6}, \quad \dfrac{x}{x^2-6x+9}, \quad \dfrac{2x-1}{x^2-2x-8}$

CAUTION Remember that the numerator *and* denominator of a fraction must be multiplied by the *same* expression to obtain an equal fraction.

Example 7 Change $\frac{2}{3}$ to a fraction having a denominator of 12.

$$\frac{2}{3} = \frac{?}{12}$$

Solution If we factor 12 into prime factors, we get

$$12 = 2 \cdot 2 \cdot 3.$$

We can then rewrite the problem as

$$\frac{2}{3} = \frac{?}{2 \cdot 2 \cdot 3}.$$

This form shows clearly that the denominator of the original fraction has been multiplied by $2 \cdot 2$. Hence the numerator must also be multiplied by $2 \cdot 2$. Thus

$$\frac{2}{3} = \frac{\boxed{2 \cdot 2} \cdot 2}{\boxed{2 \cdot 2} \cdot 3} = \frac{8}{12}.$$

Example 8 Change $\frac{3x + 1}{2x^2 + 11x + 12}$ to a fraction having a denominator of $(2x + 3)(x + 4)(x - 2)$.

$$\frac{3x + 1}{2x^2 + 11x + 12} = \frac{?}{(2x + 3)(x + 4)(x - 2)}$$

Solution First we factor the original denominator $(2x^2 + 11x + 12)$ into $(2x + 3)(x + 4)$.

$$\frac{3x + 1}{(2x + 3)(x + 4)} = \frac{?}{(2x + 3)(x + 4)(x - 2)}$$

This form shows clearly that the denominator of the original fraction has been multiplied by $(x - 2)$. Thus the original numerator must also be multiplied by $(x - 2)$.

$$\frac{3x + 1}{(2x + 3)(x + 4)} = \frac{(3x + 1)\boxed{(x - 2)}}{(2x + 3)(x + 4)\boxed{(x - 2)}} = \frac{3x^2 - 5x - 2}{(2x + 5)(x + 4)(x - 2)}$$

Notice that the factors of the numerator have been multiplied, but the denominator is left in factored form.

EXERCISE 5-3-3

Find the missing numerator in each fraction.

1. $\frac{1}{x+2} = \frac{?}{(x+2)(x-3)}$

2. $\frac{x+1}{2x+1} = \frac{?}{(2x+1)(x-2)}$

3. $\frac{2x+3}{x-4} = \frac{?}{4x-16}$

4. $\frac{x-3}{x+5} = \frac{?}{x^2+9x+20}$
(Hint: Factor the second denominator.)

5. $\frac{2x-1}{3x+4} = \frac{?}{3x^2-14x-24}$

6. $\frac{2x-5}{3x+4} = \frac{?}{9x^2+24x+16}$

7. $\frac{x-5}{x^2+2x-15} = \frac{?}{(x+5)(x-3)(x+2)}$

8. $\frac{2x+1}{x^2-3x-28} = \frac{?}{(x-7)(x+3)(x+4)}$

9. $\dfrac{x+7}{x^2-6x+5} = \dfrac{?}{2(x+7)(x-5)(x-1)}$

10. $\dfrac{3x+1}{2x^2-5x-3} = \dfrac{?}{(2x+1)(x-3)(x+2)}$

11. $\dfrac{3x+2}{x^2+4x-12} = \dfrac{?}{(2x-3)(x+6)(x-2)}$

12. $\dfrac{2x-1}{3x^2-16x+5} = \dfrac{?}{3(3x-1)(2x+1)(x-5)}$

5–4 ADDING ALGEBRAIC FRACTIONS

Upon completion of this section you should be able to:

1. Determine the LCD of algebraic fractions to be added.
2. Apply the five-step rule for adding algebraic fractions.

In the previous section you learned to find the least common denominator of two or more fractions and also to change a fraction to an equal fraction with a different denominator. We will now use these skills to add algebraic fractions.

To add two or more algebraic fractions:

Step 1 Find the LCD (least common denominator) of all fractions involved.
Step 2 Change each fraction to an equal fraction having as its denominator the least common denominator, and place the sum of the numerators over the common denominator.
Step 3 Clear all parentheses in the numerator.
Step 4 Combine like terms in the numerator.
Step 5 Simplify the answer by factoring and reducing to lowest terms if possible.

Example 1 Add: $\frac{3x}{y} + \frac{2x}{4z} + \frac{3}{8}$

Solution **Step 1** The LCD for y, $4z$, and 8 is $8yz$.

Step 2 Since

$$\frac{3x}{y} = \frac{24xz}{8yz}$$

$$\frac{2x}{4z} = \frac{4xy}{8yz}$$

and

$$\frac{3}{8} = \frac{3yz}{8yz}$$

we can write

$$\frac{3x}{y} + \frac{2x}{4z} + \frac{3}{8} = \frac{24xz}{8yz} + \frac{4xy}{8yz} + \frac{3yz}{8yz}$$

$$= \frac{24xz + 4xy + 3yz}{8yz}.$$

Step 3 Not necessary in this example.
Step 4 No like terms to combine.
Step 5 The answer is already in reduced form.

Example 2 Add: $\frac{x + 3y}{12} + \frac{x - 2y}{18}$

Solution **Step 1** The LCD for 12 and 18 is 36.

Step 2 $$\frac{x + 3y}{12} + \frac{x - 2y}{18} = \frac{\boxed{3}(x + 3y)}{\boxed{3}(12)} + \frac{\boxed{2}(x - 2y)}{\boxed{2}(18)}$$

$$= \frac{3(x + 3y) + 2(x - 2y)}{36}$$

Step 3 $$= \frac{3x + 9y + 2x - 4y}{36}$$

Step 4 $$= \frac{5x + 5y}{36}$$

Step 5 The numerator and denominator have no common factor, so the answer is in reduced form.

Example 3 Add: $\frac{x - 4}{x^2 - 2x - 15} + \frac{3}{x^2 + 5x + 6}$

Solution **Step 1** Since $x^2 - 2x - 15 = (x + 3)(x - 5)$ and $x^2 + 5x + 6 = (x + 3)(x + 2)$, the least common denominator is $(x + 3)(x - 5)(x + 2)$.

Step 2 $$\frac{x - 4}{(x + 3)(x - 5)} + \frac{3}{(x + 3)(x + 2)} = \frac{(x - 4)(x + 2) + 3(x - 5)}{(x + 3)(x - 5)(x + 2)}$$

(Note that in this example, instead of taking each fraction separately and changing to the least common denominator, we have put both new numerators over the *common* denominator. Since it is the *common* denominator, there is no advantage in writing the fractions separately.)

Step 3 $$= \frac{x^2 - 2x - 8 + 3x - 15}{(x + 3)(x - 5)(x + 2)}$$

Step 4 $= \dfrac{x^2 + x - 23}{(x + 3)(x - 5)(x + 2)}$

Step 5 Since $x^2 + x - 23$ does not factor, the answer is reduced.

Example 4 Add: $\dfrac{2}{x - 5} + \dfrac{x - 23}{x^2 - x - 20}$

Solution **Step 1** The LCD is $(x - 5)(x + 4)$.

Step 2 $\dfrac{2}{x - 5} + \dfrac{x - 23}{(x - 5)(x + 4)} = \dfrac{2(x + 4) + x - 23}{(x - 5)(x + 4)}$

Step 3 $= \dfrac{2x + 8 + x - 23}{(x - 5)(x + 4)}$

Step 4 $= \dfrac{3x - 15}{(x - 5)(x + 4)}$

Step 5 $= \dfrac{3\cancel{(x - 5)}}{\cancel{(x - 5)}(x + 4)} = \dfrac{3}{x + 4}$

EXERCISE 5-4-1

Add.

1. $\dfrac{1}{x} + \dfrac{1}{y}$

2. $\dfrac{1}{x} + \dfrac{1}{y} + \dfrac{1}{z}$

3. $\dfrac{1}{3} + \dfrac{2}{x + 5}$

4. $\dfrac{2}{x} + \dfrac{1}{x + 1}$

5. $\dfrac{1}{x + 2} + \dfrac{2}{x^2 - x - 6}$

6. $\dfrac{2}{x + 1} + \dfrac{4x}{x^2 + 4x + 3}$

7. $\dfrac{2}{x + 2} + \dfrac{1}{x - 5}$

8. $\dfrac{5}{x + 4} + \dfrac{2x - 13}{x^2 + x - 12} + \dfrac{1}{x - 3}$

9. $\dfrac{3}{x^2 - 9} + \dfrac{2}{x^2 + 6x + 9}$

10. $\dfrac{6}{5x + 10} + \dfrac{2x}{x^2 - 3x - 10}$

11. $\dfrac{1}{x + 2} + \dfrac{1}{x + 1} + \dfrac{1}{x - 3}$

12. $\dfrac{2}{x - 2} + \dfrac{3}{x + 1} + \dfrac{x - 8}{x^2 - x - 2}$

13. $\dfrac{3}{x + 1} + \dfrac{2}{x - 3} + \dfrac{5}{x + 4}$

14. $\dfrac{x + 3}{x - 6} + \dfrac{x - 1}{x^2 - 2x - 24}$

15. $\dfrac{x + 2}{x + 5} + \dfrac{x - 3}{x^2 + 3x - 10}$

16. $\dfrac{x + 3}{x^2 - x - 2} + \dfrac{2x - 1}{x^2 + 2x - 8}$

17. $\dfrac{2x}{x^2 + 6x + 9} + \dfrac{x - 1}{x^2 - 9}$

Sometimes more than one operation may appear in a problem. Recall the order of operations and the meaning of parentheses in the following problems.

18. $\frac{3}{x} + \frac{5}{y} \cdot \frac{4y}{3x}$

19. $\frac{1}{x^2 - x - 2} \div \frac{1}{x + 1} + \frac{3}{x - 2}$

20. $\frac{1}{x^2 - x - 2} \div \left(\frac{1}{x + 1} + \frac{3}{x - 2}\right)$

5–5 SUBTRACTING ALGEBRAIC FRACTIONS

OBJECTIVES

Upon completion of this section you should be able to:

1. Find the LCD of algebraic fractions to be subtracted.
2. Apply the five-step rule for subtracting algebraic fractions.

Subtraction of fractions, like addition, can only be performed if the fractions have the same denominator. When the denominators are the same, the difference is found by placing the difference of the numerators over the common denominator. In symbols this is expressed as

$$\frac{a}{c} - \frac{b}{c} = \frac{a - b}{c}.$$

If the denominators are not the same, a common denominator must be found and each fraction changed to an equal fraction having the common denominator. As was true in addition, our work can be simplified by using the *least* common denominator.

RULE

To subtract one fraction from another:

Step 1 Find the LCD (least common denominator) of the two fractions.
Step 2 Change each fraction to an equal fraction having as its denominator the least common denominator and place the difference of the numerators over the common denominator.
Step 3 Clear all parentheses in the numerator.
Step 4 Combine like terms in the numerator.
Step 5 Simplify the answers by factoring and reduce to lowest terms if possible.

Example 1 Subtract: $\dfrac{x+3}{x+4} - \dfrac{5}{x+2}$

Solution **Step 1** The least common denominator is $(x+4)(x+2)$.

Step 2 $$\frac{x+3}{x+4} - \frac{5}{x+2} = \frac{(x+3)(x+2)}{(x+4)(x+2)} - \frac{5(x+4)}{(x+2)(x+4)}$$
$$= \frac{(x+3)(x+2) - 5(x+4)}{(x+4)(x+2)}$$

Step 3 $$= \frac{x^2 + 5x + 6 - 5x - 20}{(x+4)(x+2)}$$

Step 4 $$= \frac{x^2 - 14}{(x+4)(x+2)}$$

Step 5 The answer is in simplified form.

Example 2 Subtract: $\dfrac{2}{x-3} - \dfrac{x+7}{x^2 - x - 6}$

Solution **Step 1** The least common denominator is $(x-3)(x+2)$.

Step 2 $$\frac{2}{x-3} - \frac{x+7}{x^2 - x - 6} = \frac{2(x+2) - (x+7)}{(x-3)(x+2)}$$

CAUTION Note that in this step the numerator of the second fraction is placed in parentheses. This is necessary since the subtraction sign will change each sign of the factor $(x + 7)$, giving $-x - 7$.

Step 3 $$= \frac{2x + 4 - x - 7}{(x-3)(x+2)}$$

Step 4 $$= \frac{x-3}{(x-3)(x+2)}$$

Step 5 $$= \frac{1}{x+2}$$

Example 3 Subtract: $\dfrac{x+1}{x^2-x-6} - \dfrac{x+2}{2x-6}$

Solution **Step 1** The least common denominator is $2(x-3)(x+2)$.

Step 2 $\dfrac{x+1}{(x-3)(x+2)} - \dfrac{x+2}{2(x-3)} = \dfrac{2(x+1)-(x+2)(x+2)}{2(x-3)(x+2)}$

Step 3 $= \dfrac{2x+2-x^2-4x-4}{2(x-3)(x+2)}$

CAUTION Note that the subtraction sign in the numerator changed every sign in the product $(x+2)(x+2)$. A good practice is to perform the multiplication $(x+2)(x+2) = x^2+4x+4$ and then change each sign. Don't attempt to multiply and change the signs at the same time or mistakes are likely to result.

Step 4 $= \dfrac{-x^2-2x-2}{2(x-3)(x+2)}$

Step 5 The answer can be left as it is, or factoring a -1 from the numerator could change it to

$$= -\frac{x^2+2x+2}{2(x-3)(x+2)}.$$

Either answer is correct.

EXERCISE 5-5-1

Subtract.

1. $\dfrac{3}{5} - \dfrac{4}{x}$

2. $\dfrac{2}{x} - \dfrac{4}{y}$

3. $\dfrac{x}{x+1} - \dfrac{2}{3}$

4. $\dfrac{5}{6} - \dfrac{2x}{3x-12}$

5. $\dfrac{4}{x} - \dfrac{6}{x+3}$

6. $\dfrac{1}{x+2} - \dfrac{1}{x+3}$

7. $\dfrac{2x}{x-3} - \dfrac{4}{x+5}$

8. $\dfrac{x}{x+1} - \dfrac{4}{x^2-1}$

9. $\dfrac{6}{x+2} - \dfrac{x}{x^2-x-6}$

10. $\dfrac{2x}{x-1} - \dfrac{3x+2}{x^2+3x-4}$

11. $\dfrac{5}{x+3} - \dfrac{4x-13}{x^2+x-6}$

12. $\dfrac{x+2}{x-5} - \dfrac{5x+31}{x^2-2x-15}$

13. $\dfrac{x+1}{3x^2-12} - \dfrac{x-3}{x+2}$

14. $\dfrac{x+4}{x^2-x-2} - \dfrac{x}{x^2+x-6}$

15. $\dfrac{2x}{x^2-1} - \dfrac{2x+4}{x^2+4x-5}$

Combine. Recall the order of operations and the meaning of parentheses.

16. $\dfrac{2}{x+5} + \dfrac{1}{x-4} - \dfrac{3}{x+3}$

17. $\dfrac{x}{x^2-4x-5} - \dfrac{2}{x+1} + \dfrac{2x-1}{x^2-7x+10}$

18. $\dfrac{1}{x^2 - x - 6} \div \dfrac{1}{x+2} - \dfrac{1}{x-3}$

19. $\dfrac{1}{x^2 - x - 6} \div \left(\dfrac{1}{x+2} - \dfrac{1}{x-3}\right)$

20. $\dfrac{2}{x-1} \cdot \dfrac{3x-3}{x+5} + \dfrac{x+1}{x^2+4x-5} \div \dfrac{2}{x-1} - \dfrac{1}{2}$

5-6 SIMPLIFYING COMPLEX FRACTIONS

OBJECTIVES

Upon completion of this section you should be able to:

1. Identify a complex fraction.
2. Simplify a complex fraction.

A fraction that has either a numerator or denominator, or both, composed of fractions is called a **complex fraction.**

Examples of complex fractions are

$$\frac{\frac{a}{b}}{\frac{c}{d}}, \quad \frac{\frac{x}{y} + 1}{x}, \quad \frac{\frac{1}{x} + \frac{1}{y}}{\frac{a}{b} + \frac{c}{d}}.$$

There are two methods that can be used to change a complex fraction to a simple fraction. The answer must then be put in simplified form.

RULE FOR METHOD 1
To simplify a complex fraction:

Step 1 Find the LCD (least common denominator) of all fractions in the expression.

Step 2 Multiply the numerator and denominator of the complex fraction by the LCD.

Step 3 If necessary, simplify the resulting fraction.

Example 1 Simplify: $\dfrac{\dfrac{1}{x}+\dfrac{1}{y}}{\dfrac{1}{x}-\dfrac{1}{y}}$

Solution **Step 1** The least common denominator of all the fractions is xy.

Step 2
$$\frac{\dfrac{1}{x}+\dfrac{1}{y}}{\dfrac{1}{x}-\dfrac{1}{y}} = \frac{xy\left(\dfrac{1}{x}+\dfrac{1}{y}\right)}{xy\left(\dfrac{1}{x}-\dfrac{1}{y}\right)}$$
$$= \frac{xy\left(\dfrac{1}{x}\right)+xy\left(\dfrac{1}{y}\right)}{xy\left(\dfrac{1}{x}\right)-xy\left(\dfrac{1}{y}\right)}$$
$$= \frac{y+x}{y-x}$$

Step 3 The answer is simplified.

Example 2 Simplify: $\dfrac{\dfrac{1}{a^2}-\dfrac{1}{b^2}}{a+b}$

Solution **Step 1** The least common denominator is a^2b^2.

Step 2
$$\frac{\dfrac{1}{a^2}-\dfrac{1}{b^2}}{a+b} = \frac{a^2b^2\left(\dfrac{1}{a^2}-\dfrac{1}{b^2}\right)}{a^2b^2(a+b)}$$
$$= \frac{b^2-a^2}{a^2b^2(a+b)}$$
$$= \frac{(b-a)\cancel{(b+a)}}{a^2b^2\cancel{(a+b)}}$$
$$= \frac{b-a}{a^2b^2}$$

Example 3 Simplify: $\dfrac{\dfrac{3}{x^2 + 6x + 8} + \dfrac{1}{x + 2}}{\dfrac{2}{x + 4} - 1}$

Solution **Step 1** After factoring the first denominator $x^2 + 6x + 8$ into $(x + 2)(x + 4)$, we find the LCD is $(x + 2)(x + 4)$.

Step 2

$$\frac{\dfrac{3}{(x + 2)(x + 4)} + \dfrac{1}{x + 2}}{\dfrac{2}{x + 4} - 1} = \frac{(x + 2)(x + 4)\left(\dfrac{3}{(x + 2)(x + 4)} + \dfrac{1}{x + 2}\right)}{(x + 2)\,(x + 4)\left(\dfrac{2}{x + 4} - 1\right)}$$

$$= \frac{3 + (x + 4)}{2(x + 2) - (x + 2)(x + 4)}$$

$$= \frac{3 + x + 4}{2x + 4 - x^2 - 6x - 8}$$

$$= \frac{x + 7}{-x^2 - 4x - 4}$$

Step 3 The answer is in simplest form but could be changed to $-\dfrac{x + 7}{x^2 + 4x + 4}$.

RULE FOR METHOD 2

To simplify a complex fraction:

Step 1 Perform the operations necessary to change both numerator and denominator to single fractions.

Step 2 Divide the numerator by the denominator.

Example 4 Simplify: $\dfrac{\dfrac{1}{x} + \dfrac{1}{y}}{\dfrac{1}{x} - \dfrac{1}{y}}$

Solution **Step 1**

$$\frac{1}{x} + \frac{1}{y} = \frac{(1)y + (1)x}{xy} = \frac{y + x}{xy} \leftarrow \text{Numerator}$$

$$\frac{1}{x} - \frac{1}{y} = \frac{(1)y - (1)x}{xy} = \frac{y - x}{xy} \leftarrow \text{Denominator}$$

Step 2

$$\frac{y + x}{xy} \div \frac{y - x}{xy} = \frac{y + x}{\cancel{xy}} \cdot \frac{\cancel{xy}}{y - x} = \frac{y + x}{y - x}$$

Example 5 Simplify: $\dfrac{\dfrac{1}{a^2} - \dfrac{1}{b^2}}{a + b}$

Solution **Step 1** $\dfrac{1}{a^2} - \dfrac{1}{b^2} = \dfrac{(1)b^2 - (1)a^2}{a^2b^2} = \dfrac{b^2 - a^2}{a^2b^2}$

Step 2
$$\frac{b^2 - a^2}{a^2b^2} \div (a + b) = \frac{b^2 - a^2}{a^2b^2} \cdot \frac{1}{a + b}$$
$$= \frac{(b - a)\cancel{(b + a)}}{a^2b^2} \cdot \frac{1}{\cancel{(a + b)}}$$
$$= \frac{b - a}{a^2b^2}$$

Example 6 Simplify: $\dfrac{\dfrac{3}{x^2 + 6x + 8} + \dfrac{1}{x + 2}}{\dfrac{2}{x + 4} - 1}$

Solution **Step 1**
$$\frac{3}{x^2 + 6x + 8} + \frac{1}{x + 2} = \frac{3 + 1(x + 4)}{(x + 2)(x + 4)}$$
$$= \frac{x + 7}{(x + 2)(x + 4)}$$

$$\frac{2}{x + 4} - 1 = \frac{2 - 1(x + 4)}{x + 4} = \frac{-x - 2}{x + 4}$$

Step 2
$$\frac{x + 7}{(x + 2)(x + 4)} \div \frac{-x - 2}{x + 4} = \frac{x + 7}{(x + 2)\cancel{(x + 4)}} \cdot \frac{\cancel{x + 4}}{-x - 2}$$
$$= \frac{x + 7}{(x + 2)(-x - 2)}$$
$$= \frac{x + 7}{-1(x + 2)(x + 2)}$$
$$= -\frac{x + 7}{(x + 2)^2}$$

EXERCISE 5–6–1

Simplify using either method.

1. $\dfrac{\dfrac{1}{a}}{\dfrac{1}{a^2}}$

2. $\dfrac{\dfrac{1}{a}}{\dfrac{1}{b}}$

3. $\dfrac{\dfrac{1}{x}}{y}$

4. $\dfrac{\dfrac{1}{x} + \dfrac{1}{y}}{xy}$

5. $\dfrac{\dfrac{1}{a} + \dfrac{1}{b}}{a + b}$

6. $\dfrac{\dfrac{1}{x} + \dfrac{1}{y}}{\dfrac{1}{y} + \dfrac{1}{z}}$

7. $\dfrac{1 + \dfrac{1}{a}}{ab}$

8. $\dfrac{\dfrac{1}{a} + \dfrac{1}{b}}{a^2 - b^2}$

9. $\dfrac{2 - \dfrac{1}{x}}{2x - 1}$

10. $\dfrac{\dfrac{1}{3} + \dfrac{1}{a}}{\dfrac{1}{b}}$

11. $\dfrac{\dfrac{a}{b} - \dfrac{b}{a}}{a + b}$

12. $\dfrac{\dfrac{1}{x + 2} - \dfrac{1}{x}}{2}$

13. $\dfrac{\dfrac{1}{x^2} - \dfrac{1}{y^2}}{\dfrac{1}{x^2y^2}}$

14. $\dfrac{\dfrac{1}{x^2 - 1} + 1}{\dfrac{1}{x + 1} - \dfrac{1}{x - 1}}$

15. $\dfrac{\dfrac{1}{x - 3} + \dfrac{1}{x^2 - 2x - 3}}{\dfrac{1}{x + 1}}$

16. $\dfrac{\dfrac{1}{a+b} - \dfrac{1}{a-b}}{\dfrac{2b}{a^2 - b^2}}$

17. $\dfrac{\dfrac{1}{a} + \dfrac{1}{a+b}}{\dfrac{2}{a+b} + 1}$

18. $\dfrac{2 + \dfrac{1}{x-2}}{\dfrac{1}{x+3} - 1}$

19. $\dfrac{\dfrac{1}{x+y} - \dfrac{1}{y}}{\dfrac{1}{x^2 - y^2}}$

20. $\dfrac{\dfrac{1}{2x} - \dfrac{1}{6y}}{\dfrac{1}{3y} + \dfrac{1}{4x}}$

21. $\dfrac{\dfrac{6}{x^2 + 3x - 10} - \dfrac{1}{x-2}}{\dfrac{1}{x-2} + 1}$

22. $\dfrac{\dfrac{1}{x+3} + \dfrac{6}{x^2 - x - 12}}{1 - \dfrac{1}{x+3}}$

23. $\dfrac{\dfrac{1}{x^2 + 4x + 3} + \dfrac{1}{x-2}}{\dfrac{1}{x+3} + \dfrac{1}{x^2 - x - 2}}$

5-7 FRACTIONAL EQUATIONS

OBJECTIVES
Upon completion of this section you should be able to:
1. Apply the six-step rule for solving fractional equations.
2. Determine when a fractional equation does not have a solution.

In chapter 2 we solved many equations containing fractions. However, all of those fractions were numerical fractions. In this section we will solve equations containing fractions that are algebraic. The skills you learned in chapter 2 and in the previous sections of this chapter will be used as you solve these equations.

An equation with an unknown in the denominator of at least one of its fractions is called a **fractional equation.**

$$\frac{1}{x} + \frac{1}{x-1} = \frac{2}{x}$$

is an example of a fractional equation.

A STEP-BY-STEP PROCEDURE FOR SOLVING FRACTIONAL EQUATIONS

Step 1 Find the LCD (least common denominator) of all fractions in the equation.

Step 2 Multiply each term of the equation by the common denominator from step 1. This step will eliminate all denominators.

Step 3 Combine similar terms on each side of the equation.

Step 4 Add or subtract numbers on both sides of the equation to get the unknowns on one side and the numbers on the other.

Step 5 Divide both sides of the equation by the coefficient of the unknown and simplify.

Step 6 Substitute the result of step 5 in the original equation to check the solution.

Note: Step 6 is especially necessary when solving fractional equations. Sometimes the answer will not check even though an error has not been made in the procedure. This happens when the solution of the equation in step 2 results in a denominator of the original equation equal to zero. In such cases the equation has no solution.

Example 1 Solve for x: $\frac{2}{x} + \frac{3}{2} = 2$

Solution

Step 1 The least common denominator is $2x$.

Step 2 $\boxed{2x}\left(\frac{2}{x}\right) + \boxed{2x}\left(\frac{3}{2}\right) = \boxed{2x}(2)$

$$4 + 3x = 4x$$

Step 3 There are no similar terms to combine.

Step 4 $-x = -4$

Step 5 $x = 4$

Step 6 We check in the original equation.

$$\frac{2}{x} + \frac{3}{2} = 2$$

$$\frac{2}{4} + \frac{3}{2} \quad \Big| \quad 2$$

$$\frac{1}{2} + \frac{3}{2}$$

$$\frac{4}{2}$$

$$2$$

The answer checks so the solution is $x = 4$.

Example 2 Solve for x: $\dfrac{2x}{x+1} + \dfrac{1}{3x-2} = 2$

Solution **Step 1** The least common denominator is $(3x - 2)(x + 1)$.

Step 2 $\boxed{(3x-2)(x+1)}\left(\dfrac{2x}{x+1}\right) + \boxed{(3x-2)(x+1)}\left(\dfrac{1}{3x-2}\right) = 2\boxed{(3x-2)(x+1)}$

$$(3x-2)(2x) + (x+1)(1) = 2(3x^2 + x - 2)$$

$$6x^2 - 4x + x + 1 = 6x^2 + 2x - 4$$

Step 3 $6x^2 - 3x + 1 = 6x^2 + 2x - 4$

Step 4 $-5x = -5$

Note that the terms involving x^2 cancelled. If this did not occur, we would be unable, at this time, to solve the resulting equation. Equations involving x^2 are called **quadratic equations** and will be studied later.

Step 5 $x = 1$

Step 6 $\dfrac{2x}{x+1} + \dfrac{1}{3x-2} = 2$

$$\frac{2(1)}{(1)+1} + \frac{1}{3(1)-2} \quad \Big| \quad 2$$

$$\frac{2}{2} + \frac{1}{1}$$

$$1 + 1$$

$$2$$

The answer checks so the solution is $x = 1$.

Example 3 Solve for x: $\dfrac{2}{x^2 - 1} - \dfrac{1}{x + 1} = \dfrac{1}{x - 1}$

Solution **Step 1** The least common denominator is $(x + 1)(x - 1)$.

Step 2
$$\boxed{(x + 1)(x - 1)}\left(\frac{2}{x^2 - 1}\right) - \boxed{(x + 1)(x - 1)}\left(\frac{1}{x + 1}\right) = \boxed{(x + 1)(x - 1)}\left(\frac{1}{x - 1}\right)$$
$$2 - (x - 1) = x + 1$$
$$2 - x + 1 = x + 1$$

Step 3 $3 - x = x + 1$

Step 4 $-2x = -2$

Step 5 $x = 1$

Step 6
$$\frac{2}{x^2 - 1} - \frac{1}{x + 1} = \frac{1}{x - 1}$$
$$\frac{2}{(1)^2 - 1} - \frac{1}{(1) + 1} \quad \Big| \quad \frac{1}{(1) - 1}$$
$$\frac{2}{0} - \frac{1}{2} \quad \Big| \quad \frac{1}{0}$$

We observe here that $\frac{2}{0}$ and $\frac{1}{0}$ are meaningless expressions. Thus $x = 1$ cannot be a solution of the original equation. We conclude that there is no solution to this equation.

CAUTION Be careful to distinguish between adding fractions and solving fractional equations. Confusion may arise because the least common denominator is used in both processes.

Example 4 Add: $\dfrac{2}{x} + \dfrac{x}{x + 1}$

Solution We must first find the LCD of the fractions and then convert each to a fraction having the LCD as its denominator.

$$\frac{2}{x} \cdot \boxed{\frac{x + 1}{x + 1}} + \frac{x}{x + 1} \cdot \boxed{\frac{x}{x}} = \frac{2x + 2 + x^2}{x(x + 1)}$$

The result is a *fraction* that represents the sum of the two original fractions.

Example 5 Solve: $\dfrac{2}{x} + \dfrac{x}{x + 1} = 1$

Solution We must first find the LCD and then multiply it by each term in the equation.

$$\boxed{x(x + 1)}\frac{2}{x} + \boxed{x(x + 1)}\frac{x}{x + 1} = \boxed{x(x + 1)}(1)$$
$$2x + 2 + x^2 = x^2 + x$$
$$x = -2$$

The result is a *value* that yields a true statement when substituted for x in the original equation.

Example 6 The numerator of the fraction $\frac{13}{27}$ is increased by an amount so that the value of the resulting fraction is $\frac{8}{9}$. By what amount was the numerator increased?

Solution If we let x represent the number added to the numerator then we have

$$\frac{13 + x}{27} = \frac{8}{9}.$$

Solving for x, we use the least common denominator, 27, to multiply each term.

$$\frac{\cancel{27}(13 + x)}{\cancel{27}} = \frac{\overset{3}{\cancel{27}}(8)}{\cancel{9}}$$

$$13 + x = 24$$

$$x = 11$$

Check:

$$\frac{13 + x}{27} = \frac{8}{9}$$

$$\frac{13 + 11}{27} \;\Big|\; \frac{8}{9}$$

$$\frac{24}{27}$$

$$\frac{8}{9}$$

Therefore 11 is the amount by which the numerator was increased.

Finding the solution to an applied problem often requires the use of a fractional equation. **Work problems** fall into this category.

Example 7 Bill can clean the outside of a travel trailer in 6 hours. Bill and his helper working together can do the job in 4 hours. How long would it take his helper working alone to clean the trailer?

Solution The equation needed to solve this problem is based on the amount of work done in one unit of time.

	Hours to Complete the Job	Amount of Job Completed in One Hour
Bill	6	$\frac{1}{6}$
Helper	x	$\frac{1}{x}$
Together	4	$\frac{1}{4}$

$$\begin{pmatrix}\text{Amount Bill}\\ \text{completed in}\\ \text{one hour}\end{pmatrix} + \begin{pmatrix}\text{Amount helper}\\ \text{completed in}\\ \text{one hour}\end{pmatrix} = \begin{pmatrix}\text{Amount both}\\ \text{completed in}\\ \text{one hour}\end{pmatrix}$$

$$\frac{1}{6} + \frac{1}{x} = \frac{1}{4}$$

The LCD is $12x$.

$$12x\left(\frac{1}{6}\right) + 12x\left(\frac{1}{x}\right) = 12x\left(\frac{1}{4}\right)$$

$$2x + 12 = 3x$$

$$x = 12$$

Checking this value in the original equation, we have

$$\frac{1}{6} + \frac{1}{x} = \frac{1}{4}$$

$$\frac{1}{6} + \frac{1}{12} \;\Big|\; \frac{1}{4}$$

$$\frac{2 + 1}{12}$$

$$\frac{3}{12}$$

$$\frac{1}{4}$$

It would therefore take the helper 12 hours to clean the trailer alone.

EXERCISE 5-7-1

Solve.

1. $\dfrac{5}{x} = \dfrac{1}{2}$

2. $\dfrac{1}{x} + \dfrac{1}{2x} = 3$

3. $\dfrac{2}{x} + \dfrac{1}{3} = \dfrac{1}{2x} - 1$

4. $\dfrac{x + 1}{x} - \dfrac{x - 2}{2x} = 1$

5. $\dfrac{1}{x} = \dfrac{3}{2x}$

6. $\dfrac{4}{x} + \dfrac{1}{2x} = \dfrac{9}{4}$

7. $\dfrac{3}{x - 1} = 1$

8. $\dfrac{x}{x - 2} = \dfrac{4}{3}$

9. $\dfrac{2}{x} = \dfrac{1}{x + 1}$

10. $\dfrac{x}{x + 1} = \dfrac{x - 1}{x - 2}$

11. $\dfrac{1}{x + 1} = \dfrac{2}{1 - x^2}$

12. $\dfrac{x + 1}{x^2 + 2x - 3} + \dfrac{1}{x - 1} = \dfrac{1}{x + 3}$

13. $\dfrac{x + 2}{x^2 + 7x} = \dfrac{1}{x + 3}$

14. $\dfrac{1}{x^2 - 3x} = \dfrac{2}{x^2 - 9}$

15. $\dfrac{x}{x+1} + \dfrac{1}{2x+1} = 1$

16. $\dfrac{3}{x+5} = 1 - \dfrac{x-4}{2x+10}$

17. $\dfrac{x}{x+4} - \dfrac{x}{x-4} = \dfrac{x+18}{x^2-16}$

18. $\dfrac{2}{x-1} + \dfrac{x}{x+1} = 1 - \dfrac{1}{x^2-1}$

19. $\dfrac{2}{3x+6} = \dfrac{1}{6} - \dfrac{1}{2x+4}$

20. $\dfrac{x+1}{x+5} - \dfrac{2x+1}{x-2} = \dfrac{5-x^2}{x^2+3x-10}$

21. Three divided by a given number is equal to four divided by three more than the given number. Find the number.

22. In a fraction the numerator is three less than the denominator. If one is added to both the numerator and denominator, the value of the resulting fraction is $\dfrac{5}{6}$. Find the original fraction.

23. The numerator of a fraction is four less than the denominator. If the numerator is increased by three and the denominator is increased by seven, the resulting fraction has the same value as the original fraction. Find the original fraction.

24. The sum of the reciprocals of two consecutive even integers is equal to seven divided by four times the first number. Find the numbers.

25. A camera utilizes the formula from optics $\frac{1}{f} = \frac{1}{a} + \frac{1}{b}$, where f represents the focal length of the lens, a represents the distance of the object from the lens, and b represents the distance of the image from the lens. If a lens has a focal length of 12 centimeters, how far from the lens will the image appear when the object is 36 centimeters from the lens?

26. If an electric circuit is wired in parallel with two resistors, the total resistance R (measured in ohms) is given by the formula $\frac{1}{R} = \frac{1}{r_1} + \frac{1}{r_2}$, where r_1 and r_2 represent the resistances of the two resistors. If $R = 300$ ohms and $r_2 = 750$ ohms, find r_1.

27. An experienced bricklayer and his apprentice can build a wall together in 3 hours. It would take the apprentice 12 hours to do the job alone. How long would it take the experienced bricklayer to do the job alone?

28. Mower A can cut a field in 6 hours and mower B can cut the same field in 2 hours. How long would it take both mowers working together to cut the field?

29. Two pumps working together can fill a swimming pool in 4 hours. If pump A can fill it alone in 5 hours, how long would it take pump B to fill it alone?

30. A drain can empty a tank in 4 hours. Pump A can fill the tank in 6 hours and pump B can fill it in 8 hours. If both pumps are working together and the drain is accidentally left open, how long will it take to fill the tank?

SUMMARY

The number in brackets refers to the section of the chapter that discusses that concept.

Terminology

- The **fundamental principle of fractions** is

$$\frac{a}{b} = \frac{ax}{bx}, x \neq 0. \qquad [5\text{–}1]$$

- A fraction is in **simplified form** or **reduced form** if the numerator and denominator have no prime factors in common. [5–1]
- A **common denominator** for two or more fractions is an expression that contains all factors of the denominators of each fraction. [5–3]
- A **least common denominator** (LCD) contains the minimum number of factors to be a common denominator. [5–3]
- A fraction that has either a numerator or denominator, or both, composed of fractions is called a **complex fraction.** [5–6]

Rules and Procedures

Simplifying Fractions

- To simplify a fraction factor both the numerator and denominator into prime factors, then cancel all common factors. [5–1]

Multiplication of Fractions

- To multiply algebraic fractions: [5–2]
  STEP 1 Factor all numerators and denominators into prime factors.
  STEP 2 Cancel any factor that is common to a numerator and a denominator.
  STEP 3 Multiply the remaining factors in the numerator and place this product over the product of the remaining factors in the denominator.

Division of Fractions

- To divide fractions invert the divisor and multiply. [5–2]

Least Common Denominator

- To find the least common denominator (LCD) of two or more fractions: [5–3]
  STEP 1 Factor each denominator into prime factors.
  STEP 2 Write the first denominator as the proposed common denominator.
  STEP 3 By inspection, determine which factors of the second denominator are not already in the proposed denominator and include them.
  STEP 4 Repeat step 3 for each fraction.

The resulting expression is the *least* common denominator.

Addition and Subtraction of Fractions

- To add or subtract fractions change all fractions so they have a common denominator and combine the numerators. Reduce all results. [5–4]

Simplifying Complex Fractions

- Method 1. To simplify a complex fraction: [5–6]
  STEP 1 Find the LCD (least common denominator) of all fractions in the expression.
  STEP 2 Multiply the numerator and denominator of the complex fraction by the LCD.
  STEP 3 If necessary, simplify the resulting fraction.
- Method 2. To simplify a complex fraction: [5–6]
  STEP 1 Perform the operations necessary to change both numerator and denominator to single fractions.
  STEP 2 Divide the numerator by the denominator.

Fractional Equations

- To solve fractional equations: [5–7]
  STEP 1 Find the LCD of all fractions in the equation.
  STEP 2 Multiply each term of both sides of the equation by the LCD from step 1.
  STEP 3 Combine similar terms on both sides of the equation.
  STEP 4 Add or subtract numbers on both sides of the equation to obtain the unknown on one side and the numbers on the other.
  STEP 5 Divide both sides of the equation by the coefficient of the unknown and simplify.
  STEP 6 Substitute the result of step 5 in the original equation to check the solution.

REVIEW

Reduce.

1. $\dfrac{3x + 45}{12}$

2. $\dfrac{2x + 8}{3x + 12}$

3. $\dfrac{x^2 - 2x - 15}{x^2 + 3x - 40}$

4. $\dfrac{x^2 - 49}{x^2 + 14x + 49}$

5. $\dfrac{6x + 30}{10x^2 + 40x - 50}$

Multiply.

6. $\dfrac{3x + 6}{x} \cdot \dfrac{5x}{21x + 42}$

7. $\dfrac{x^2 - 25}{x^2 + 7x + 10} \cdot \dfrac{x^2 + x - 2}{x^2 - 9x + 20}$

8. $\dfrac{3x^2 - 10x - 8}{3x^2 - 11x - 4} \cdot \dfrac{12x^2 + 7x + 1}{6x^2 + x - 2}$

9. $\dfrac{x^2 - 16}{2x^2 - 5x - 12} \cdot \dfrac{4x^2 + 8x + 3}{2x^2 + 7x - 4}$

10. $\dfrac{x^2 + 3x + 2}{x^2 + 5x + 6} \cdot \dfrac{x^2 - 2x - 15}{x^2 - 3x - 10} \cdot \dfrac{2x^2 - x - 10}{x^2 + 5x + 4}$

Divide.

11. $\dfrac{2x - 6}{3x} \div \dfrac{x^2 - 2x - 3}{x^2 + x}$

12. $\dfrac{x + 1}{x^2 - 5x - 6} \div \dfrac{x^2 - 1}{x - 6}$

13. $\dfrac{x^2 + 7x + 10}{x^2 + 4x - 5} \div \dfrac{x^2 + 7x + 12}{x^2 + 2x - 3}$

14. $\dfrac{9x^2 + 18x + 8}{6x^2 + 19x + 10} \div \dfrac{12x^2 + 13x - 4}{8x^2 + 10x - 3}$

15. $\dfrac{x^2 - 16}{x^2 + 5x + 4} \div \dfrac{x^2 - 8x + 16}{x^2 - 1}$

Find the least common denominator.

16. $\dfrac{3}{x}, \quad \dfrac{2}{x + 2}$

17. $\dfrac{4}{x^2 + 3x}, \quad \dfrac{x}{x + 3}$

18. $\dfrac{3}{x^2 - 5x}, \quad \dfrac{x - 1}{x^2 - 4x - 5}$

19. $\dfrac{4}{x + 1}, \quad \dfrac{1}{x^2 - 4}, \quad \dfrac{5}{x + 2}$

20. $\dfrac{2x}{x^2 - 1}, \quad \dfrac{3}{x^2 - x - 2}$

Add.

21. $\dfrac{3}{x} + \dfrac{2}{x + 2}$

22. $\dfrac{4}{x^2 + 3x} + \dfrac{x}{x + 3}$

23. $\dfrac{3}{x^2 - 5x} + \dfrac{x - 1}{x^2 - 4x - 5}$

24. $\dfrac{4}{x + 1} + \dfrac{1}{x^2 - 4} + \dfrac{5}{x + 2}$

25. $\frac{2x}{x^2 - 1} + \frac{3}{x^2 - x - 2}$

Subtract.

26. $\frac{3}{4} - \frac{x}{x + 5}$

27. $\frac{3}{x + 1} - \frac{2}{x - 2}$

28. $\frac{5}{x - 1} - \frac{2x + 6}{x^2 + 2x - 3}$

29. $\frac{2x}{x^2 - 4} - \frac{x + 4}{x^2 + 4x - 12}$

30. $\frac{1}{x + 2} + \frac{1}{x + 1} - \frac{2}{x - 7}$

Simplify.

31. $\dfrac{\dfrac{1}{a} + \dfrac{1}{b}}{a + b}$

32. $\dfrac{1 - \dfrac{1}{a}}{a - 1}$

33. $\dfrac{\dfrac{1}{x} + \dfrac{1}{y}}{\dfrac{1}{x + y}}$

34. $\dfrac{\dfrac{1}{x + 3} - 1}{\dfrac{1}{x^2 - 9}}$

35. $\dfrac{\dfrac{-13}{x^2 - 2x - 35} - \dfrac{1}{x + 5}}{\dfrac{1}{x + 5} + 1}$

Solve.

36. $\frac{3}{2x} = \frac{1}{x} + \frac{1}{2}$

37. $\frac{x}{x+7} = \frac{x-1}{x}$

38. $\frac{1}{x-2} + \frac{1}{x+2} = \frac{4}{x^2-4}$

39. $\frac{1}{x+1} - \frac{1}{2} = \frac{1}{3x+3}$

40. $\frac{x+1}{x+2} - \frac{x-3}{x+5} = \frac{4}{x^2+7x+10}$

41. In the electrical formula $\frac{1}{R} = \frac{1}{r_1} + \frac{1}{r_2} + \frac{1}{r_3}$, R, r_1, r_2, and r_3 represent resistances (measured in ohms). Find r_2 if $R = 5$ ohms, $r_1 = 40$ ohms, and $r_3 = 8$ ohms.

42. A painter can paint a house alone in 12 hours. His helper can paint the same house alone in 15 hours. How long will it take both of them working together to paint the house?

CHAPTER 5

PRACTICE TEST

1. Simplify: $\dfrac{5x+15}{20}$

1. ______________________

2. Simplify: $\dfrac{x^2-2x-15}{x^2-25}$

2. ______________________

3. Multiply: $\dfrac{x^2+8x+15}{x^2+x-6}\cdot\dfrac{x^2+3x-10}{x^2+4x-5}$

3. ______________________

4. Divide: $\dfrac{2x+8}{3x}\div\dfrac{x^2+3x-4}{x^2-x}$

4. ______________________

5. Find the LCD: $\dfrac{4}{x},\quad \dfrac{2}{x-7}$

5. ______________________

6. Find the LCD: $\dfrac{2}{x + 6}$, $\dfrac{5}{x^2 - 36}$, $\dfrac{1}{x + 4}$

6. ____________________

7. Combine: $\dfrac{3}{x + 1} + \dfrac{2}{x}$

7. ____________________

8. Combine: $\dfrac{3x}{x + 2} - \dfrac{x + 3}{x^2 + 6x + 8}$

8. ____________________

9. Simplify: $\dfrac{1 + \dfrac{3}{a}}{a + 3}$

9. ____________________

10. Simplify: $\dfrac{1 - \dfrac{3}{x + 3}}{\dfrac{1}{x^2 - 9}}$

10. ____________________

11. Solve: $\dfrac{3}{x + 3} + \dfrac{1}{x - 1} = \dfrac{8}{x^2 + 2x - 3}$

11. ______________________

12. Water tap A can fill a washtub in 2 minutes. Water tap B can fill the same tub in 4 minutes. If both taps are open, how long will it take to fill the tub?

12. ______________________

CHAPTER 6

SURVEY

The following questions refer to material discussed in this chapter. Work as many problems as you can and check your answers with the answer section in the back of the book. The results will direct you to the sections of the chapter in which you need to work. If you answer all questions correctly, you have a good understanding of the material contained in this chapter.

1. Simplify: $(xy^2)^4$ **1.** ______________

2. Simplify: $\left(\frac{x^2}{2}\right)^3\left(\frac{4}{x^3}\right)^2$ **2.** ______________

Simplify using only positive exponents.

3. $\dfrac{x^{-2}y^{-5}}{x^{-4}y}$ **3.** ______________

4. $(a^{-1} + b^{-1})^{-2}$ **4.** ______________

5. Write the number 3,500,000 in scientific notation. **5.** ______________

6. Evaluate: $\sqrt[3]{-64}$ **6.** ______________

7. Evaluate: $(-27)^{2/3}$ **7.** ______________

8. Simplify: $\sqrt[3]{x^9y^6}$

8. ______________________________

9. Simplify: $\sqrt[5]{-64x^8y^{16}}$

9. ______________________________

Multiply and simplify.

10. $\sqrt{5}(3\sqrt{2} + \sqrt{3}) + \sqrt{2}(\sqrt{3} - 2\sqrt{5})$

10. ______________________________

11. $(2\sqrt{3} + 5\sqrt{7})(2\sqrt{3} - 5\sqrt{7})$

11. ______________________________

12. Simplify: $\dfrac{3}{\sqrt[3]{x}}$

12. ______________________________

13. Simplify: $\dfrac{4}{\sqrt{5} + \sqrt{2}}$

13. ______________________________

Exponents and Radicals

When we go beyond the four basic operations of addition, subtraction, multiplication, and division, a knowledge of radicals becomes necessary. In this chapter you will study techniques of simplifying radicals, the laws of exponents, and the relationship that exists between radicals and exponents.

6–1 THE LAWS OF EXPONENTS

OBJECTIVES

Upon completion of this section you should be able to:

1. State the five laws of exponents.
2. Apply one or more of the laws to simplify an expression.

In chapter 1 we defined the positive whole number exponent and gave the first two laws of exponents.

FIRST LAW OF EXPONENTS $x^a \cdot x^b = x^{a+b}$

SECOND LAW OF EXPONENTS

$$\frac{x^a}{x^b} = x^{a-b}, \text{ if } a > b$$

$$\frac{x^a}{x^b} = \frac{1}{x^{b-a}}, \text{ if } a < b$$

$$\frac{x^a}{x^b} = 1, \text{ if } a = b$$

Remember that in these laws the bases must be the same.

EXERCISE 6-1-1

As a review of the first two laws of exponents, do the following exercises. If the law does not apply, explain why.

1. $x^3 \cdot x^5$

2. $x^2 \cdot x^4$

3. $x \cdot x^6 \cdot x^3$

4. $x^2 \cdot y^3$

5. $x^3 \cdot y \cdot x \cdot y^2$

6. $\frac{x^5}{x^3}$

7. $\frac{x^4}{x}$

8. $\frac{x^2}{y}$

9. $\frac{x^3}{x^4}$

10. $\frac{x^2y^3}{x^3y}$

There are three more laws of exponents.

THIRD LAW OF EXPONENTS To find a power of a power multiply exponents.

$$(x^a)^b = x^{ab}$$

Example 1 $(x^5)^2 = x^{5 \cdot 2} = x^{10}$

Example 2 $(x^3)^3 = x^{3 \cdot 3} = x^9$

Example 3 $(2^3)^2 = 2^{3 \cdot 2} = 2^6$

Example 4 $[(x^2)^3]^2 = x^{2 \cdot 3 \cdot 2} = x^{12}$

FOURTH LAW OF EXPONENTS Raising the product of factors to a power is the same as raising each factor to the power and then finding the product.

$$(xy)^a = x^a y^a$$

Example 5 $(xy)^3 = x^3y^3$

Example 6 $[(2)(3)]^2 = (2)^2(3)^2$

Example 7 $(x^2y^3)^2 = x^4y^6$

Example 8 $[2(x^2y^3)^2]^3 = [2x^4y^6]^3 = 2^3x^{12}y^{18} = 8x^{12}y^{18}$

Example 9 $(-x^4y^3)^2 = (-1)^2x^8y^6 = x^8y^6$

Example 10 $(-2x^3y)^3 = (-2)^3x^9y^3 = -8x^9y^3$

CAUTION Do not attempt to use the fourth law of exponents on sums or differences since

$$(x + y)^2 = x^2 + 2xy + y^2$$
$$\text{NOT } x^2 + y^2.$$

You can never distribute an exponent over a sum or difference.

FIFTH LAW OF EXPONENTS Raising a quotient to a power is the same as raising the numerator and denominator to the power and then finding the quotient.

$$\left(\frac{x}{y}\right)^a = \frac{x^a}{y^a}$$

Example 11 $\left(\frac{x}{y}\right)^4 = \frac{x^4}{y^4}$

Example 12 $\left(\frac{3}{2}\right)^2 = \frac{3^2}{2^2}$

Example 13 $\left(\frac{x^2}{y^3}\right)^4 = \frac{x^8}{y^{12}}$

Example 14 $\left(\frac{-3x}{2y^2}\right)^3 = \frac{(-3)^3x^3}{2^3y^6} = \frac{-27x^3}{8y^6}$

Many problems involving exponents will require a combination of the laws of exponents.

Example 15 $2xy(y^3)^2 = 2xy(y^6) = 2xy^7$

Example 16 $[(x^2y)(3x^3y)]^2 = [3x^5y^2]^2 = 9x^{10}y^4$

Example 17 $(3xy)^2(2x^2y^3)^3 = (9x^2y^2)(8x^6y^9) = 72x^8y^{11}$

Example 18 $\left(\frac{x^4}{4}\right)^2\left(\frac{2}{x^2y}\right)^3 = \left(\frac{x^8}{16}\right)\left(\frac{8}{x^6y^3}\right) = \frac{x^2}{2y^3}$

Example 19 $\left[\frac{(xy)^2}{3}\right]^2\left[\frac{2xy}{3}\right] = \left[\frac{x^2y^2}{3}\right]^2\left[\frac{2xy}{3}\right] = \left(\frac{x^4y^4}{9}\right)\left(\frac{2xy}{3}\right) = \frac{2x^5y^5}{27}$

EXERCISE 6–1–2

Apply either one or a combination of the five laws of exponents.

1. $(x^2)^5$

2. $(xy)^3$

3. $(x^2y)^4$

4. $\left(\frac{x}{y}\right)^2$

5. $\left(\frac{x^2}{y^3}\right)^3$

6. $\left(\frac{x}{2}\right)^3$

7. $(2x^2)^4$

8. $(3xy^2)^3$

9. $\left(\frac{2}{x^2y}\right)^3$

10. $[(x^2)^3]^4$

11. $\left(\frac{xy^2}{x^2y}\right)^2$

12. $\left(\frac{2x^3}{xy^2}\right)^3$

13. $[2(xy)^2]^3$

14. $\left(\frac{1}{3x^2}\right)^2$

15. $[(2x^3y)^2]^3$

16. $\frac{(x^2)^3}{(x^3)^2}$

17. $\frac{(-3x)^3}{9x}$

18. $\frac{-4x^2}{(2x)^3}$

19. $\frac{-5xy^3}{(-5x)^2}$

20. $\frac{(2x^2y^3)^3}{(2x^3y^2)^2}$

21. $2x^2y(x^2y)^3$

22. $(2xy)^3(2x^2y)^2$

23. $(-x^2y)^3(-2xy^3)^2$

24. $(x^2yz^2)^3(-2xy^3z)^2$

25. $[2(xy)^2]^5[3(x^2y)^3]^2$

26. $\left(\frac{x}{y}\right)^3\left(\frac{2x}{y}\right)^4$

27. $\left(\frac{-2}{x}\right)^3\left(\frac{x}{2}\right)^2$

28. $\left(\frac{x^2y}{3}\right)\left(\frac{3}{xy^2}\right)^2$

29. $\left(\frac{-2x}{5y^2}\right)^3\left(\frac{5y}{4x^2}\right)^2$

30. $\left(\frac{x^3}{8}\right)^2\left(\frac{4}{x^2}\right)^3$

31. $(6xy^2)(-4x^2y)(5x^3y^2)$

32. $\left(\frac{10}{x^5}\right)\left(\frac{x^4}{5}\right)^2\left(\frac{5}{2x^2}\right)$

6–2 NEGATIVE AND ZERO EXPONENTS AND SCIENTIFIC NOTATION

OBJECTIVES

Upon completion of this section you should be able to:

1. Define zero and negative exponents.
2. Use these definitions to simplify expressions.
3. Write a given number in scientific notation.
4. Change a number from scientific notation to one without exponents.

The set of numbers used as exponents in our discussion thus far has been the set of positive integers. This is the only set that can be used when exponents are defined as they were in chapter 1. In this section, however, we would like to expand this set to include all integers (positive, negative, and zero) as exponents. This will, of course, require further definitions. These new definitions must be consistent with the system. Furthermore, we will expect all of the laws of exponents as well as other previously known facts to still be true.

DEFINITION

If $x \neq 0$, then $x^0 = 1$. (Any number, except zero, when raised to the zero power is 1.)

Example 1 $5^0 = 1$

Example 2 $10^0 = 1$

Example 3 $(xyz)^0 = 1$, if $x, y, z \neq 0$

Example 4 $\left(\dfrac{5x^2}{2y^3}\right)^0 = 1$, if $x, y \neq 0$

To check the consistency of the definition consider the product $x^0 \cdot x^5$, if $x \neq 0$.

$$x^0 \cdot x^5 = x^{0+5} = x^5$$

Since we know that the product of x^5 and a number will be x^5 only if that number is 1, $x^0 = 1$ is consistent in this instance.

You may wish to verify that this definition is consistent with the other laws of exponents.

Another definition becomes necessary if we are to allow negative numbers to be used as exponents.

DEFINITION

$x^{-a} = \dfrac{1}{x^a}$, if $x \neq 0$ and $a > 0$. (A negative exponent represents the reciprocal.)

Example 5 $x^{-2} = \dfrac{1}{x^2}$

Example 6 $x^{-2}y^{-3} = \dfrac{1}{x^2y^3}$

Example 7 $3^{-2} = \dfrac{1}{3^2} = \dfrac{1}{9}$

Example 8 $5^{-1} = \dfrac{1}{5}$

Example 9 $\dfrac{1}{x^{-2}} = \dfrac{1}{\dfrac{1}{x^2}} = x^2$

CAUTION The negative sign of the exponent *never* affects the sign of the expression.

As mentioned earlier, the preceding definitions are consistent with the laws of exponents. That is, negative and zero exponents follow the same laws for positive exponents.

LAW 1

$x^ax^b = x^{a+b}$

Example 10 $x^{-2}x^{-3} = x^{[(-2)+(-3)]} = x^{-5} = \dfrac{1}{x^5}$

Example 11 $x^5x^{-2} = x^{[5+(-2)]} = x^3$

LAW 2

The law in section 6–1 could now be shortened to one rule stated as

$$\frac{x^a}{x^b} = x^{a-b}$$

without regard to the relative size of *a* and *b*.

Example 12 $\dfrac{x^5}{x^2} = x^{5-2} = x^3$

Example 13 $\dfrac{x^3}{x^7} = x^{3-7} = x^{-4} = \dfrac{1}{x^4}$

Example 14 $\dfrac{x^3}{x^3} = x^{3-3} = x^0 = 1$

LAW 3
$(x^a)^b = x^{ab}$

Example 15 $(x^{-2})^3 = x^{(-2)(3)} = x^{-6} = \frac{1}{x^6}$

LAW 4
$(xy)^a = x^a y^a$

Example 16 $(xy)^{-2} = x^{-2}y^{-2} = \frac{1}{x^2y^2}$

LAW 5
$\left(\frac{x}{y}\right)^a = \frac{x^a}{y^a}$

Example 17 $\left(\frac{x}{y}\right)^{-3} = \frac{x^{-3}}{y^{-3}} = \frac{y^3}{x^3}$

We can now work with expressions involving both positive and negative exponents.

Example 18 Rewrite and simplify $\frac{a^2a^{-8}}{a^{-4}}$ using only positive exponents.

Solution Using the first law, we write

$$\frac{a^2a^{-8}}{a^{-4}} = \frac{a^{[2+(-8)]}}{a^{-4}} = \frac{a^{-6}}{a^{-4}}.$$

Using the second law, we write

$$\frac{a^{-6}}{a^{-4}} = a^{-6-(-4)} = a^{-2}.$$

Using the definition of negative exponents, we write

$$a^{-2} = \frac{1}{a^2}.$$

When a factor is moved from numerator to denominator or from denominator to numerator, the sign of the exponent is changed. Thus example 18 could be worked as follows.

$$\frac{a^2a^{-8}}{a^{-4}} = \frac{a^2a^4}{a^8} = \frac{a^6}{a^8} = \frac{1}{a^2}$$

This shortcut can only be used with factors, never with terms.

Example 19 Rewrite $(-3a^2)^{-4}(b^{-1})^{-3}$ using only positive exponents. Using the fourth law of exponents, we can write

$$(-3)^{-4}(a^2)^{-4}(b^{-1})^{-3}.$$

Using the third law of exponents, we obtain

$$(-3)^{-4}a^{-8}b^3.$$

The definition of negative exponents enables us to write

$$\left(\frac{1}{(-3)^4}\right)\left(\frac{1}{a^8}\right)(b^3)$$

or

$$\frac{b^3}{81a^8}.$$

Example 20 Rewrite $\frac{y^{-1} - x^{-1}}{x - y}$ using only positive exponents.

Using the definition of negative exponents, we write

$$\frac{\frac{1}{y} - \frac{1}{x}}{x - y}.$$

Here we have a complex fraction. To simplify it we need to multiply both the numerator and denominator by xy (the LCD of the fractions contained in the complex fraction). Thus

$$\frac{xy\left(\frac{1}{y} - \frac{1}{x}\right)}{xy(x - y)}.$$

Simplifying, we obtain

$$\frac{x - y}{xy(x - y)} = \frac{1}{xy}.$$

EXERCISE 6-2-1

Rewrite using only positive exponents and simplify.

1. x^{-3} **2.** $(xy)^{-5}$ **3.** $x^{-3}y^{-5}$

4. 2^{-3}

5. $\left(\frac{a}{b}\right)^{-5}$

6. $\frac{1}{3^{-2}}$

7. $(ab)^{-1}$

8. $(a + b)^{-1}$

9. $a^{-1} + b^{-1}$

10. $x^{-2}y^4z^{-1}$

11. $(x^3)^{-2}$

12. $(x^{-2}y^4)^{-4}$

13. $\frac{x^{-3}}{x^5}$

14. x^5x^{-3}

15. $\frac{x^{-2}x^{-6}}{x^7}$

16. $(x^2y^{-3})^{-2}$

17. $(2x^2)^{-3}$

18. $\frac{5^3 \cdot 5^0}{5^2}$

19. $\frac{2^{-5} \cdot 2^0}{2^{-2}}$

20. $\frac{x^{-3}y^2}{x^{-5}y^{-1}}$

21. $\frac{x^3y^{-7}}{x^3y^0}$

22. $\left(\dfrac{3^{-1} \cdot 3^{0}}{3^{5}}\right)^{0}$

23. $(-2x^{3})^{-3}(3x^{-1})^{2}$

24. $(2x^{2}y^{-3})^{-3}(-2x^{-3}y^{2})^{2}$

25. $\left(\dfrac{x^{2}y^{-1}}{x^{-5}}\right)\left(\dfrac{x^{-1}y^{3}}{x^{5}}\right)^{-3}$

26. $\left(\dfrac{2^{-3}}{x^{4}}\right)^{-2}\left(\dfrac{-2x^{3}}{x^{-1}}\right)^{3}$

27. $\dfrac{x^{-1} + y^{-1}}{xy}$

28. $(x^{-1} + 2^{-2})^{-1}$

29. $(x^{-1} + y^{-1})^{-2}$

30. $\dfrac{x^{-1} + y^{-1}}{x + y}$

Exponents are used in many fields of science to write numbers in what is called scientific notation. If a number is either very large or very small, this method of expressing it keeps it from being cumbersome and can make computations easier. Many calculators utilize scientific notation when numbers are too large or too small for the display.

> A number is in **scientific notation** if it is expressed as the product of a power of 10 and a number equal to or greater than 1 and less than 10.

Example 21 3.6×10^{3} is in scientific notation.

$$3.6 \times 10^{3} = 3{,}600$$

Example 22 The earth is approximately 93,000,000 miles from the sun. Express this number in scientific notation.

Answer

$$93{,}000{,}000 = 9.3 \times 10^{7}$$

Notice that the definition is very explicit. The product must be of a number equal to or greater than 1 and less than 10, and a power of 10.

$$93{,}000{,}000 = .93 \times 10^{8}$$

but this is *not* scientific notation because .93 is not equal to or greater than 1.

Example 23 Write 2.5×10^6 without using exponents.

Answer

$$2.5 \times 10^6 = 2,500,000$$

Note that multiplying by 10^6 moves the decimal point six places to the right.

Example 24 Write 2.5×10^{-6} without using exponents.

Answer

$$2.5 \times 10^{-6} = .0000025$$

Note that multiplying by 10^{-6} moves the decimal point six places to the left.

Example 25 Write .0000000345 in scientific notation.

Solution Immediately we see that part of the answer must be 3.45 (equal to or greater than 1 and less than 10 always gives one nonzero digit to the left of the decimal point).

If we now look at 3.45 we must ask, "What power of 10 will return the decimal point to its original position?" Counting, we get eight places to the left, so

$$.0000000345 = 3.45 \times 10^{-8}.$$

EXERCISE 6-2-2

State whether or not the given number is in scientific notation.

1. 3.6×10^5

2. $.05 \times 10^8$

3. 2.78×10^{16}

4. 54.1×10^6

5. 8.2×10^{-3}

6. 7.8×10^{-4}

7. 25×10^{-3}

8. $.645 \times 10^4$

9. 5×10^8

10. 10×10^7

Write each number in scientific notation.

11. 5,000

12. 346,000,000

13. .000000235

14. 23,000,000,000,000

15. .0000000052

16. 5,280

17. 68

18. 728

19. 728,000

20. 728,000,000

Write each number without using exponents.

21. 3.201×10^5

22. 7.28×10^{-6}

23. 6.23×10^{23}

24. 1.07×10^{-9}

25. 5.02×10^{10}

26. 3.58×10^{-1}

27. 5.3762×10^2

28. 4.07×10^{-5}

29. 3.6×10^0

30. 9.9×10^1

31. Mars is approximately 49,000,000 miles from Earth. Express this distance in scientific notation.

32. An enormous cloud of hydrogen gas eight million miles in diameter was discovered around the comet Bennet by NASA in 1970. Express this distance in scientific notation.

33. A light year (the distance light travels in a year) is approximately 6.0×10^{12} miles. Express this distance without exponents.

34. The mass of the Earth is 1.3×10^{25} pounds. Express this without exponents.

35. An "Angstrom" is a unit of length. One Angstrom is approximately 1×10^{-8} cm. Express this length without exponents.

36. The diameter of the nucleus of an average atom is approximately 3.5×10^{-12} cm. Express this distance without exponents.

37. A red blood cell is approximately .001 cm in diameter. Express this length in scientific notation.

38. One electron volt produces approximately .0000000000016 ergs of energy. Express this number in scientific notation.

39. The Rubik's Cube puzzle has over 43,000,000,000,000,000,000 color combinations. Express this number in scientific notation.

40. The Milky Way galaxy has a diameter of approximately 590,000,000,000,000,000 miles. Express this distance in scientific notation.

6–3 RADICALS

Upon completion of this section you should be able to:

1. Give the principal root of a number.
2. Use a calculator or a table to find the square root of a number.

The symbol $\sqrt{}$ is called the **radical sign.** An expression containing the radical sign is called a **radical.** The radical $\sqrt[5]{32}$ is read "the principal fifth root of 32." The number under the radical sign (32 in this case) is called the **radicand** and the integer 5 is called the **index** and indicates the **root.** If the index is 2, it is usually omitted. Hence, both $\sqrt{8}$ and $\sqrt[2]{8}$ indicate the principal square root of 8. If the radical sign is used to indicate any root other than the square root, the index must be written. The index number is always a positive integer greater than 1.

Example 1 $\sqrt[4]{24}$ means the principal fourth root of 24.

Example 2 $\sqrt[3]{24}$ means the principal cube root of 24.

Example 3 $\sqrt{24}$ means the principal square root of 24.

> **DEFINITION**
> x is an ***n*th root** of y if $x^n = y$.

Example 4 3 is a square root of 9 since $3^2 = 9$.

Example 5 -3 is a square root of 9 since $(-3)^2 = 9$.

Example 6 2 is a fourth root of 16 since $2^4 = 16$.

Example 7 -2 is a fourth root of 16 since $(-2)^4 = 16$.

Notice in the preceding examples that 9 has two square roots, 3 and -3. It can be shown by more advanced methods that every number has two square roots, three cube roots, four fourth roots, and so on. Not all of these roots, however, are in the set of real numbers.

We use the radical sign to indicate the *principal* root of a number.

> **DEFINITION**
> The **principal *n*th root** of x ($\sqrt[n]{x}$) is
> a. zero if $x = 0$,
> b. positive if $x > 0$,
> c. negative if $x < 0$ and n is an odd integer,
> d. not a real number if $x < 0$ and n is an even integer.

Example 8 $\sqrt{16} = 4$

Example 9 $\sqrt[3]{8} = 2$

Example 10 $\sqrt[3]{-8} = -2$

Example 11 $\sqrt[4]{-16}$ is not a real number.

EXERCISE 6-3-1

Evaluate. If not a real number, so state.

1. $\sqrt{4}$ **2.** $\sqrt{25}$ **3.** $\sqrt[3]{27}$

4. $\sqrt{1}$ **5.** $\sqrt[3]{-1}$ **6.** $\sqrt{9}$

7. $\sqrt{100}$

8. $\sqrt[3]{-27}$

9. $\sqrt[4]{16}$

10. $\sqrt{-169}$

11. $\sqrt[5]{-32}$

12. $\sqrt[3]{125}$

13. $\sqrt{225}$

14. $\sqrt[9]{-1}$

15. $\sqrt{144}$

16. $\sqrt[3]{-216}$

17. $\sqrt{625}$

18. $\sqrt[6]{-64}$

19. $\sqrt[87]{-1}$

20. $\sqrt[7]{-128}$

One of the most common uses of radicals is in finding the square root of a number. In the previous set of exercises we found square roots such as $\sqrt{4}$, $\sqrt{9}$, $\sqrt{100}$, and so on. Each of these values was a whole number. However, we will not always obtain a whole number or even an exact value as the principal square root of a given number. For example, consider $\sqrt{3}$. There is no exact value that when multiplied by itself yields 3 as a result. In such a case we must find an approximation of the square root of the number.

You may remember from arithmetic how to find the approximation of a square root of a number. To simplify our work we will use a square root table or a calculator. If you have a calculator with a square root key, you will want to use it in the following exercises. Most calculators require the user to enter the number first and then push the square root key. If you do not have such a calculator, the square root table on the inside back cover will aid you. Read the following instructions carefully on how to use the table.

To find the square root of a number between 1 and 200, simply read the value directly from the table. For example, if we want to find $\sqrt{3}$, we find 3 in the first column and look for the proper entry in the square root column. In this case, the value is 1.732. Thus $(1.732)^2 \approx 3$. Suppose we want to find $\sqrt{5.6}$ to three decimal places. The actual value 5.6 does not appear in the table. We observe, however, that the value of $\sqrt{5.6}$ is between $\sqrt{5}$ and $\sqrt{6}$, or between 2.236 and 2.449. The most accurate method of finding the square root of a number not in the table is to use a calculator having a square root key on it. Using such a calculator, we find $\sqrt{5.6} = 2.366$ rounded to three decimal places.

EXERCISE 6–3–2

Using either a calculator or the table of square roots, evaluate each to three decimal places.

1. $\sqrt{15}$ **2.** $\sqrt{96}$ **3.** $\sqrt{8.3}$

4. $\sqrt{120}$ **5.** $\sqrt{21.4}$ **6.** $\sqrt{2{,}300}$

7. $\sqrt{268}$ **8.** $\sqrt{279}$ **9.** $\sqrt{472}$

10. $\sqrt{110}$ **11.** $\sqrt{1{,}075}$ **12.** $\sqrt{143}$

6–4 FRACTIONAL EXPONENTS

OBJECTIVES

Upon completion of this section you should be able to:

1. State the meaning of a fractional exponent.
2. Convert expressions with fractional exponents to radical form.
3. Convert expressions with radicals to exponential form.
4. Evaluate expressions having fractional exponents.

Having discussed radicals in the previous section, we are now ready to give meaning to an exponent that is a fraction.

DEFINITION

$x^{a/b} = \sqrt[b]{x^a}$ or $(\sqrt[b]{x})^a$

Note that in the fractional exponent the denominator represents the index number and the numerator represents a power.

This definition is consistent with the laws of exponents. For example,

$$x^{1/2} = \sqrt{x}.$$

Since

$$x^{1/2}x^{1/2} = x^{1/2 + 1/2} = x,$$

we see that the definition is consistent in this instance.

Example 1 $x^{2/3} = \sqrt[3]{x^2}$

Example 2 $a^{4/5} = \sqrt[5]{a^4}$

Example 3 $2^{1/2} = \sqrt{2}$

Example 4 $x^{-1/3} = \dfrac{1}{\sqrt[3]{x}}$

Example 5 $8^{2/3} = \sqrt[3]{8^2} = (\sqrt[3]{8})^2 = 2^2 = 4$

Example 6 $(-32)^{2/5} = (\sqrt[5]{-32})^2 = (-2)^2 = 4$

Example 7 $-32^{2/5} = -(\sqrt[5]{32})^2 = -(2)^2 = -4$

EXERCISE 6–4–1

Change from exponential to radical form.

1. $x^{1/3}$
2. $x^{-1/2}$
3. $a^{3/4}$
4. $6^{1/2}$
5. $x^{-2/3}$
6. $2^{2/5}$
7. $(-3)^{2/7}$
8. $(a^{-1})^{2/3}$
9. $(ab)^{-2/3}$
10. $(2x)^{2/3}$

Change from radical to exponential form.

11. $\sqrt{x}$
12. $\sqrt[5]{x^2}$
13. $\sqrt[4]{a^3}$

14. $(\sqrt[3]{a})^2$

15. $\dfrac{1}{\sqrt{x}}$

16. $(\sqrt[5]{x})^4$

17. $(\sqrt[3]{ab})^2$

18. $\sqrt[5]{\sqrt{x}}$

19. $(\sqrt[7]{x^2})^2$

20. $\dfrac{1}{(\sqrt[5]{x})^2}$

Evaluate.

21. $4^{1/2}$

22. $9^{1/2}$

23. $(-27)^{2/3}$

24. $-27^{2/3}$

25. $(64)^{2/3}$

26. $\left(\dfrac{1}{8}\right)^{-1/3}$

27. $\left(-\dfrac{1}{8}\right)^{-2/3}$

28. $-\dfrac{1}{8}^{-2/3}$

29. $(-1)^{4/5}$

30. $\left(\dfrac{64}{27}\right)^{-2/3}$

31. $(\sqrt[3]{27})^2$

32. $(\sqrt[5]{-1})^3$

33. $(\sqrt[5]{-1})^2$

34. $(\sqrt[4]{16})^2$

35. $(\sqrt{25})^3$

36. $\sqrt{144^2}$

37. $\sqrt[3]{(-2)^3}$

38. $\sqrt[4]{(53)^4}$

39. $\sqrt[5]{(-21)^5}$

40. $\sqrt{\sqrt{16}}$

6–5 SIMPLIFICATION OF RADICALS

OBJECTIVE

Upon completion of this section you should be able to simplify radicals by removing perfect roots.

In the previous section we established the relationship between radicals and fractional exponents. In preparation for operations on radicals we turn our attention to changing a radical to simplest form. In this section we will assume all variables represent positive numbers.

A radical is in **simplest form** when no power of a factor of the radicand is equal to or greater than the index number. For example, $\sqrt{48}$ may be simplified since

$$\sqrt{48} = \sqrt{(4^2)(3)}$$

and the factor 4 is raised to a power equal to the index number. We may then write

$$\sqrt{(4^2)(3)} = \sqrt{4^2}\sqrt{3}$$

since the fourth law of exponents gives

$$[(4^2)(3)]^{1/2} = (4^2)^{1/2}(3)^{1/2}.$$

Thus

$$\begin{aligned} \sqrt{48} &= \sqrt{(4^2)(3)} \\ &= \sqrt{4^2}\sqrt{3} \\ &= 4\sqrt{3} \end{aligned}$$

which is the simplified form.

Example 1 Simplify: $2\sqrt{75}$

Solution The question to ask is, "Does 75 have a factor that is raised to the second or greater power?" To answer this question we need only look for factors of 75 that are perfect squares.

$$\begin{aligned} 2\sqrt{75} &= 2\sqrt{(25)(3)} \\ &= 2\sqrt{(5^2)(3)} \\ &= 2\sqrt{5^2}\sqrt{3} \\ &= 10\sqrt{3} \end{aligned}$$

Example 2 Simplify: $\sqrt{8x^3y^4}$

Solution

$$\begin{aligned} \sqrt{8x^3y^4} &= \sqrt{(2^2)(2)(x^2)(x)(y^4)} \\ &= \sqrt{2^2}\sqrt{2}\sqrt{x^2}\sqrt{x}\sqrt{y^4} \\ &= 2xy^2\sqrt{2x} \end{aligned}$$

Example 3 Simplify: $3\sqrt[3]{16x^4y^5}$

Solution

$$\begin{aligned} 3\sqrt[3]{16\,x^4y^5} &= 3\sqrt[3]{(2^3)(2)(x^3)(x)(y^3)(y^2)} \\ &= 6xy\sqrt[3]{2xy^2} \end{aligned}$$

EXERCISE 6-5-1

Simplify.

1. $\sqrt{8}$
2. $\sqrt{18}$
3. $4\sqrt{27}$
4. $3\sqrt{125}$
5. $\sqrt[3]{16}$
6. $\sqrt[3]{54}$
7. $\sqrt{x^3}$
8. $\sqrt[3]{x^7}$
9. $\sqrt[3]{x^6y^9}$
10. $\sqrt{25x^3}$
11. $\sqrt{18x^5}$
12. $\sqrt{27x^3y^5}$

13. $2\sqrt{8xy^6}$

14. $\sqrt{32x^5y^7}$

15. $5\sqrt[3]{16x^5}$

16. $\sqrt[3]{54x^4y^6}$

17. $\sqrt[5]{64x^{11}}$

18. $\sqrt{16x^8}$

19. $\sqrt{18x^5y^3}$

20. $\sqrt[3]{-8x^7y^6}$

21. $\sqrt[4]{162x^9y^5}$

22. $\sqrt[3]{64x^4y^6}$

23. $\sqrt[5]{-64x^{10}y^2}$

24. $2xy\sqrt{144x^{10}y^5}$

25. $\sqrt[3]{81x^{14}y^7}$

26. $\sqrt[5]{64x^7y^{12}}$

6–6 OPERATIONS WITH RADICALS

Upon completion of this section you should be able to:

1. Combine similar radicals.
2. Multiply two radicals.
3. Multiply expressions containing radicals.

Since all radicals can be written in exponential form and we already have rules for adding, subtracting, multiplying, and dividing algebraic expressions containing exponents, it is not really necessary to have separate rules for these operations with radicals. However, it is sometimes more convenient to operate with the radicals so we will develop such rules. In this section we will develop addition, subtraction, and multiplication rules. Division will be discussed in the next section.

Let us look first at the problem of simplifying.

Example 1 Simplify: $2\sqrt{x} + 3\sqrt{x}$

Solution Changing to exponents, we have

$$2\sqrt{x} + 3\sqrt{x} = 2(x)^{1/2} + 3(x)^{1/2}.$$

We see a common factor of $x^{1/2}$. Thus

$$2(x)^{1/2} + 3(x)^{1/2} = (2 + 3)(x)^{1/2} = 5x^{1/2} = 5\sqrt{x}.$$

Hence $$2\sqrt{x} + 3\sqrt{x} = 5\sqrt{x}.$$

This example, plus the facts we already know about operating on algebraic expressions, leads us to the following definition and rule.

DEFINITION

Similar (or like) radicals are simplified radicals that have the same radicand and index number.

RULE

Only similar radicals can be added or subtracted.

Example 2 Simplify: $4\sqrt[3]{2} + 3\sqrt[3]{2} - \sqrt[3]{2}$

Solution $4\sqrt[3]{2} + 3\sqrt[3]{2} - \sqrt[3]{2} = (4 + 3 - 1)\sqrt[3]{2} = 6\sqrt[3]{2}$

Example 3 Simplify: $3\sqrt{2} + \sqrt{50} + \sqrt{32}$

Solution We first simplify $\sqrt{50}$ and $\sqrt{32}$ obtaining

$$\begin{aligned} 3\sqrt{2} + \sqrt{50} + \sqrt{32} &= 3\sqrt{2} + 5\sqrt{2} + 4\sqrt{2} \\ &= (3 + 5 + 4)\sqrt{2} \\ &= 12\sqrt{2}. \end{aligned}$$

Example 4 Simplify: $\sqrt[3]{a^4} + \sqrt[3]{8a^4} - \sqrt[6]{4a^2}$

Solution Again we first simplify the radicals, obtaining

$$\begin{aligned} \sqrt[3]{a^4} + \sqrt[3]{8a^4} - \sqrt[6]{4a^2} &= a\sqrt[3]{a} + 2a\sqrt[3]{a} - \sqrt[3]{2a} \\ &= 3a\sqrt[3]{a} - \sqrt[3]{2a}. \end{aligned}$$

Note that $\sqrt[3]{a}$ and $\sqrt[3]{2a}$ are not similar radicals and therefore cannot be combined.

EXERCISE 6-6-1

Simplify by combining similar radicals.

1. $\sqrt{3} + 4\sqrt{3} - 2\sqrt{3}$

2. $5\sqrt{2} - 7\sqrt{2} + \sqrt{2}$

3. $3\sqrt{5} - 2\sqrt{3} + 4\sqrt{5} - \sqrt{3}$

4. $2\sqrt{6} + 3\sqrt[3]{6} - \sqrt{6} + 2\sqrt[3]{6}$

5. $2\sqrt{2} - 3\sqrt[3]{2} - \sqrt{8}$

6. $\sqrt{8} - 2\sqrt{18} + \sqrt{50}$

7. $\sqrt{9} + \sqrt{18} - \sqrt{36}$

8. $\sqrt[3]{16} - \sqrt[3]{54} + \sqrt[3]{3}$

9. $\sqrt{50} + \sqrt[3]{2} - \sqrt{32} - \sqrt[3]{16}$

10. $2\sqrt{8} - 3\sqrt{2} + \sqrt{32}$

11. $2\sqrt{12} - 5\sqrt{3} + \sqrt{27}$

12. $3\sqrt{8} - 4\sqrt{18} + \sqrt{27}$

Multiplication with radicals can also be accomplished by changing the radicals to exponential form and using the laws of exponents. However, the following examples will show that working with radical forms may be more convenient.

RULE
$\sqrt[n]{a}\sqrt[n]{b} = \sqrt[n]{ab}$

This rule is an application of the fourth law of exponents. Since

$$\sqrt[n]{a} = a^{1/n} \text{ and } \sqrt[n]{b} = b^{1/n}$$

then

$$\sqrt[n]{a}\sqrt[n]{b} = a^{1/n}b^{1/n} = (ab)^{1/n} = \sqrt[n]{ab}.$$

Example 5 $\sqrt{3}\sqrt{7} = \sqrt{(3)(7)} = \sqrt{21}$

Example 6 $(2\sqrt{5})(3\sqrt{2}) = (2)(3)\sqrt{(5)(2)} = 6\sqrt{10}$

Example 7 $\sqrt[3]{4}\sqrt[3]{2} = \sqrt[3]{8} = 2$

Example 8 $\sqrt[5]{x^2}\sqrt[5]{x^4} = \sqrt[5]{x^6} = x\sqrt[5]{x}$

EXERCISE 6-6-2

Simplify.

1. $\sqrt{2}\sqrt{5}$
2. $\sqrt[3]{3}\sqrt[3]{11}$
3. $(2\sqrt{7})(3\sqrt{2})$
4. $\sqrt{8}\sqrt{2}$
5. $(4\sqrt{a})(2\sqrt{a^3})$
6. $(\sqrt[3]{16})(2\sqrt[3]{5})$
7. $\sqrt[4]{x^3}\sqrt[4]{x^2}$
8. $(3\sqrt{12a})(2\sqrt{18a})$

9. $(4\sqrt{20x^3})(-\sqrt{50x})$

10. $\sqrt[3]{54a^2}\sqrt[3]{-16a^2}$

The second rule can be expanded to simplify the following examples.

Example 9 Multiply: $2\sqrt{2}(3\sqrt{3} + 4\sqrt{5})$

Solution Here we multiply each term in the parentheses by $2\sqrt{2}$. Hence

$$2\sqrt{2}(3\sqrt{3} + 4\sqrt{5}) = 6\sqrt{6} + 8\sqrt{10}.$$

Example 10 Simplify: $2\sqrt[3]{3}(\sqrt[3]{7} + \sqrt[3]{5}) - 4\sqrt[3]{5}(\sqrt[3]{3} + \sqrt[3]{2})$

Solution

$$\begin{aligned} 2\sqrt[3]{3}(\sqrt[3]{7} + \sqrt[3]{5}) - 4\sqrt[3]{5}(\sqrt[3]{3} + \sqrt[3]{2}) &= 2\sqrt[3]{21} + 2\sqrt[3]{15} - 4\sqrt[3]{15} - 4\sqrt[3]{10} \\ &= 2\sqrt[3]{21} - 2\sqrt[3]{15} - 4\sqrt[3]{10} \end{aligned}$$

EXERCISE 6–6–3

Multiply and simplify.

1. $\sqrt{2}(\sqrt{3} + \sqrt{5})$

2. $\sqrt{3}(2\sqrt{2} - \sqrt{7})$

3. $3\sqrt{3}(2\sqrt{5} - 3\sqrt{7})$

4. $\sqrt{2}(3\sqrt{3} - 2\sqrt{2})$

5. $5\sqrt{3}(2\sqrt{6} + \sqrt{15})$

6. $\sqrt[3]{2}(4\sqrt[3]{4} - 2\sqrt[3]{32})$

7. $4\sqrt{3}(\sqrt{6} - \sqrt{3} + \sqrt{18})$

8. $2\sqrt{2x}(5\sqrt{2x} - \sqrt{6x^3})$

9. $\sqrt{3}(\sqrt{2} + \sqrt{7}) + \sqrt{2}(\sqrt{5} - \sqrt{3})$

10. $\sqrt{2}(\sqrt{6} + \sqrt{2}) - \sqrt{3}(\sqrt{6} + \sqrt{3})$

11. $\sqrt{2}(3\sqrt{6} - \sqrt{10}) - 2\sqrt{3}(\sqrt{15} + 2\sqrt{12})$

12. $2\sqrt{5}(2\sqrt{3} - \sqrt{10}) - \sqrt{3}(3\sqrt{5} - \sqrt{6})$

13. $\sqrt[3]{3}(2\sqrt[3]{9} + \sqrt[3]{2}) + 3\sqrt[3]{2}(\sqrt[3]{3} - 3\sqrt[3]{4})$

14. $2\sqrt[3]{5}(\sqrt[3]{50} - \sqrt[3]{2}) - \sqrt[3]{2}(2\sqrt[3]{5} - \sqrt[3]{4})$

The methods of multiplying polynomial expressions that were covered in chapter 4 can also be transferred to the multiplication of some radical expressions.

Example 11 Expand: $(\sqrt{3} + 4\sqrt{2})(2\sqrt{3} - \sqrt{2})$

Solution In chapter 4 we learned that

$$(x + 4y)(2x - y) = 2x^2 + 7xy - 4y^2.$$

Using the same pattern,

$$\begin{aligned}(\sqrt{3} + 4\sqrt{2})(2\sqrt{3} - \sqrt{2}) &= 2(\sqrt{3})^2 + 7(\sqrt{3})(\sqrt{2}) - 4(\sqrt{2})^2\\ &= 6 + 7\sqrt{6} - 8\\ &= 7\sqrt{6} - 2.\end{aligned}$$

Example 12 Expand: $(3\sqrt{2} + 5\sqrt{3})(3\sqrt{2} - 5\sqrt{3})$

Solution Using the fact that $(a + b)(a - b) = a^2 - b^2$, we have

$$\begin{aligned}(3\sqrt{2} + 5\sqrt{3})(3\sqrt{2} - 5\sqrt{3}) &= (3\sqrt{2})^2 - (5\sqrt{3})^2 \\ &= 9(2) - 25(3) \\ &= -57.\end{aligned}$$

Example 13 Expand: $(4\sqrt{5} + \sqrt{6})^2$

Solution Using the fact that $(x + y)^2 = x^2 + 2xy + y^2$, we obtain

$$\begin{aligned}(4\sqrt{5} + \sqrt{6})^2 &= (4\sqrt{5})^2 + 2(4\sqrt{5})(\sqrt{6}) + (\sqrt{6})^2 \\ &= 80 + 8\sqrt{30} + 6 \\ &= 86 + 8\sqrt{30}.\end{aligned}$$

EXERCISE 6-6-4

Expand.

1. $(\sqrt{2} + \sqrt{5})(\sqrt{3} + \sqrt{2})$

2. $(\sqrt{6} + \sqrt{2})(\sqrt{3} - \sqrt{2})$

3. $(2\sqrt{3} - \sqrt{5})(3\sqrt{5} - \sqrt{2})$

4. $(3\sqrt{6} + 2\sqrt{3})(2\sqrt{3} - \sqrt{2})$

5. $(\sqrt{3} + \sqrt{5})(\sqrt{3} - \sqrt{5})$

6. $(2\sqrt{2} + \sqrt{6})(2\sqrt{2} - \sqrt{6})$

7. $(2\sqrt{2} - 5\sqrt{7})(2\sqrt{2} + 5\sqrt{7})$

8. $(2\sqrt{3} + \sqrt{2})(2\sqrt{3} - 5\sqrt{2})$

9. $(\sqrt{3} + \sqrt[3]{2})(2\sqrt{6} - 3\sqrt{10})$

10. $(\sqrt[3]{2} - 2\sqrt{3})(\sqrt{2} + 3\sqrt{6})$

11. $(\sqrt{2} + 3\sqrt{5})^2$

12. $(3\sqrt{5} - 4)^2$

13. $(2\sqrt{3} - 5\sqrt{7})^2$

6–7 SIMPLIFYING RADICAL EXPRESSIONS

OBJECTIVES

Upon completion of this section you should be able to:

1. Simplify fractions having a single radical term in the denominator.
2. Simplify fractions having a denominator composed of two terms, at least one of which is a radical.

In section 6–5 we learned to simplify radicals. We now expand and use that skill as we simplify algebraic expressions that contain radicals.

An algebraic expression containing radicals is in simplest form when

1. each individual radical is in simplest form and
2. no radical appears in the denominator of a fraction.

Simplifying the individual radicals was discussed in section 6–5, so we will concern ourselves here with the second requirement. Removing all radicals from the denominator of an expression is referred to as **rationalizing the denominator.**

Problems of rationalizing the denominator fall into two categories:

1. those that have only one term in the denominator and
2. those that have two or more terms in the denominator.

Two different techniques are used so we will discuss these separately.

Example 1 Simplify: $\dfrac{1}{\sqrt{3}}$

Solution Our problem here is to rename the given fraction as another fraction with no radical in the denominator. We know, of course, that multiplying the numerator and denominator of any fraction by the same nonzero number will yield an equal fraction. We must multiply by a number that will give a perfect square in the denominator. In this case the number is $\sqrt{3}$. Thus

$$\begin{aligned}\frac{1}{\sqrt{3}} &= \frac{1}{\sqrt{3}} \cdot \frac{\sqrt{3}}{\sqrt{3}} \\ &= \frac{\sqrt{3}}{\sqrt{9}} \\ &= \frac{\sqrt{3}}{3}.\end{aligned}$$

Example 2 Simplify: $\dfrac{3}{\sqrt{8}}$

Solution Here the requirement is to find a number that when multiplied by $\sqrt{8}$ yields the square root of a perfect square. Since $(\sqrt{8})(\sqrt{2}) = \sqrt{16}$, the desired number is $\sqrt{2}$. Hence

$$\begin{aligned}\frac{3}{\sqrt{8}} &= \frac{3}{\sqrt{8}} \cdot \frac{\sqrt{2}}{\sqrt{2}} \\ &= \frac{3\sqrt{2}}{\sqrt{16}} \\ &= \frac{3\sqrt{2}}{4}.\end{aligned}$$

Example 3 Simplify: $\dfrac{2}{\sqrt[3]{5}}$

Solution We must now find a number that when multiplied by $\sqrt[3]{5}$ yields the cube root of a perfect cube. Since

$$(\sqrt[3]{5})(\sqrt[3]{25}) = \sqrt[3]{125},$$

the desired number is $\sqrt[3]{25}$. Thus

$$\begin{aligned}\frac{2}{\sqrt[3]{5}} &= \frac{2}{\sqrt[3]{5}} \cdot \frac{\sqrt[3]{25}}{\sqrt[3]{25}} \\ &= \frac{2\sqrt[3]{25}}{\sqrt[3]{125}} \\ &= \frac{2\sqrt[3]{25}}{5}.\end{aligned}$$

Example 4 Simplify: $\dfrac{x}{\sqrt[3]{x^2y}}$

Solution Remember that perfect cube exponents are multiples of three.

$$\begin{aligned}\frac{x}{\sqrt[3]{x^2y}} &= \frac{x}{\sqrt[3]{x^2y}} \cdot \frac{\sqrt[3]{xy^2}}{\sqrt[3]{xy^2}}\\ &= \frac{x\sqrt[3]{xy^2}}{\sqrt[3]{x^3y^3}}\\ &= \frac{x\sqrt[3]{xy^2}}{xy}\\ &= \frac{\sqrt[3]{xy^2}}{y}\end{aligned}$$

Example 5 Simplify: $\dfrac{3}{\sqrt[5]{16x^3y^2}}$

Solution

$$\begin{aligned}\frac{3}{\sqrt[5]{16x^3y^2}} &= \frac{3}{\sqrt[5]{16x^3y^2}} \cdot \frac{\sqrt[5]{2x^2y^3}}{\sqrt[5]{2x^2y^3}}\\ &= \frac{3\sqrt[5]{2x^2y^3}}{\sqrt[5]{32x^5y^5}}\\ &= \frac{3\sqrt[5]{2x^2y^3}}{2xy}\end{aligned}$$

Example 6 Simplify: $\sqrt{\dfrac{x+y}{x}}$

Solution Remember that $\left(\dfrac{a}{b}\right)^n = \dfrac{a^n}{b^n}$. Thus

$$\begin{aligned}\sqrt{\frac{x+y}{x}} &= \frac{\sqrt{x+y}}{\sqrt{x}}\\ &= \frac{\sqrt{x+y}}{\sqrt{x}} \cdot \frac{\sqrt{x}}{\sqrt{x}}\\ &= \frac{\sqrt{x(x+y)}}{\sqrt{x^2}}\\ &= \frac{\sqrt{x^2+xy}}{x}.\end{aligned}$$

Example 7 Simplify: $\dfrac{a-b}{\sqrt{a+b}}$

Solution

$$\begin{aligned}\frac{a-b}{\sqrt{a+b}} &= \frac{a-b}{\sqrt{a+b}} \cdot \frac{\sqrt{a+b}}{\sqrt{a+b}}\\ &= \frac{(a-b)\sqrt{a+b}}{\sqrt{(a+b)^2}}\\ &= \frac{(a-b)\sqrt{a+b}}{a+b}\end{aligned}$$

EXERCISE 6–7–1

Simplify.

1. $\frac{1}{\sqrt{2}}$

2. $\frac{1}{\sqrt{5}}$

3. $\frac{1}{\sqrt{7}}$

4. $\frac{2}{\sqrt{3}}$

5. $\frac{1}{\sqrt{12}}$

6. $\frac{3}{\sqrt{20}}$

7. $\frac{1}{\sqrt[3]{2}}$

8. $\frac{1}{\sqrt[3]{4}}$

9. $\frac{1}{\sqrt[5]{8}}$

10. $\frac{1}{\sqrt[5]{9}}$

11. $\frac{1}{\sqrt[3]{x^2}}$

12. $\frac{2}{\sqrt[5]{x^3}}$

13. $\dfrac{3}{\sqrt[3]{2x}}$

14. $\dfrac{x}{\sqrt[3]{x}}$

15. $\dfrac{2y}{\sqrt[3]{4x^2y}}$

16. $\dfrac{2xy}{\sqrt[5]{8x^3y^2}}$

17. $\sqrt{\dfrac{1}{2}}$

18. $\sqrt{\dfrac{3}{5}}$

19. $\sqrt{\dfrac{1}{8}}$

20. $\sqrt{\dfrac{x + 2}{x}}$

21. $\dfrac{1}{\sqrt{x + 3}}$

22. $\sqrt{\dfrac{x}{x + 1}}$

23. $\sqrt{\dfrac{8}{x - 2}}$

24. $\dfrac{x + 3}{\sqrt{x - 5}}$

25. $\dfrac{x + y}{\sqrt{x + y}}$

When the denominator of a fraction contains two or more terms and one or more of these contains a radical, the problem of rationalizing the denominator becomes more complex. We will discuss only the simplest type of these problems, which are those problems with two terms, in which one or both contain radicals with the index of two. Problems containing radicals with higher indices and more than two terms in the denominator are left for more advanced courses.

Example 8 Simplify: $\dfrac{3}{\sqrt{5}+1}$

Solution $(\sqrt{5}+1)$ is a binomial. Multiplication of a binomial by another binomial will yield at least a trinomial in all cases except the special case of the product of the sum and difference of the same two numbers.

$$(a-b)(a+b)=a^2-b^2$$

If $(\sqrt{5}+1)$ is multiplied by any number other than $(\sqrt{5}-1)$, a middle term still containing a radical will result. Therefore, to rationalize the denominator in this example we will multiply numerator and denominator by $(\sqrt{5}-1)$. Thus

$$\begin{aligned}\frac{3}{\sqrt{5}+1} &= \frac{3}{(\sqrt{5}+1)}\cdot\frac{(\sqrt{5}-1)}{(\sqrt{5}-1)}\\ &= \frac{3(\sqrt{5}-1)}{\sqrt{25}-1}\\ &= \frac{3\sqrt{5}-3}{4}.\end{aligned}$$

RULE

To simplify an algebraic expression having a binomial of the form $a\sqrt{b}+c\sqrt{d}$ in the denominator, where $\sqrt{b}$ and $\sqrt{d}$ are not both rational, multiply the numerator and denominator by $a\sqrt{b}-c\sqrt{d}$.

$(a\sqrt{b}+c\sqrt{d})$ and $(a\sqrt{b}-c\sqrt{d})$ are sometimes referred to as **conjugates.** Their product will always result in a rational number. Therefore the denominator of the resulting fraction will not contain a radical.

Example 9 Simplify: $\dfrac{5}{\sqrt{2}+\sqrt{3}}$

Solution

$$\begin{aligned}\frac{5}{\sqrt{2}+\sqrt{3}} &= \frac{5}{(\sqrt{2}+\sqrt{3})}\cdot\frac{(\sqrt{2}-\sqrt{3})}{(\sqrt{2}-\sqrt{3})}\\ &= \frac{5(\sqrt{2}-\sqrt{3})}{2-3}\\ &= \frac{5(\sqrt{2}-\sqrt{3})}{-1}\\ &= -5(\sqrt{2}-\sqrt{3})\end{aligned}$$

or

$$= 5\sqrt{3}-5\sqrt{2}$$

Example 10 Simplify: $\dfrac{\sqrt{2} - 1}{\sqrt{2} + 1}$

Solution

$$\begin{aligned}\frac{\sqrt{2} - 1}{\sqrt{2} + 1} &= \frac{(\sqrt{2} - 1)}{(\sqrt{2} + 1)} \cdot \frac{(\sqrt{2} - 1)}{(\sqrt{2} - 1)}\\ &= \frac{2 - 2\sqrt{2} + 1}{2 - 1}\\ &= 3 - 2\sqrt{2}\end{aligned}$$

Example 11 Simplify: $\dfrac{2\sqrt{x} + y}{3\sqrt{x} - \sqrt{y}}$

Solution

$$\begin{aligned}\frac{2\sqrt{x} + y}{3\sqrt{x} - \sqrt{y}} &= \frac{(2\sqrt{x} + y)}{(3\sqrt{x} - \sqrt{y})} \cdot \frac{(3\sqrt{x} + \sqrt{y})}{(3\sqrt{x} + \sqrt{y})}\\ &= \frac{6x + 3y\sqrt{x} + 2\sqrt{xy} + y\sqrt{y}}{9x - y}\end{aligned}$$

EXERCISE 6–7–2

Simplify.

1. $\dfrac{1}{\sqrt{2} + 1}$

2. $\dfrac{1}{\sqrt{3} - 2}$

3. $\dfrac{2}{\sqrt{5} + 1}$

4. $\dfrac{4}{3 - \sqrt{5}}$

5. $\dfrac{6}{\sqrt{x} + y}$

6. $\dfrac{3}{\sqrt{2} + \sqrt{3}}$

7. $\dfrac{1}{\sqrt{3} - \sqrt{5}}$

8. $\dfrac{\sqrt{2}}{\sqrt{2} + 1}$

9. $\dfrac{\sqrt{2}}{\sqrt{6} - 2}$

10. $\dfrac{\sqrt{3}+1}{\sqrt{3}-1}$

11. $\dfrac{a-1}{\sqrt{a}-1}$

12. $\dfrac{\sqrt{3}+\sqrt{2}}{\sqrt{3}-\sqrt{2}}$

13. $\dfrac{5}{\sqrt{7}-\sqrt{2}}$

14. $\dfrac{2\sqrt{3}-5}{3\sqrt{3}+1}$

15. $\dfrac{3\sqrt{2}+\sqrt{3}}{\sqrt{2}-2\sqrt{3}}$

16. $\dfrac{\sqrt{x}}{\sqrt{x}+\sqrt{y}}$

17. $\dfrac{2\sqrt{x}-y}{3\sqrt{x}+y}$

18. $\dfrac{3\sqrt{x}+\sqrt{y}}{\sqrt{x}-2\sqrt{y}}$

CHAPTER

6 SUMMARY

The number in brackets refers to the section of the chapter that discusses that concept.

Terminology

- A **zero exponent** is defined as $x^0 = 1$, if $x \neq 0$. [6–2]
- A **negative exponent** is defined as $x^{-a} = \dfrac{1}{x^a}$, if $x \neq 0$. [6–2]
- A number is in **scientific notation** if it is expressed as the product of a number equal to or greater than 1 but less than 10, and a power of 10. [6–2]
- The symbol $\sqrt{}$ is a **radical sign.** [6–3]
- In the expression $\sqrt[n]{x}$, x is the **radicand** and n is the **index.** [6–3]
- If $x^n = y$, then x is an ***n*th root** of y. [6–3]
- The **principal *n*th root** of x ($\sqrt[n]{x}$) is [6–3]
  a. zero if $x = 0$,
  b. positive if $x > 0$,
  c. negative if $x < 0$ and n is an odd integer,
  d. not a real number if $x < 0$ and n is an even integer.
- A radical is in **simplest form** when no power of a factor of the radicand is greater than or equal to the index number. [6–5]
- **Similar radicals** are simplified radicals that have the same radicand and index number. [6–6]

Rules and Procedures

Laws of Exponents

1. $x^a x^b = x^{a+b}$ [6–1]
2. $\dfrac{x^a}{x^b} = x^{a-b}$ [6–1]
3. $(x^a)^b = x^{ab}$ [6–1]
4. $(xy)^a = x^a y^a$ [6–1]
5. $\left(\dfrac{x}{y}\right)^a = \dfrac{x^a}{y^a}$ [6–1]

Fractional Exponents

- To change a fractional exponent to radical form use

$$x^{a/b} = \sqrt[b]{x^a} \text{ or } (\sqrt[b]{x})^a. \quad [6\text{–}4]$$

Radicals

- Only similar radicals can be added or subtracted. [6–6]
- To multiply radicals use

$$\sqrt[n]{a}\sqrt[n]{b} = \sqrt[n]{ab}. \quad [6\text{–}6]$$

- To simplify an algebraic expression containing radicals: [6–7]
  Step 1 Simplify each radical in the expression.
  Step 2 Rationalize all denominators in the expression.
- To simplify an algebraic expression having a binomial of the form $(\sqrt{a} \pm \sqrt{b})$ in the denominator multiply the numerator and denominator by a binomial having the same terms but whose second term has the opposite sign. [6–7]

CHAPTER 6 REVIEW

Apply the laws of exponents.

1. $(2x^2y^3)^5$
2. $\left(\dfrac{x}{y^2}\right)^3$
3. $\left(\dfrac{2x^3}{3y}\right)^3$
4. $[-3(x^2y)^2]^3[2xy^2]^2$
5. $\left(\dfrac{-2}{x^2}\right)^3\left(\dfrac{x}{4}\right)^2$

Simplify.

6. x^8x^{-2}
7. $\dfrac{x^2 + 3^0}{x^{-1}}$
8. $(x^3)^{-4}(x^{-2})^3$

9. $\left(\frac{2x}{y}\right)^{-2}\left(\frac{x^3}{y^2}\right)^3$

10. $\frac{a^{-1} - b^{-1}}{a - b}$

11. Write 54,200,000 in scientific notation.

12. Write .0003921 in scientific notation.

13. Write 3.2×10^{-5} without exponents.

14. Write 7.1×10^{6} without exponents.

15. The operating capacity of the Cowans Ford Dam nuclear power plant in North Carolina is 1,180,000,000 watts. Express this number in scientific notation.

Evaluate.

16. $\sqrt{16}$

17. $\sqrt[3]{-125}$

18. $\sqrt[5]{243}$

19. $\sqrt{175}$

20. $\sqrt{408}$

21. $27^{2/3}$

22. $\left(\frac{4}{9}\right)^{-1/2}$

23. $\left(-\frac{125}{8}\right)^{-2/3}$

24. $-64^{-2/3}$

25. $\sqrt[3]{(216)^2}$

Simplify.

26. $\sqrt{50}$

27. $\sqrt[3]{32}$

28. $\sqrt[5]{x^6y^{12}}$

29. $\sqrt{75x^3y^6}$

30. $\sqrt[3]{-40x^3y^7}$

31. $\sqrt{75} + 3\sqrt{24} - 4\sqrt{27}$

32. $\sqrt{2a^3} - 3\sqrt{4a^2} + 5a\sqrt{288a}$

33. $(2\sqrt{12x^3})(3\sqrt{48x^5})$

34. $\sqrt{5}(3\sqrt{20} - 2\sqrt{3})$

35. $(2\sqrt{6} - \sqrt{2})^2$

36. $\dfrac{1}{\sqrt{18}}$

37. $\dfrac{3}{\sqrt[3]{4}}$

38. $\dfrac{a - b}{\sqrt{a - b}}$

39. $\dfrac{x - 2}{\sqrt{x} - \sqrt{2}}$

40. $\dfrac{\sqrt{3} - 1}{2\sqrt{3} + 3}$

CHAPTER

6 PRACTICE TEST

Simplify (do not leave negative exponents in your answer).

1. $[(x^2y^3)^4]^3$ — 1. ____________

2. $\left(\frac{2x}{y^2}\right)^3\left(\frac{y}{2x^2}\right)^5$ — 2. ____________

3. x^6x^{-2} — 3. ____________

4. $\frac{x^4 + 2^0}{x^{-2}}$ — 4. ____________

5. $(a^2)^{-3}(a^{-5})^2$ — 5. ____________

6. $\sqrt{98}$ — 6. ____________

7. $\sqrt[5]{x^6y^{12}}$

7. ______________________

8. $\sqrt[3]{-27ab^8}$

8. ______________________

9. $\sqrt{18} + 5\sqrt{50} - 4\sqrt{8}$

9. ______________________

10. $\sqrt{7}(2\sqrt{28} - \sqrt{2})$

10. ______________________

11. $\dfrac{1}{\sqrt{5}}$

11. ______________________

12. $\dfrac{5}{\sqrt{2} - \sqrt{x}}$

12. ______________________

13. $\frac{1}{\sqrt[3]{2}}$

13. ______________________

14. $(\sqrt{5} + \sqrt{3})^2$

14. ______________________

15. Write .00000561 in scientific notation.

15. ______________________

16. Write 3.8×10^8 without exponents.

16. ______________________

Evaluate.

17. $8^{2/3}$

17. ______________________

18. $\left(\frac{49}{25}\right)^{-1/2}$

18. ______________________

19. $\sqrt[3]{-216}$

19. ____________________

20. $\sqrt[3]{(125)^2}$

20. ____________________

CHAPTER

SURVEY

The following questions refer to material discussed in this chapter. Work as many problems as you can and check your answers with the answer section in the back of the book. The results will direct you to the sections of the chapter in which you need to work. If you answer all questions correctly, you have a good understanding of the material contained in this chapter.

1. Solve: $3x^2 + 11x - 4 = 0$ **1.** ____________

2. Solve: $3x^2 = 5x$ **2.** ____________

3. Solve: $5x^2 - 10 = 0$ **3.** ____________

4. Solve by completing the square: $5x^2 + 3 = 20x$ **4.** ____________

5. Use the quadratic formula to solve: $3x^2 + 8x = 5$ **5.** ____________

6. Give the nature of the roots by evaluating the discriminant: $3x^2 - 7x + 5 = 0$ **6.** ____________

7. Divide: $(5 + 3i) \div (1 - 2i)$ **7.** ____________

8. Solve: $3x^4 - 8x^2 + 4 = 0$ **8.** ____________

9. Solve: $\sqrt{x + 2} + \sqrt{2x + 11} = 4$

9. ______________________

10. One number is five greater than another number. The sum of their squares is 97. Find the numbers.

10. ______________________

Quadratic Equations

The equations discussed in chapter 2 were first-degree equations in one unknown or variable. Of course, there are equations with higher degree variables. The technique discussed in chapter 2 will not suffice to work these. In this chapter we will concentrate on equations of second-degree. A knowledge of exponents, radicals, and factoring is necessary for an understanding of equations of degrees greater than one.

7–1 QUADRATICS SOLVED BY FACTORING

OBJECTIVES

Upon completion of this section you should be able to:

1. Write a quadratic equation in standard form.
2. Solve a quadratic equation by factoring.

The degree of a polynomial equation in one unknown is the highest degree of the unknown in any term. For instance, $3x^2 + x - 1 = 0$ is a second-degree equation and $x^5 + 3x^2 = 8$ is a fifth-degree equation. Equations of degree one (discussed in chapter 2) are called **linear equations.** Equations of degree two are called **quadratic equations.** In this chapter we will discuss methods of solving quadratic equations.

Every equation of degree two can be put in the form

$$ax^2 + bx + c = 0,$$

where a, b, and c are real numbers and $a \neq 0$. This is called the **standard form** of a quadratic equation.

Example 1 Write $5x + 2 = 7 - 3x^2$ in standard form.

Solution Since we want the equation in the form $ax^2 + bx + c = 0$, we need to place all terms on the left. By subtracting 7 from both sides and adding $3x^2$ to both sides we obtain $3x^2 + 5x - 5 = 0$, which is in standard form.

EXERCISE 7-1-1

Write each equation in standard form.

1. $x^2 = 3x - 2$
2. $2x = 4 - x^2$
3. $6x^2 + 1 = 5x$
4. $4x^2 = 3x$
5. $2x^2 + 7 = 5x + 6$
6. $4x = 10 + 3x - 6x^2$
7. $x^2 + 5x = 2 - 4x^2$
8. $5x^2 - 10 + 3x = 2x + 8 + 2x^2$

A very important property of real numbers is that if a product is zero, then at least one of the factors must be zero. In symbols this is written

if $AB = 0$, then $A = 0$ or $B = 0$.

This property is the basis for the first method of solving quadratic equations.

Example 2 Solve: $x^2 + x - 6 = 0$

Solution **Step 1** Factor $x^2 + x - 6$ to give

$$(x + 3)(x - 2) = 0.$$

In this form we have a product of two factors as zero. This means that at least one factor is zero. We now find the values of x that will make these factors zero.

Step 2 Set each factor equal to zero.

$$x + 3 = 0 \text{ or } x - 2 = 0$$

Step 3 Solve the resulting equations.

$$x + 3 = 0 \quad \text{or} \quad x - 2 = 0$$
$$x = -3 \quad \text{or} \quad x = 2$$

Either of these values for x will make one of the factors zero. Therefore the product will be zero. Since the factored form is equivalent to the original equation, these two values are solutions of the equation.

Step 4 It is always a good idea to check the solutions in the original equation.

If $x = -3$,
$$x^2 + x - 6 = 0$$
$$(-3)^2 + (-3) - 6 \mid 0$$
$$9 - 3 - 6$$
$$0$$

If $x = 2$,
$$x^2 + x - 6 = 0$$
$$2^2 + 2 - 6 \mid 0$$
$$4 + 2 - 6$$
$$0$$

Thus we see that -3 and 2 are solutions to the equation. Sometimes it is helpful to write the solution in set notation. The **solution set** for this problem is $\{-3,2\}$.

Example 3 Solve: $5x^2 - 3x - 2 = 0$

Solution

Step 1 $(5x + 2)(x - 1) = 0$

Step 2 $5x + 2 = 0$ or $x - 1 = 0$

Step 3 $5x = -2$ or $x = 1$

$$x = -\frac{2}{5}$$

Step 4 Checking both values in the original equation, we find that the solution set is $\left\{-\frac{2}{5}, 1\right\}$.

Example 4 Solve: $4x^2 - 2x - 6 = 0$

Solution

Step 1 $2(2x - 3)(x + 1) = 0$

Step 2 Here we have three factors and since we know that $2 \neq 0$, then

$$2x - 3 = 0 \text{ or } x + 1 = 0.$$

Step 3 $2x = 3$ or $x = -1$

$$x = \frac{3}{2}$$

Step 4 Checking these values, we find the solution set to be $\left\{-1, \frac{3}{2}\right\}$.

Example 5 Solve: $\dfrac{x}{2} = \dfrac{5}{2} - \dfrac{3}{x}$

Solution This is not a quadratic equation as written. Multiplying both sides of such equations by a common denominator will often result in a quadratic equation. In this case the lowest common denominator is $2x$. Thus

$$\boxed{2x}\left(\frac{x}{2}\right) = \boxed{2x}\left(\frac{5}{2}\right) - \boxed{2x}\left(\frac{3}{x}\right)$$

$$x^2 = 5x - 6$$

$$x^2 - 5x + 6 = 0.$$

We now have a quadratic equation that can be solved by factoring.

Step 1 $(x - 3)(x - 2) = 0$
Step 2 $x - 3 = 0$ or $x - 2 = 0$
Step 3 $x = 3$ or $x = 2$
Step 4 Checking these values in the original equation, we find the solution set to be $\{2,3\}$.

In solving equations such as those found in example 5 it is extremely important to check the solutions using the original equation. Multiplying both sides of an equation by a variable will create an equation that is not equivalent to the original, and all solutions of this new equation may not be solutions to the original. However, any solution(s) of the original equation will be found in the solution set of the new equation.

EXERCISE 7–1–2

Solve by factoring.

1. $x^2 + 3x + 2 = 0$

2. $x^2 + 8x + 15 = 0$

3. $x^2 + 8x + 7 = 0$

4. $x^2 - 4x + 3 = 0$

5. $x^2 + 2x = 8$

6. $x^2 = 3x + 10$

7. $x^2 + 18 = 9x$

8. $x^2 + 8 = 6x$

9. $x^2 + 4x - 21 = 0$

10. $x^2 - 5x - 24 = 0$

11. $x^2 + 2x = 3$

12. $x^2 = 2x + 15$

13. $x^2 + 2x + 1 = 0$

14. $x^2 - 6x + 9 = 0$

15. $x^2 + 4x = 0$

16. $x^2 = 6x$

17. $2x^2 = x$

18. $3x^2 + 2x = 0$

19. $x + 7 + \frac{10}{x} = 0$

20. $x = 2 + \frac{35}{x}$

21. $2x^2 + 5x + 3 = 0$

22. $6x^2 + 5x + 1 = 0$

23. $10x^2 + 19x + 6 = 0$

24. $2x^2 + 5x = 12$

25. $6x^2 - 13x = 5$

26. $24x^2 + 66x = 63$

27. $10x^2 + 19x - 15 = 0$

28. $6x^2 - 13x + 6 = 0$

29. $x + \frac{17}{6} + \frac{5}{3x} = 0$

30. $5x + \frac{4}{3x} = \frac{23}{3}$

7–2 INCOMPLETE QUADRATICS

Upon completion of this section you should be able to:

1. Identify an incomplete quadratic equation.
2. Solve an incomplete quadratic equation.

If, when an equation is placed in standard form $ax^2 + bx + c = 0$, either $b = 0$ or $c = 0$, the equation is an **incomplete quadratic.**

Example 1 $5x^2 - 10 = 0$ is an incomplete quadratic since the middle term is missing and therefore $b = 0$.

When you encounter an incomplete quadratic with $c = 0$ (third term missing), it can be solved by factoring.

Example 2 Solve for x if $3x^2 - 2x = 0$.

Solution Factor $x(3x - 2) = 0$.

Then

$$x = 0 \quad \text{or} \quad 3x - 2 = 0$$
$$3x = 2$$
$$x = \frac{2}{3}$$

The solution set is $\left\{0, \frac{2}{3}\right\}$.

Notice that if the c term is missing, you can always factor x from the other terms. This means that in all such equations zero will be one of the solutions.

Example 3 Solve for x if $7x^2 + 14x = 0$.

Solution

$$7x(x + 2) = 0$$
$$7x = 0 \quad \text{or} \quad x + 2 = 0$$
$$x = 0 \quad \text{or} \quad x = -2$$

The solution set is $\{-2, 0\}$.

An incomplete quadratic with the b term missing must be solved by another method, since factoring will be possible only in special cases.

Example 4 Solve for x if $x^2 - 12 = 0$.

Solution Since $x^2 - 12$ has no common factor and is not the difference of squares, it cannot be factored into rational factors. But, from previous observations, we have the following theorem.

> THEOREM If $A^2 = B$, then $A = \pm\sqrt{B}$.

Using this theorem, we have

$$x^2 - 12 = 0$$
$$x^2 = 12$$
$$x = \pm\sqrt{12}.$$

Since all answers should be left in simplified form, we will simplify $\pm\sqrt{12}$ and obtain $x = \pm 2\sqrt{3}$. Notice that this gives two solutions, $+2\sqrt{3}$ and $-2\sqrt{3}$. We can write the solutions using set notation as $\{2\sqrt{3}, -2\sqrt{3}\}$ or simply as $\{\pm 2\sqrt{3}\}$.

Example 5 Solve for x if $2x^2 - 10 = 0$.

Solution

$$2x^2 - 10 = 0$$
$$2x^2 = 10$$
$$x^2 = 5$$
$$x = \pm\sqrt{5}$$

The solution set is $\{\pm\sqrt{5}\}$.

Example 6 Solve for x if $x^2 + 25 = 0$.

Solution

$$x^2 + 25 = 0$$
$$x^2 = -25$$

No real solution

Note that in this example we have the square of a number equal to a negative number. This can never be true in the real number system and therefore we have no real solution.

EXERCISE 7–2–1

Solve.

1. $x^2 + 3x = 0$ **2.** $x^2 - 5x = 0$ **3.** $2x^2 - 3x = 0$

4. $3x^2 + x = 0$

5. $x^2 = 8x$

6. $x^2 + x = 0$

7. $3x^2 = 5x$

8. $4x^2 + 12x = 0$

9. $5x^2 = 10x$

10. $2x^2 + 3x = 7x$

11. $x^2 = 4$

12. $x^2 = 9$

13. $x^2 = 5$

14. $x^2 = 13$

15. $x^2 = 20$

16. $x^2 = 32$

17. $x^2 - 14 = 0$

18. $x^2 + 16 = 0$

19. $x^2 + 6 = 42$

20. $5x^2 = 45$

21. $2x^2 = 32$

22. $3x^2 = 108$

23. $2x^2 = 4$

24. $3x^2 = 9$

25. $2x^2 = 16$

26. $3x^2 = 54$

27. $5x^2 - 15 = 0$

28. $4x^2 + 3 = 103$

29. $6x^2 + 19 = -5$

30. $x = \frac{147}{3x}$

7–3 SOLVING QUADRATIC EQUATIONS BY COMPLETING THE SQUARE

OBJECTIVES

Upon completion of this section you should be able to:

1. Supply the coefficient of the x term that will make a trinomial a perfect square.
2. Supply the value of the constant term that will make a trinomial a perfect square.
3. Apply the seven-step rule to solve a quadratic equation by completing the square.

Consider the quadratic equation $x^2 - 6x + 4 = 0$. Attempting to solve this equation by the methods discussed so far leads to a dead end. We are not dealing with an incomplete quadratic, but neither can we factor the trinomial expression on the left side of the equation.

Since not all quadratics are factorable as polynomials having rational terms, it is necessary to have another method for finding solutions. The method of completing the square involves a process based on the form of the perfect square trinomial discussed in chapter 4.

$$(x + a)^2 = x^2 + 2ax + a^2$$

Example 1 $(x + 3)^2 = x^2 + 6x + 9$

Example 2 $(x + 5)^2 = x^2 + 10x + 25$

Example 3 $(x - 4)^2 = x^2 - 8x + 16$

From these examples we can make the following observations.

1. If the coefficient of x^2 is 1, then the coefficient of x is twice the square root of the third term.
2. The sign of the middle term in the expansion is the same as the sign of the second term in the binomial being squared.

Example 4 Supply the coefficient of x in $x^2 + (\ )x + 81$ that will make the expression a perfect square trinomial.

Solution From observation 1 the coefficient of x should be twice the square root of 81. Therefore since $\sqrt{81} = 9$ and $2(9) = 18$, the desired trinomial is

$$x^2 + (18)x + 81.$$

Note that $x^2 + 18x + 81 = (x + 9)^2$.

EXERCISE 7-3-1

Supply the coefficient of x that will make the trinomials perfect squares.

1. $x^2 + (\quad)x + 36$ **2.** $x^2 + (\quad)x + 49$ **3.** $x^2 + (\quad)x + 100$

4. $x^2 + (\quad)x + 1$ **5.** $x^2 + (\quad)x + 4$ **6.** $x^2 - (\quad)x + 64$

7. $x^2 - (\quad)x + 144$ **8.** $x^2 + (\quad)x + 169$

We might also state the same observation in a slightly different manner:

In the trinomial expansion of $x^2 + 2ax + a^2$ the last term is a perfect square number and it is the square of one-half the coefficient of x (one-half of $2a$ is a and $\left[\frac{1}{2}(2a)\right]^2$ is a^2).

Example 5 Find c such that $x^2 + 6x + c$ will be a perfect square.

Solution Since the coefficient of x is 6, the required number is $\left[\frac{1}{2}(6)\right]^2 = (3)^2 = 9$.

Thus $x^2 + 6x + 9$ is a perfect square trinomial that factors into $(x + 3)^2$. This process is known as **completing the square.**

EXERCISE 7-3-2

Find c in each of the following such that the resulting trinomial will be a perfect square. Then factor the perfect square trinomial.

1. $x^2 + 8x + c$

2. $x^2 + 10x + c$

3. $x^2 + 22x + c$

4. $x^2 - 4x + c$

5. $x^2 - 24x + c$

6. $x^2 - 6x + c$

7. $x^2 + 5x + c$

8. $x^2 - x + c$

We are now prepared to solve quadratic equations by completing the square. We will use the following step-by-step method.

STEP-BY-STEP PROCEDURE FOR SOLVING A QUADRATIC EQUATION BY COMPLETING THE SQUARE

Step 1 Arrange the equation in standard form.

Step 2 If the coefficient of x^2 is not 1, divide each side of the equation by that coefficient.

Step 3 Rearrange the equation so that the x^2 and x terms are on the left side of the equation and the numerical term is on the right.

Step 4 Complete the square on the left and keep the equation balanced by adding the square of one-half of the coefficient of the x term to both sides.

Step 5 Factor the completed square.

Step 6 Take the square root of both sides of the equation remembering that if $A^2 = B$, then $A = \pm \sqrt{B}$.

Step 7 Solve for the two values of x by using the $(+)$ sign and then the $(-)$ sign of the right-hand term.

Example 6 Solve $x^2 - 15 = -2x$ by completing the square.

Solution

Step 1 $x^2 + 2x - 15 = 0$

Step 2 The coefficient of x^2 is already 1.

Step 3 $x^2 + 2x = 15$

Step 4 Since $\left[\frac{1}{2}(2)\right]^2 = 1$, we add 1 to both sides.

$$x^2 + 2x + 1 = 15 + 1$$

Step 5 $(x + 1)^2 = 16$

Step 6 $x + 1 = \pm\sqrt{16}$

$x + 1 = \pm 4$

Step 7 $x + 1 = 4$ or $x + 1 = -4$

$x = 3$ $\qquad$ $x = -5$

The solution set is $\{-5,3\}$.

Example 7 Solve $2x^2 + 3x - 5 = 0$ by completing the square.

Solution

Step 1 $2x^2 + 3x - 5 = 0$

Step 2 $x^2 + \frac{3}{2}x - \frac{5}{2} = 0$

Step 3 $x^2 + \frac{3}{2}x = \frac{5}{2}$

Step 4 Since $\left[\left(\frac{1}{2}\right)\left(\frac{3}{2}\right)\right]^2 = \frac{9}{16}$, we add $\frac{9}{16}$ to both sides.

$$x^2 + \frac{3}{2}x + \frac{9}{16} = \frac{5}{2} + \frac{9}{16}$$

Step 5 $\left(x + \frac{3}{4}\right)^2 = \frac{49}{16}$

Step 6 $x + \frac{3}{4} = \pm\sqrt{\frac{49}{16}}$

$x + \frac{3}{4} = \pm\frac{7}{4}$

Step 7 $x + \frac{3}{4} = \frac{7}{4}$ or $x + \frac{3}{4} = -\frac{7}{4}$

$x = \frac{4}{4}$ $\qquad$ $x = -\frac{10}{4}$

$x = 1$ $\qquad$ $x = -\frac{5}{2}$

The solution set is $\left\{-\frac{5}{2},1\right\}$.

Example 8 Solve $x^2 - 6x + 4 = 0$ by completing the square.

Solution

$$x^2 - 6x + 4 = 0$$
$$x^2 - 6x = -4$$
$$x^2 - 6x + 9 = -4 + 9$$
$$(x - 3)^2 = 5$$
$$x - 3 = \pm\sqrt{5}$$
$$x - 3 = \sqrt{5} \quad \text{or} \quad x - 3 = -\sqrt{5}$$
$$x = 3 + \sqrt{5} \qquad x = 3 - \sqrt{5}$$

Since $\sqrt{5}$ is not rational, it is sufficient to leave it in this form. The solution set is therefore $\{3 - \sqrt{5}, 3 + \sqrt{5}\}$.

Example 9 Solve $2x^2 + x + 3 = 0$ by completing the square.

Solution

$$2x^2 + x + 3 = 0$$
$$x^2 + \frac{1}{2}x + \frac{3}{2} = 0$$
$$x^2 + \frac{1}{2}x = -\frac{3}{2}$$
$$x^2 + \frac{1}{2}x + \frac{1}{16} = -\frac{3}{2} + \frac{1}{16}$$
$$\left(x + \frac{1}{4}\right)^2 = -\frac{23}{16}$$

In this step we see that the square of a number is $-\frac{23}{16}$. The square of a real number is never negative, therefore we have *no real solution.*

EXERCISE 7-3-3

Solve by completing the square.

1. $x^2 + 4x - 5 = 0$

2. $x^2 + 6x - 16 = 0$

3. $x^2 - 3 = 2x$

4. $x^2 - 8x + 7 = 0$

5. $x^2 + 20 = 4x$

6. $x^2 + 3x - 1 = 0$

7. $x^2 - 3 = 5x$

8. $x^2 - 2 = x$

9. $x^2 + x + 5 = 0$

10. $2x^2 + 3x - 2 = 0$

11. $2x^2 - 5x + 2 = 0$

12. $3x^2 + 6x - 4 = 0$

13. $5x^2 + 3 = 20x$

14. $2x^2 + 3x + 4 = 0$

15. $2x^2 + \frac{1}{2} = 2x$

16. $\frac{2}{3}x^2 + 2x - \frac{1}{2} = 0$

7–4 SOLVING QUADRATIC EQUATIONS BY FORMULA

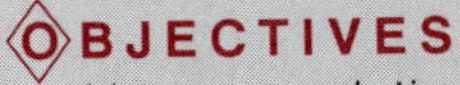

Upon completion of this section you should be able to:

1. State the quadratic formula.
2. Solve quadratic equations using the quadratic formula.

In a sense, a formula is the solution of all problems of a particular type. Since all quadratic equations can be put in the form $ax^2 + bx + c = 0$, the solution of this general equation in standard form will yield a formula for the solutions of any quadratic equation.

We will solve the general quadratic equation by the method of completing the square.

Step 1 $ax^2 + bx + c = 0$

Step 2 Divide by a.

$$x^2 + \frac{b}{a}x + \frac{c}{a} = 0$$

Step 3 $x^2 + \frac{b}{a}x = -\frac{c}{a}$

Step 4 Since $\left[\left(\frac{1}{2}\right)\left(\frac{b}{a}\right)\right]^2 = \frac{b^2}{4a^2}$, we will add $\frac{b^2}{4a^2}$ to each side.

$$x^2 + \frac{b}{a}x + \frac{b^2}{4a^2} = \frac{b^2}{4a^2} - \frac{c}{a}$$

Step 5

$$\left(x + \frac{b}{2a}\right)^2 = \frac{b^2 - 4ac}{4a^2}$$

Step 6

$$x + \frac{b}{2a} = \pm\sqrt{\frac{b^2 - 4ac}{4a^2}}$$

$$x + \frac{b}{2a} = \pm\frac{\sqrt{b^2 - 4ac}}{2a}$$

$$x = -\frac{b}{2a} \pm \frac{\sqrt{b^2 - 4ac}}{2a}$$

$$x = \frac{-b \pm \sqrt{b^2 - 4ac}}{2a}$$

Thus $x = \frac{-b + \sqrt{b^2 - 4ac}}{2a}$ or $x = \frac{-b - \sqrt{b^2 - 4ac}}{2a}$.

RULE

To solve a quadratic equation by using the quadratic formula first rearrange the equation in standard form, then substitute the values of a, b, and c into the formula. (Note that when an equation is in standard form, a represents the coefficient of x^2, b represents the coefficient of x, and c represents the numerical term.)

Note: You should now memorize the **quadratic formula.**

$$x = \frac{-b \pm \sqrt{b^2 - 4ac}}{2a}$$

Example 1 Use the quadratic formula to solve $5x^2 + 3x - 8 = 0$.

Solution The equation is already in standard form, thus $a = 5$, $b = 3$, $c = -8$. Substituting these values in the formula, we obtain

$$x = \frac{-3 \pm \sqrt{(3)^2 - 4(5)(-8)}}{2(5)}$$

$$x = \frac{-3 \pm \sqrt{9 + 160}}{10}$$

$$x = \frac{-3 \pm \sqrt{169}}{10}$$

$$x = \frac{-3 \pm 13}{10}$$

$$x = \frac{-3 + 13}{10} \quad \text{or} \quad x = \frac{-3 - 13}{10}$$

$$x = \frac{10}{10} \qquad x = \frac{-16}{10}$$

$$x = 1 \qquad x = -\frac{8}{5}.$$

The solution set is $\left\{-\frac{8}{5}, 1\right\}$.

Example 2 Solve $x^2 - 4x + 12 = 0$ by using the formula you have memorized.

Solution Again we see that the equation is in standard form so $a = 1$, $b = -4$, $c = 12$.

$$x = \frac{-(-4) \pm \sqrt{(-4)^2 - 4(1)(12)}}{2(1)}$$

$$x = \frac{4 \pm \sqrt{16 - 48}}{2}$$

$$x = \frac{4 \pm \sqrt{-32}}{2}$$

Since $\sqrt{-32}$ is not a real number (that is, we cannot find a real number whose square is -32), the equation has no real solution.

Example 3 Solve $x^2 - 4x - 4 = 0$ using the formula.

Solution $a = 1$, $b = -4$, $c = -4$.

$$x = \frac{4 \pm \sqrt{16 - 4(1)(-4)}}{2(1)}$$

$$x = \frac{4 \pm \sqrt{32}}{2}$$

$$x = \frac{4 \pm 4\sqrt{2}}{2}$$

$$x = \frac{2(2 \pm 2\sqrt{2})}{2}$$

$$x = 2 \pm 2\sqrt{2}$$

The solution set is $\{2 - 2\sqrt{2}, 2 + 2\sqrt{2}\}$.

EXERCISE 7-4-1

Solve by the quadratic formula. Leave all solutions in simplest form.

1. $x^2 + 2x - 15 = 0$

2. $x^2 - 9x + 20 = 0$

3. $5x^2 - 7x - 6 = 0$

4. $6x^2 = x + 2$

5. $x^2 + 3x + 1 = 0$

6. $2x^2 = x + 3$

7. $2x^2 + 1 = 3x$

8. $4x^2 + 5x + 1 = 0$

9. $3x^2 = 2x + 1$

10. $x^2 - 3x + 4 = 0$

11. $3x^2 + 5x + 2 = 0$

12. $5x^2 + 7x + 1 = 0$

13. $2x^2 - 6x + 3 = 0$

14. $x^2 + 2x = 7$

15. $x^2 + 25 = 10x$

16. $2x^2 - 3x + 5 = 0$

17. $9x^2 + 12x + 4 = 0$

18. $3x^2 + 8x = 5$

19. $\frac{5}{6}x^2 - x + \frac{1}{3} = 0$

20. $\frac{3}{10}x^2 + \frac{2}{5}x - \frac{1}{2} = 0$

7–5 THE NATURE OF THE ROOTS OF A QUADRATIC EQUATION

BJECTIVES

Upon completion of this section you should be able to:

1. Compute the value of the discriminant $b^2 - 4ac$.
2. Determine the nature of the roots of a quadratic equation from the value of the discriminant.

When we examine the solution of the general quadratic equation (the formula), we discover some important facts.

$$x = \frac{-b \pm \sqrt{b^2 - 4ac}}{2a}$$

First we notice that the radicand $b^2 - 4ac$ will determine whether or not we have solutions that are real numbers. If $b^2 - 4ac$ is negative, and we are working with real numbers, we have no solution.

The radicand $b^2 - 4ac$ is called the **discriminant** of a quadratic equation because it determines the nature of the roots (solutions).

The discriminant $b^2 - 4ac$ can give further information about the roots. Since $\sqrt{b^2 - 4ac}$ is added to give one solution and subtracted to give the other, then the only way both solutions could be the same is for the discriminant to be zero. This leads us to the second property.

If $b^2 - 4ac = 0$, the solutions of the equation are equal.

Both solutions of a quadratic equation can be the same only if the quadratic is a perfect square trinomial.

If $b^2 - 4ac$ is positive, we will have two different solutions for the equation. Furthermore, if the value of $b^2 - 4ac$ is positive and not a perfect square (such as 4, 9, 25, and so on), the solution must contain a radical and is therefore irrational.

If the solution to a quadratic equation does not contain a radical, the roots can be obtained by factoring. Thus the value of $b^2 - 4ac$ becomes a sure test for the factorability of a trinomial.

The following table summarizes the facts discussed.

Roots of a Quadratic

Value of the Discriminant	Nature of the Roots
$b^2 - 4ac < 0$	No real roots
$b^2 - 4ac = 0$	Real and equal roots (The trinomial is a perfect square.)
$b^2 - 4ac > 0$ but not a perfect square	Irrational and unequal roots Two different roots that contain a radical
$b^2 - 4ac > 0$ and a perfect square	Unequal and rational roots Two different roots containing no radical (The trinomial is factorable.)

EXERCISE 7-5-1

Compute $b^2 - 4ac$ for each of the following and give the nature of their roots.

1. $2x^2 + 5x + 3 = 0$

2. $3x^2 - 7x - 1 = 0$

3. $5x^2 - 3x + 1 = 0$

4. $x^2 - 8x + 11 = 0$

5. $x^2 + 4x + 4 = 0$

6. $4x^2 - 20x + 25 = 0$

7. $x^2 + 6x - 9 = 0$

8. $x^2 - 12x - 12 = 0$

9. $x^2 - 4x + 4 = 0$

10. $x^2 - 3x + 1 = 0$

11. $2x^2 - 7x + 3 = 0$

12. $x^2 - 3x + 5 = 0$

13. $4x^2 + 4x + 1 = 0$

14. $2x^2 + x + 1 = 0$

15. $3x^2 + 5x + 2 = 0$

16. $2x^2 + 5x - 3 = 0$

17. $x^2 + 5x + 7 = 0$

18. $x^2 - 11x + 31 = 0$

19. $2x^2 - 7x + 4 = 0$

20. $3x^2 - 8x + 4 = 0$

7-6 COMPLEX NUMBERS

Upon completion of this section you should be able to:

1. Add, subtract, multiply, and divide complex numbers.
2. Solve a quadratic equation for complex roots.

In preceding sections many problems have had "no real solution" or "no real roots." To the observant students this may have implied that there is a set of numbers other than the real numbers in which such equations do have solutions. This is, in fact, the case.

To introduce this new set of numbers we define the imaginary unit "i" as the square root of negative 1.

DEFINITION

$i = \sqrt{-1}$ or $i^2 = -1$

Accepting this definition of i makes it possible to find values for the square roots of negative numbers, or at least to indicate such values.

Example 1 Find the value of $\sqrt{-4}$.

Solution

$$\begin{aligned}\sqrt{-4} &= \sqrt{(-1)(4)}\\ &= \sqrt{-1}\,\sqrt{4}\\ &= i\sqrt{4}\\ &= 2i\end{aligned}$$

To check $\sqrt{-4} = 2i$ by the definition of the square root, we evaluate $(2i)^2$.

$$\begin{aligned}(2i)^2 &= (2i)(2i)\\ &= (2)(2)(i)(i)\\ &= 4i^2\end{aligned}$$

But since $i^2 = -1$,

then $4i^2 = 4(-1) = -4$.

Example 2 Find the value of $\sqrt{-10}$.

Solution

$$\begin{aligned}\sqrt{-10} &= \sqrt{(-1)(10)}\\ &= \sqrt{-1}\,\sqrt{10}\\ &= i\sqrt{10}\end{aligned}$$

The answer is left in this form since $\sqrt{10}$ is not rational.

A number such as $2i$ or $i\sqrt{10}$ is called an **imaginary number.**

EXERCISE 7-6-1

Express as imaginary numbers.

1. $\sqrt{-16}$ **2.** $\sqrt{-25}$ **3.** $\sqrt{-30}$ **4.** $\sqrt{-7}$

5. $\sqrt{-50}$ **6.** $\sqrt{-2}$ **7.** $\sqrt{-49}$ **8.** $\sqrt{-100}$

9. $\sqrt{-18}$ **10.** $\sqrt{-27}$

DEFINITION
The indicated sum of a real number and an imaginary number, $(a + bi)$ where a and b are real, is called a **complex number.**

It should first be noted that the set of real numbers is a subset of the set of complex numbers. A real number x can be expressed as $x + 0i$ and by the definition is complex. Also, the set of imaginary numbers is a subset of the set of complex numbers since an imaginary number bi can be expressed as $0 + bi$.

Complex numbers can be added, subtracted, multiplied, divided, raised to powers, and so on, as you would expect of any set of numbers. Rules for these operations follow.

To add or subtract complex numbers combine the real parts of each number and the imaginary parts of each number. (Note that this follows a previous rule that only like terms can be combined.)

Example 3 Add: $(7 + 6i) + (3 + 2i)$

Solution

$$\begin{aligned}(7 + 6i) + (3 + 2i) &= (7 + 3) + (6i + 2i) \\ &= 10 + 8i\end{aligned}$$

Example 4 Add: $(3 + 4i) + (2 - 6i)$

Solution

$$\begin{aligned}(3 + 4i) + (2 - 6i) &= (3 + 2) + (4i - 6i) \\ &= 5 - 2i\end{aligned}$$

To multiply complex numbers we use the distributive law.

Example 5 Multiply: $5(3 + 4i)$

Solution

$$5(3 + 4i) = 15 + 20i$$

When both complex numbers are in the form $a + bi$, we use the FOIL method developed in chapter 4.

Example 6 Multiply: $(3 + 2i)(4 + 6i)$

Solution

$$\begin{aligned}(3 + 2i)(4 + 6i) &= 12 + 26i + 12i^2 \\ &= 12 + 26i + 12(-1) \\ &= 26i\end{aligned}$$

Example 7 Find: $(3 + i)^2$

Solution

$$\begin{aligned}(3 + i)^2 &= (3 + i)(3 + i) \\ &= 9 + 6i + i^2 \\ &= 9 + 6i - 1 \\ &= 8 + 6i\end{aligned}$$

The next operation we will discuss is division. Division is always defined as multiplying by the inverse. This is actually rationalizing the denominator since i is just a symbol for a radical.

Example 8 Divide: $(3 + 4i) \div (2 + i)$

Solution

$$(3 + 4i) \div (2 + i) = \frac{3 + 4i}{2 + i}$$

If we write this as $\frac{3 + 4\sqrt{-1}}{2 + \sqrt{-1}}$, you will recognize it as a type of problem found in chapter 6 (exercise 6–7–2). We should multiply the numerator and denominator by $2 - i$ (known as the *conjugate* of $2 + i$).

$$\begin{aligned}\frac{3 + 4i}{2 + i} \cdot \frac{2 - i}{2 - i} &= \frac{6 + 5i - 4i^2}{4 - i^2} \\ &= \frac{10 + 5i}{5}\end{aligned}$$

This solution can be reduced to

$$\frac{10 + 5i}{5} = \frac{5(2 + i)}{5} = 2 + i.$$

EXERCISE 7–6–2

Perform the operations.

1. $(2 + 3i) + (5 + 4i)$ **2.** $(6 + 5i) + (1 - 6i)$ **3.** $(5 - 4i) + 2(4 + 2i)$

4. $(5 - 2i) - (8 - i)$

5. $(11 + 7i) - (3 + 4i)$

6. $(11 + i) - 3(2 - 5i)$

7. $(2 + 3i)(1 + 4i)$

8. $(x + iy)(x - iy)$

9. $(5 + 4i)^2$

10. $(6 - 5i)^2$

11. $(2 + 3i) \div (1 - 2i)$

12. $5 \div (4 + 3i)$

13. $2i \div (6 + i)$

14. $(3 + i) \div i$

15. $(1 - 4i) \div (2 - 3i)$

The most common use of complex numbers at this level of algebra is in expressing solutions to quadratic equations that have no real solution.

Example 9 Solve and check: $x^2 - 2x + 5 = 0$

Solution Since the expression on the left will not factor, we will use the formula with $a = 1$, $b = -2$, and $c = 5$.

$$\begin{aligned} x &= \frac{2 \pm \sqrt{(-2)^2 - 4(1)(5)}}{2} \\ &= \frac{2 \pm \sqrt{-16}}{2} \\ &= \frac{2 \pm 4i}{2} \\ &= 1 \pm 2i \end{aligned}$$

Check: $x = 1 + 2i$.

$$\begin{array}{r|l} x^2 - 2x + 5 & = 0 \\ (1 + 2i)^2 - 2(1 + 2i) + 5 & 0 \\ 1 + 4i + 4i^2 - 2 - 4i + 5 & \\ 4 + 4i^2 & \\ 4 + 4(-1) & \\ 0 & \end{array}$$

Check: $x = 1 - 2i$.

$$\begin{array}{r|l} x^2 - 2x + 5 & = 0 \\ 1 - 4i + 4i^2 - 2 + 4i + 5 & 0 \\ 4 + 4i^2 & \\ 4 + 4(-1) & \\ 0 & \end{array}$$

We see that both solutions check.
Thus the solution set is $\{1 \pm 2i\}$.

EXERCISE 7–6–3

Solve and check.

1. $x^2 - 4x + 8 = 0$

2. $2x^2 - 2x + 1 = 0$

3. $8x^2 - 4x + 1 = 0$

4. $x^2 + x + 1 = 0$

5. $x^2 - x + 1 = 0$

6. $x^2 - 2x + 4 = 0$

7. $3x^2 + x + 1 = 0$

8. $x^2 - 4x + 5 = 0$

9. $5x^2 - 2x + 3 = 0$

10. $x^2 + 3x + 4 = 0$

7–7 EQUATIONS QUADRATIC IN FORM

OBJECTIVES

Upon completion of this section you should be able to:

1. Determine if an equation is quadratic in form.
2. Solve equations that are quadratic in form.

An equation is **quadratic in form** if a suitable substitution for the unknown can be found so that the resulting equation is a quadratic equation.

For example,

$$x^4 - 5x^2 + 6 = 0$$

is quadratic in form since the substitution $u = x^2$ would give

$$u^2 - 5u + 6 = 0,$$

which is a quadratic equation in the variable u. Such equations may often be solved by solving the resulting quadratic equation and then using these solutions and the substitution to obtain solutions to the original equation.

Example 1 Solve: $x^4 - 5x^2 + 6 = 0$

Solution Let $u = x^2$.

Then

$$\begin{aligned} u^2 - 5u + 6 &= 0 \\ (u - 3)(u - 2) &= 0 \\ u - 3 = 0 \quad \text{or} \quad u - 2 &= 0 \\ u = 3 \qquad\qquad u &= 2. \end{aligned}$$

Since $u = x^2$, we have

$$\begin{aligned} x^2 = 3 \quad &\text{or} \quad x^2 = 2 \\ x = \pm\sqrt{3} \quad &\quad\; x = \pm\sqrt{2}. \end{aligned}$$

Checking these values in the original equation, we find they all satisfy the equation. Therefore the solution set is $\{\sqrt{3}, -\sqrt{3}, \sqrt{2}, -\sqrt{2}\}$. Note that this fourth-degree equation has four roots. An equation of integral degree "n" will have n roots.

Example 2 Solve: $2x^{1/2} - x^{1/4} - 6 = 0$

Solution Let $u = x^{1/4}$.

Then

$$\begin{aligned} u^2 - u - 6 &= 0 \\ (2u + 3)(u - 2) &= 0 \\ 2u + 3 = 0 \quad &\text{or} \quad u - 2 = 0 \\ 2u = -3 \quad & \quad\quad u = 2 \\ u = -\frac{3}{2}. & \end{aligned}$$

Since $u = x^{1/4}$, then

$$x^{1/4} = -\frac{3}{2} \quad \text{or} \quad x^{1/4} = 2.$$

Raising both sides to the fourth power, we obtain

$$(x^{1/4})^4 = \left(-\frac{3}{2}\right)^4 \quad \text{or} \quad (x^{1/4})^4 = 2^4$$

$$x = \frac{81}{16} \qquad\qquad x = 16.$$

Checking these values, we have

$$\text{if } x = \frac{81}{16},\ 2x^{1/2} - x^{1/4} - 6 = 0$$

$$\begin{array}{r|l} 2\left(\frac{81}{16}\right)^{1/2} - \left(\frac{81}{16}\right)^{1/4} - 6 & 0 \\ 2\left(\frac{9}{4}\right) - \frac{3}{2} - 6 & \\ \frac{9}{2} - \frac{3}{2} - 6 & \\ 3 - 6 & \\ -3. & \end{array}$$

We see that $\frac{81}{16}$ is *not* a solution.

$$\text{If } x = 16,\ 2x^{1/2} - x^{1/4} - 6 = 0$$

$$\begin{array}{r|l} 2(16)^{1/2} - (16)^{1/4} - 6 & 0 \\ 8 - 2 - 6 & \\ 0. & \end{array}$$

Therefore 16 is a solution and the solution set is $\{16\}$.

EXERCISE 7–7–1

Solve for real or complex roots.

1. $x^4 - 5x^2 + 4 = 0$

2. $x^4 - 10x^2 + 9 = 0$

3. $x^4 - 6x^2 + 8 = 0$

4. $x^4 - 7x^2 + 12 = 0$

5. $x^4 - 7x^2 + 10 = 0$

6. $2x^4 - 7x^2 + 3 = 0$

7. $3x^4 - 8x^2 + 4 = 0$

8. $x^4 - 3x^2 - 4 = 0$

9. $x^4 - 7x^2 - 18 = 0$

10. $x^4 + 8x^2 + 15 = 0$

11. $x^{1/2} - 5x^{1/4} + 6 = 0$

12. $x^{1/2} - 4x^{1/4} - 5 = 0$

13. $x - 6x^{1/2} + 8 = 0$

14. $x - 4x^{1/2} - 21 = 0$

15. $x^{2/3} + 2x^{1/3} - 3 = 0$

16. $2x^{2/3} + 5x^{1/3} + 2 = 0$

7-8 EQUATIONS WITH RADICALS

OBJECTIVES

Upon completion of this section you should be able to:

1. Solve equations involving radicals.
2. Identify extraneous roots.

In chapter 2 we found that certain operations will always yield an equivalent equation. This fact was used to solve first-degree equations. We are now faced with equations that will require something more than the four basic operations to obtain an equivalent equation. In fact, you will find that it is not always possible to obtain an equivalent equation.

Consider the two equations $x = 2$ and $x^2 = 4$. These equations certainly have something in common but it is also obvious that they are not equivalent, since 2 is the only solution to $x = 2$ but 2 and -2 are both solutions to $x^2 = 4$. If both sides of an equation are squared, the resulting equation is not always equivalent to the original. However, any solution to the original equation will be a solution of the resulting equation.

In general, if both sides of an equation are raised to the same power, the resulting equation will contain all solutions of the original equation. However, the resulting equation may also contain solutions that are *not* solutions of the original equation. These solutions are called **extraneous roots.** It is therefore necessary to check all solutions in the original equations.

A STEP-BY-STEP PROCEDURE FOR SOLVING EQUATIONS INVOLVING SQUARE ROOT RADICALS

Step 1 Isolate the radical if possible.
Step 2 Square both sides of the equation.
Step 3 If necessary, repeat steps 1 and 2 until all radicals have been eliminated.
Step 4 Solve the resulting equation.
Step 5 Check for extraneous roots.

Example 1 Solve: $x - \sqrt{2x - 5} = 4$

Solution First we isolate the radical on one side of the equation.

$$-\sqrt{2x - 5} = 4 - x$$

Squaring each side yields

$$\begin{aligned} (-\sqrt{2x - 5})^2 &= (4 - x)^2 \\ 2x - 5 &= 16 - 8x + x^2 \\ x^2 - 10x + 21 &= 0 \\ (x - 7)(x - 3) &= 0 \\ x - 7 = 0 \quad &\text{or} \quad x - 3 = 0 \\ x = 7 \quad & \qquad\quad x = 3. \end{aligned}$$

Checking, we have

$$\begin{array}{r|l} \text{if } x = 7,\ x - \sqrt{2x - 5} & = 4 \\ 7 - \sqrt{2(7) - 5} & 4 \\ -\sqrt{9} & \\ 7 - 3 & \\ 4. & \end{array}$$

Therefore 7 is a solution.

$$\begin{array}{r|l} \text{If } x = 3,\ x - \sqrt{2x - 5} & = 4 \\ 3 - \sqrt{2(3) - 5} & 4 \\ 3 - \sqrt{1} & \\ 3 - 1 & \\ 2. & \end{array}$$

Therefore 3 is *not* a solution. The solution set is $\{7\}$.

Example 2 Solve: $\sqrt{5x - 1} - \sqrt{2x} = \sqrt{x - 1}$

Solution Here we have more than one radical in the equation. Since there is a single radical isolated on one side of the equation, it will disappear when we square both sides.

$$\begin{aligned} (\sqrt{5x - 1} - \sqrt{2x})^2 &= (\sqrt{x - 1})^2 \\ 5x - 1 - 2\sqrt{5x - 1}\,\sqrt{2x} + 2x &= x - 1 \\ -2\sqrt{5x - 1}\,\sqrt{2x} &= -6x \end{aligned}$$

or

$$-2\sqrt{10x^2 - 2x} = -6x$$

Squaring each side, we obtain

$$
\begin{aligned}
(-2\sqrt{10x^2 - 2x})^2 &= (-6x)^2 \\
4(10x^2 - 2x) &= 36x^2 \\
40x^2 - 8x &= 36x^2 \\
4x^2 - 8x &= 0 \\
4x(x - 2) &= 0
\end{aligned}
$$

$$
\begin{aligned}
4x = 0 \quad &\text{or} \quad x - 2 = 0 \\
x = 0 \quad &\qquad\quad x = 2.
\end{aligned}
$$

Checking, we have

$$
\begin{array}{r|l}
\text{if } x = 0,\ \sqrt{5x - 1} - \sqrt{2x} & = \sqrt{x - 1} \\
\sqrt{5(0) - 1} - \sqrt{2(0)} & \sqrt{0 - 1} \\
\sqrt{-1} - \sqrt{0} & \sqrt{-1} \\
\sqrt{-1}. &
\end{array}
$$

Therefore 0 is a solution.

$$
\begin{array}{r|l}
\text{If } x = 2,\ \sqrt{5x - 1} - \sqrt{2x} & = \sqrt{x - 1} \\
\sqrt{5(2) - 1} - \sqrt{2(2)} & \sqrt{2 - 1} \\
\sqrt{9} - \sqrt{4} & \sqrt{1} \\
3 - 2 & 1 \\
1. &
\end{array}
$$

Therefore 2 is also a solution and the solution set is $\{0,2\}$.

CAUTION To avoid the embarrassment of accepting an extraneous root as a solution always check the solution(s) in the original equation when solving equations involving radicals.

EXERCISE 7-8-1

Solve.

1. $x + \sqrt{x} = 6$

2. $x - 2\sqrt{x} = 3$

3. $x + \sqrt{x - 1} = 7$

4. $x - \sqrt{x + 2} = 10$

5. $x + \sqrt{2x - 5} = 10$

6. $2x - \sqrt{3x - 2} = 8$

7. $x - 2\sqrt{2x + 1} = -2$

8. $5 - \sqrt{5x - 1} = x$

9. $6 + \sqrt{3x + 1} = 2x$

10. $\sqrt{x + 7} = 1 + \sqrt{2x}$

11. $\sqrt{2x - 1} + \sqrt{x + 3} = 3$

12. $\sqrt{x + 2} + \sqrt{2x + 11} = 4$

13. $\sqrt{x + 4} + \sqrt{2x + 10} = 3$

14. $\sqrt{x + 1} + \sqrt{x - 7} = \sqrt{2x}$

15. $\sqrt{2x + 5} - \sqrt{x + 2} = \sqrt{3x - 5}$

7–9 WORD PROBLEMS

OBJECTIVES

Upon completion of this section you should be able to:

1. Identify word problems that require a quadratic equation for their solution.
2. Solve word problems involving quadratic equations.

Certain types of word problems will lead to quadratic equations. As always, the solutions to such problems must be checked in the problem itself rather than in the resulting equation. This is necessary because the physical restrictions within the problem may eliminate one or more of the solutions.

Example 1 If the length of a rectangle is one unit more than twice the width, and the area is 55 square units, find the length and width.

Solution The formula for the area of a rectangle is area = length × width. Let x = width, $2x + 1$ = length.

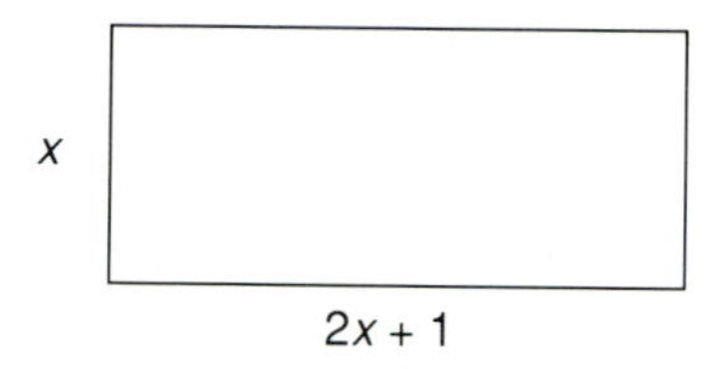

Substituting these values in the formula, we have

$$55 = (2x + 1)(x)$$
$$55 = 2x^2 + x$$
$$2x^2 + x - 55 = 0$$
$$(2x + 11)(x - 5) = 0$$
$$2x + 11 = 0 \quad \text{or} \quad x - 5 = 0$$
$$2x = -11 \qquad x = 5$$
$$x = -\frac{11}{2}.$$

At this point, we see that the solution $x = -\frac{11}{2}$ is not valid since x represents a measurement of the width and negative numbers are not used for such measurements. Therefore the solution is

$$\text{width} = x = 5$$
$$\text{length} = 2x + 1 = 11.$$

Example 2 A number added to its reciprocal is $2\frac{9}{10}$. Find the number.

Solution Let x = the number.

Then $$x + \frac{1}{x} = 2\frac{9}{10}$$

or $$x + \frac{1}{x} = \frac{29}{10}.$$

The LCD is $10x$.

$$\boxed{10x}\left[x + \frac{1}{x}\right] = \boxed{10x}\left[\frac{29}{10}\right]$$

$$10x^2 + 10 = 29x$$

$$10x^2 - 29x + 10 = 0$$

$$(5x - 2)(2x - 5) = 0$$

$$5x - 2 = 0 \quad \text{or} \quad 2x - 5 = 0$$

$$5x = 2 \qquad\qquad 2x = 5$$

$$x = \frac{2}{5} \qquad\qquad x = \frac{5}{2}$$

Both solutions check. Therefore the solution set is $\left\{\frac{5}{2}, \frac{2}{5}\right\}$.

The work problems discussed in chapter 5 were solved using first-degree equations. Such problems may sometimes result in a quadratic equation.

Example 3 Two pipes, A and B, together fill a gasoline tank truck in two hours. Pipe A alone can fill the tank in three hours less than pipe B alone. Find the time it would take pipe A to fill the tank alone.

	Time to Fill the Tank	Amount of Tank Filled in One Hour
A	$x - 3$	$\frac{1}{x - 3}$
B	x	$\frac{1}{x}$
Both	2	$\frac{1}{2}$

Recall that the equation is based on the amount of work done in one unit of time.

$$\frac{1}{x - 3} + \frac{1}{x} = \frac{1}{2}$$

$$2x + 2(x - 3) = x(x - 3)$$

$$2x + 2x - 6 = x^2 - 3x$$

$$x^2 - 7x + 6 = 0$$

$$(x - 1)(x - 6) = 0$$

$$x - 1 = 0 \quad \text{or} \quad x - 6 = 0$$

$$x = 1 \qquad\qquad x = 6$$

Notice that if we check the problem using $x = 1$, we obtain pipe B takes 1 hour and pipe A takes -2 hours. The physical properties of the problem eliminate this solution.

If $x = 6$, then $x - 3 = 3$. So pipe A can fill the tank in 3 hours.

EXERCISE 7–9–1

Solve.

1. The length of a rectangle is two more than twice its width. Find the dimensions of the rectangle if its area is 24.

2. A triangle of area 35 has an altitude which is three less than its base. Find the base and altitude.

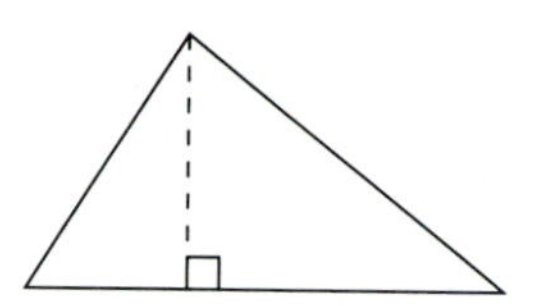

3. Find two consecutive positive odd integers whose product is 143.

4. One number is five greater than the other. Find the numbers if their product is 176.

5. Find three consecutive positive integers if the sum of their squares is 77.

6. The sum of a number and its reciprocal is $2\frac{1}{12}$. Find the number.

7. The diagonal of a rectangle is ten centimeters and the width is two centimeters less than the length. Find the dimensions of the rectangle. (Recall that the sum of the squares of two sides of a right triangle is equal to the square of the hypotenuse.)

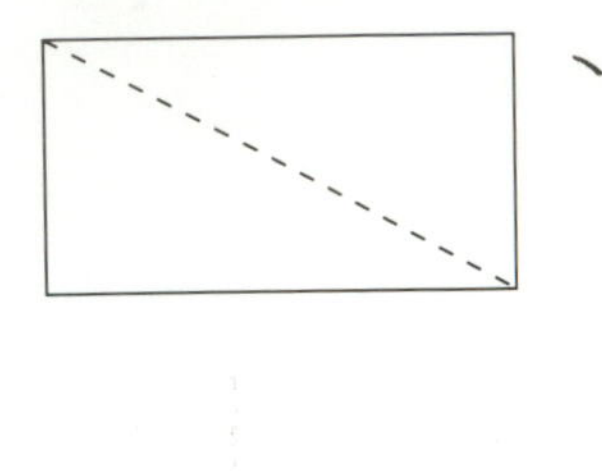

8. The sum of the reciprocals of two consecutive even integers is $\frac{9}{40}$. Find the integers.

9. One number is four greater than the other. The sum of their squares is 106. Find the numbers.

10. One number is five more than twice the other. Find the numbers if the difference of their squares is 153.

11. Find the number that is six greater than its positive square root.

12. One leg of a right triangle is two centimeters longer than the other. If the hypotenuse is ten centimeters, find the lengths of the two legs.

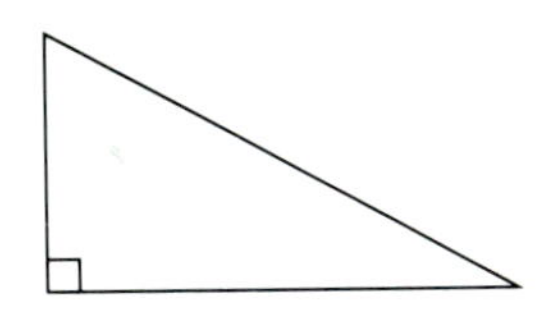

13. One side of a right triangle is four centimeters. The other side is seven centimeters less than twice the length of the hypotenuse. Find the length of the hypotenuse.

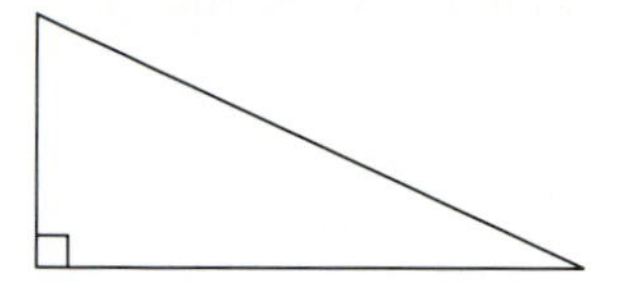

14. The area of a square is 36 square meters. If the area is to be increased by 28 square meters, how many meters should each side be increased by?

15. The length of a rectangle is twice the width. If each dimension is increased by three, the new area would be 104 square meters. Find the original dimensions.

16. A polygon of n sides has $\frac{n(n-3)}{2}$ diagonals. How many sides does a polygon have if it has 54 diagonals?

17. An object dropped from the top of a 45-meter tower falls according to the formula $s = -5t^2 + 45$, where s represents the distance of the object above the ground at any time t in seconds. How long will it take the object to reach the ground?

18. A ball is thrown upward with a velocity of 15 meters per second. The distance s of the ball above the ground in t seconds is given by $s = 15t - 5t^2$. How long will it take the ball to be within ten meters off the ground? Why are there two answers?

19. A small motorboat can travel ten kilometers per hour in still water. The boat travels eight kilometers upstream and returns in one hour and forty minutes. What is the speed of the current?

20. A merchant bought some calculators for a total price of $225.00. Each calculator was marked up $10.00 and offered for sale. All but one of the calculators were sold for a total return of $220.00. How many calculators did the merchant buy?

21. Two pumps can fill a swimming pool in three hours. It would take pump A 8 hours longer than pump B to fill the pool alone. How many hours would it take each pump to fill the pool alone?

22. A ranch hand can mend a fence in 20 hours. If his boss works with him, they can mend the fence in one hour less time than it would take for the boss to do it alone. How long would it take to mend the fence if both worked together?

SUMMARY

The number in brackets refers to the section of the chapter that discusses that concept.

Terminology

- A **quadratic equation** is a second-degree polynomial equation with one unknown. [7–1]
- The **standard form** of a quadratic equation is $ax^2 + bx + c = 0$, $a \neq 0$. [7–1]
- An **incomplete quadratic equation** is an equation of the form $ax^2 + bx + c = 0$, $a \neq 0$ and either $b = 0$ or $c = 0$. [7–2]
- **Completing the square** is a process for solving quadratic equations. [7–3]
- The **quadratic formula** is derived by completing the square on the general quadratic equation. [7–4]
- In the quadratic formula the radicand $b^2 - 4ac$ is called the **discriminant.** [7–5]
- A **complex number** may be written in the form $a + bi$, where a and b are real numbers and $i = \sqrt{-1}$. [7–6]
- An equation is **quadratic in form** if a suitable substitution for the unknown can be found so that the resulting equation is quadratic. [7–7]
- **Extraneous roots** may occur when both sides of an equation are raised to a power. [7–8]

Rules and Procedures

Solving Quadratic Equations

- The most direct and generally easiest method of finding the solutions to a quadratic equation is factoring. This method is based on the theorem if $AB = 0$, then $A = 0$ or $B = 0$. To use this theorem we put the equation in standard form, factor, and set each factor equal to zero. [7–1]

- To solve an incomplete quadratic equation use the theorem if $A^2 = B$, then $A = \pm\sqrt{B}$. [7–2]
- To solve a quadratic equation by completing the square complete the following steps. [7–3]

  STEP 1 Arrange the equation in standard form.

  STEP 2 If the coefficient of x^2 is not 1, divide each side of the equation by that coefficient.

  STEP 3 Rearrange the equation so that the x^2 and x terms are on the left side of the equation and the numerical term is on the right.

  STEP 4 Complete the square on the left and keep the equation balanced by adding the square of one-half of the coefficient of the x term to both sides.

  STEP 5 Factor the completed square.

  STEP 6 Take the square root of both sides of the equation remembering that if $A^2 = B$, then $A = \pm\sqrt{B}$.

  STEP 7 Solve for the two values of x by using the $(+)$ sign and then the $(-)$ sign of the right-hand term.
- To solve a quadratic equation by using the quadratic formula first arrange the equation in standard form $ax^2 + bx + c = 0$. Then substitute the values of a, b, and c into the formula.

$$x = \frac{-b \pm \sqrt{b^2 - 4ac}}{2a} \qquad [7\text{–}4]$$

Roots of a Quadratic Equation

- The nature of the roots of a quadratic equation is dependent on the value of the discriminant. If the value of the discriminant is greater than or equal to zero, the roots are real. If the value is negative, there are no real roots. [7–5]

Complex Numbers

- To add or subtract complex numbers combine the real parts of each number and the imaginary parts of each number. [7–6]
- To multiply or divide two complex numbers treat the numbers as binomials. [7–6]

Equations Quadratic in Form

- Since equations that are quadratic in form are quadratics in some power of the variable, we can solve for this power of the variable as if we were solving a quadratic by factoring or using the formula. We then proceed to find the solutions either by raising both sides to a power or taking a root of both sides. [7–7]

Radical Equations

- To solve equations containing square root radicals complete the following steps. [7–8]

  STEP 1 Isolate the radical if possible.

  STEP 2 Square both sides of the equation.

  STEP 3 If necessary, repeat steps 1 and 2 until all radicals have been eliminated.

  STEP 4 Solve the resulting equation.

  STEP 5 Check for extraneous roots.

REVIEW

Solve by factoring.

1. $x^2 + 10x + 21 = 0$ **2.** $x^2 + x - 30 = 0$ **3.** $x^2 - 11x + 24 = 0$

4. $2x^2 + 3x - 2 = 0$

5. $6x^2 + 7x - 3 = 0$

Solve.

6. $x^2 = 36$

7. $x^2 = 17$

8. $5x^2 = 100$

9. $2x^2 + 32 = 0$

10. $3x^2 - 27 = 0$

Solve by completing the square.

11. $x^2 + 8x + 15 = 0$

12. $x^2 - 10x + 8 = 0$

13. $x^2 + 3x - 1 = 0$

14. $3x^2 - 6x + 2 = 0$

15. $2x^2 + 5x + 4 = 0$

Solve by using the quadratic formula.

16. $x^2 + 3x - 28 = 0$

17. $x^2 - 4x + 3 = 0$

18. $x^2 - x - 1 = 0$

19. $2x^2 + 8x + 7 = 0$

20. $3x^2 - 2x - 1 = 0$

Compute $b^2 - 4ac$ for each of the following and give the nature of their roots.

21. $x^2 + 3x - 2 = 0$

22. $x^2 - 3x - 18 = 0$

23. $x^2 + 6x + 9 = 0$

24. $2x^2 - 3x + 2 = 0$

25. $3x^2 + 5x + 2 = 0$

Perform the indicated operation.

26. $(3 - 2i) + (1 + 5i)$

27. $(6 + i) - (3 - 2i)$

28. $(2 + 5i)(1 - 3i)$

29. $(7 + i) \div (2 + i)$

30. $3i \div (4 + 2i)$

Solve for real or complex roots by any method.

31. $6x^2 - 5x = 4$

32. $x^2 + 5 = 4x$

33. $9x^2 = 6x$

34. $\frac{1}{2}x^2 + x = \frac{1}{4}$

35. $2x^2 - 3x + 5 = 0$

Solve for real or complex roots.

36. $x^4 - 10x^2 + 9 = 0$

37. $x^4 + 3x^2 - 10 = 0$

38. $x^{1/2} - 7x^{1/4} + 10 = 0$

39. $x - 9x^{1/2} + 20 = 0$

40. $x^{2/3} + 3x^{1/3} - 4 = 0$

Solve for x.

41. $x + \sqrt{x} = 12$

42. $x - 3\sqrt{x} = -2$

43. $4 + \sqrt{2x} = x$

44. $\sqrt{3x + 1} + \sqrt{2x - 1} = 7$

45. $\sqrt{x + 7} - \sqrt{x + 6} = \sqrt{2x + 13}$

Solve.

46. The area of a rectangle is 105 square meters. Find the dimensions if the length is one meter longer than twice the width.

47. The sum of the squares of two consecutive odd integers is 74. Find the integers.

48. The base of a triangle is four units longer than the altitude. Find the length of the base if the area of the triangle is 96.

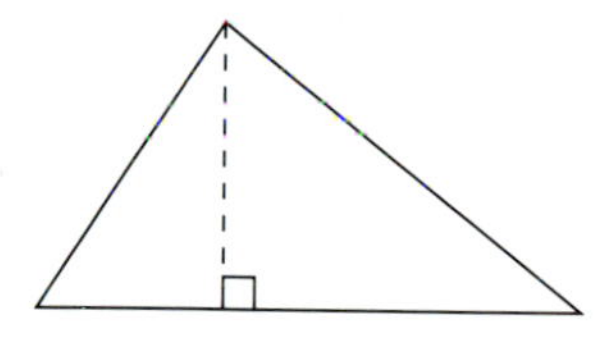

49. The sum of a number and twice its reciprocal is $\frac{27}{5}$. Find the number.

50. Bill and Bob working together can landscape a garden in 4 hours. Bill can landscape the garden by himself in 6 hours less time than it takes Bob to do it alone. How many hours would it take Bill to do the job alone?

CHAPTER 7

PRACTICE TEST

1. Complete the table. In column 2 place the value of the discriminant. In column 3 place an a, b, c, or d depending on the nature of the roots. These designations are: (a) no real roots, (b) roots are real and equal, (c) roots are irrational and unequal, and (d) roots are unequal and rational.

1. ____________________

Equation	Value of the Discriminant	Nature of the Roots
$2x^2 - 3x - 1 = 0$		
$x^2 + 6x + 9 = 0$		
$4x^2 + 3x + 1 = 0$		

2. Supply the coefficient of x that will make the trinomial a perfect square: $x^2 + (\quad)x + 49$

3. ____________________

3. Find the value of c that will make the following trinomial a perfect square: $x^2 + 10x + c$

Solve for real or complex roots:

4. ____________________

4. $x^2 = 10$

5. $3x^2 - 12 = 0$

5. ____________________

6. $x^2 - 7x = 0$

6. ____________________

7. $x^2 - 3x + 4 = 0$

7. ____________________

8. $3x^2 + 6x - 9 = 0$

8. ____________________

9. $x^2 - 4x = -1$

9. ____________________

10. $2x^2 + 5x - 3 = 0$

10. ____________________

11. $x^4 - 3x^2 - 10 = 0$

11. ____________________

12. $x - \sqrt{2x} = 12$

12. ____________________

13. $\sqrt{x+3} + \sqrt{2x-1} = 3$

13. ______________________

14. The area of a rectangle is 84 square meters. If the length is 5 meters longer than the width, find the width.

14. ______________________

CHAPTER

SURVEY

The following questions refer to material discussed in this chapter. Work as many problems as you can and check your answers with the answer section in the back of the book. The results will direct you to the sections of the chapter in which you need to work. If you answer all questions correctly, you have a good understanding of the material contained in this chapter.

1. In the following equations is y a function of x? If so, state the domain of the function.

a. $y = \dfrac{1}{\sqrt{x - 3}}$ **b.** $y^2 = x - 5$

1. a. ______________________

1. b. ______________________

2. Sketch the graph of $y = x^2 + 2x - 3$.

2. ______________________

3. Solve the following system by graphing:

$$\begin{cases} y = x^2 - 2x + 3 \\ x + y = 5 \end{cases}$$

3. ______________________

4. Find the equation, in standard form, of the line through the points $(3,-1)$ and $(-4,2)$.

4. ______________________

5. Sketch the graph of the following linear inequality:
$2x + y < 5$

5. ______________________

6. Solve the following system of inequalities by graphing:

$$\begin{cases} 3x - 2y \leq 6 \\ x + 2y \geq 4 \end{cases}$$

6. ______________________

7. Solve algebraically: $\begin{cases} 2x + 3y = 6 \\ x - 2y = -11 \end{cases}$

7. ______________________

8. Solve: $\begin{cases} 3x + 4y + z = -2 \\ y + z = 1 \\ 2x - y - z = -5 \end{cases}$

8. ______________________

Systems of Equations and Inequalities

In chapters 2 and 7 we discussed equations having one variable. In this chapter we will discuss equations having more than one variable as well as techniques for solving systems of these equations.

8–1 FUNCTIONS

OBJECTIVES

Upon completion of this section you should be able to:

1. Determine if a relation is a function.
2. Find the domain of a function.

If we are given two sets of numbers, we can establish some rule that shows a relation between elements of the first set and elements of the second set.

Example 1 Given the two sets

$$A: \quad \{1,3,5\}$$
$$B: \quad \{2,4,6,8,10\},$$

the rule "Relate a number in set A to a number in set B so that the number in set B is twice as large as the number in set A" would give

1 is related to 2
3 is related to 6
5 is related to 10.

We can express this more compactly as a set of ordered pairs

$$\{(1,2), (3,6), (5,10)\}$$

where the first number (abscissa) in each pair is from set A and the second number (ordinate) is from set B.

Example 2 Given the same two sets A and B as in example 1, if we use the rule "Relate a number in set A to a number in set B that is larger," we would obtain

1 is related to 2, 4, 6, 8, and 10
3 is related to 4, 6, 8, and 10
5 is related to 6, 8, and 10.

Writing this as a set of ordered pairs, we have

$$\{(1,2),(1,4),(1,6),(1,8),(1,10),(3,4),(3,6),(3,8),(3,10),(5,6),(5,8),(5,10)\}.$$

DEFINITION

A **relation** is a set of ordered pairs.

The set of all first numbers in the ordered pairs of a relation is called the **domain.** In example 1 the domain is $\{1,3,5\}$. It is also the domain of example 2.

The set of second numbers in the ordered pairs of a relation is called the **range.** In example 1 the range is $\{2,6,10\}$. The range in example 2 is $\{2,4,6,8,10\}$. Sometimes the numbers in the range are referred to as **images** and the range is called the **image set.**

DEFINITION

A **function** is a relation such that each element of the domain has exactly one image.

From this definition note that in example 1 we have a function, whereas in example 2 we do not have a function.

In mathematics the rule relating the elements of the domain to elements of the range is usually given as an algebraic expression. For instance, in example 1 if x represents an element of set A, then $2x$ represents the image of x in set B. This rule could be written as $y = 2x$. We note that this rule is thus expressed as an equation containing two variables, x and y. The variable to which values are assigned is called the **independent** variable and the other is called the **dependent** variable. In this case x is the independent variable and y is the dependent variable.

Given an algebraic equation in two variables, it may be important to know if one variable is a function of the other.

Example 3 Is y a function of x in the equation $y = x^2 + 2x - 5$? If so, find the domain.

Solution From the definition y is a function of x if for each value of x, there is exactly one value of y. We can determine this by substituting values for x (the independent variable) and finding the corresponding values of y (the dependent variable).

If $x = 3$, then $y = (3)^2 + 2(3) - 5 = 10$.

We note that if any number is squared and then added to twice that number, and then 5 is subtracted, we will arrive at a unique value. We conclude that y is a function of x because each x yields exactly one y.

To find the domain, we must determine the largest set of numbers that can be used for x to give valid values for y. We usually think in a "negative direction." Are there any numbers that cannot be substituted for x? Prior knowledge always enters here. In this case we know that any real number can be squared, the result can be added to twice that number, and that result can be reduced by 5. So we conclude that the domain of $y = x^2 + 2x - 5$ is the set of all real numbers.

Example 4 Is y a function of x in the equation $y = \sqrt{x}$? If so, find the domain.

Solution Again, prior knowledge must come into play. Since the radical symbol represents only the principal square root, a value substituted for x will yield only one value of y. Therefore we conclude that y is a function of x.

To determine the domain we must ask, "Does $\sqrt{x}$ yield a real number for all values of x?" Again, from prior knowledge, we know that $\sqrt{x}$ is not a real number when $x < 0$. Thus the domain is zero and all positive numbers. We can write this as $x \geq 0$.

CAUTION Be particularly alert for restrictions in the domain when a function involves radicals or fractions.

Example 5 Is y a function of x in the equation $y = \dfrac{1}{x - 2}$? If so, find the domain.

Solution If we substitute a number for x, can $\dfrac{1}{x - 2}$ have more than one value? No. Therefore y is a function of x.

Now, once again, from prior knowledge we know that zero cannot occur as the denominator of a fraction. If $x = 2$, then $\dfrac{1}{x - 2}$ becomes $\dfrac{1}{0}$, which has no meaning. The domain of this function is, therefore, the set of all real numbers except 2.

Example 6 Is y a function of x in the equation $y^2 = x$? If so, find the domain.

Solution If we let $x = 9$, we have $y^2 = 9$ and $y = \pm 3$. One value of x gave us two values of y. Therefore y is not a function of x.

EXERCISE 8–1–1

For each equation determine if y is a function of x. If so, find the domain.

1. $y = 2x$

2. $y = 3x - 1$

3. $y = x^2$

4. $y = x^2 + x - 3$

5. $y = \dfrac{1}{x - 5}$

6. $y = \dfrac{1}{2x + 1}$

7. $y^2 = x + 1$

8. $y = \dfrac{1}{\sqrt{x - 1}}$

9. $y = \dfrac{1}{x^2 - 4}$

10. $x^2 + y^2 = 4$

8–2 CARTESIAN COORDINATE SYSTEM AND GRAPHING EQUATIONS IN TWO VARIABLES

Upon completion of this section you should be able to:

1. Locate points on the Cartesian coordinate system.
2. Sketch graphs of relations by plotting solutions of the equation.

One method of uniquely naming each point in the plane is the **Cartesian coordinate system** (named for the French mathematician René Descartes, 1596–1650). This rectangular coordinate system is constructed with two mutually perpendicular real number lines in the plane that intersect at zero on each line. One line is vertical and the other is horizontal, with positive direction to the right and upward. The two lines are called **coordinate axes** and their point of intersection is called the **origin.** The horizontal axis is usually referred to as the ***x*-axis** and the vertical axis is referred to as the ***y*-axis.**

The four regions of the plane formed by the x- and y-axes are called **quadrants** and are numbered in a counterclockwise direction starting with the upper right.

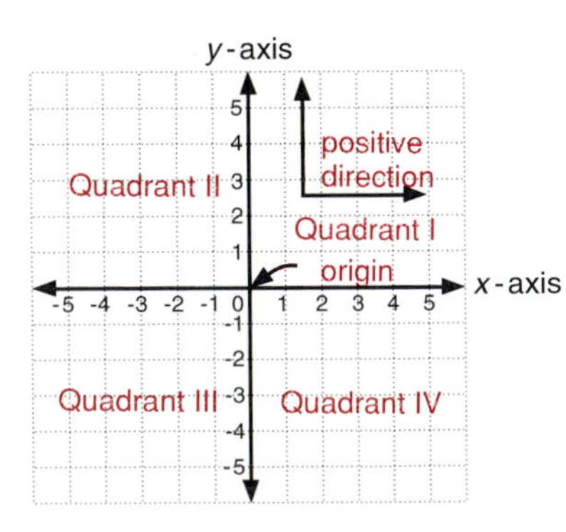

Any point in the plane can be located by moving first along the x-axis and then parallel to the y-axis. If we agree to write a pair of numbers such as (5,3) so that the first number represents the distance and direction from the origin along the x-axis and the second number represents the distance and direction parallel to the y-axis, then this **ordered pair** of numbers represents one and only one point on the plane. In the Cartesian coordinate system points on the plane are always represented by an ordered pair (x,y).

EXERCISE 8-2-1

1. Locate each of the following points on the coordinate system. Label each point with the letter and the ordered pair. A is given as an example. A: $(-3,5)$ B: $(5,2)$ C: $(5,-4)$ D: $(-4,-5)$ E: $(0,4)$ F: $(5,0)$ G: $(-3,0)$ H: $(0,-3)$

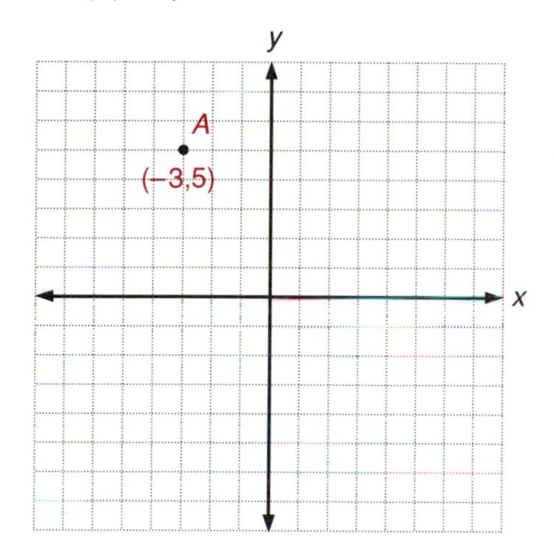

2. Note the points indicated on the coordinate system. Give the ordered pair associated with each point.

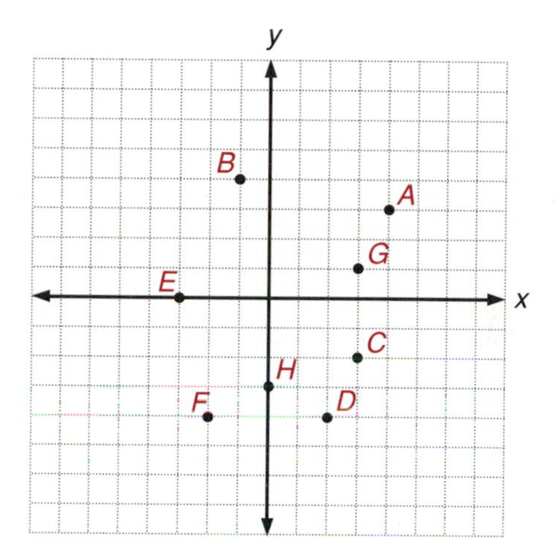

The graph of a relation between x and y can be represented on the coordinate system by expressing this relation as ordered pairs and locating the points on the plane represented by these ordered pairs. It is generally not possible to locate all points given by an algebraic equation, so we content ourselves with locating a sufficient number of points to establish a pattern and then sketch the graph.

Example 1 Sketch the graph of $y = 3x - 1$.

Solution Our first task is to find ordered pairs (x,y) that are solutions to the given equation. We accomplish this by substituting arbitrary values of x in the equation and finding the corresponding value of y.

For instance, if $x = -1$, then

$$y = 3x - 1 = 3(-1) - 1 = -4.$$

(value of x)

Therefore the ordered pair $(-1,-4)$ is a solution.

If $x = 0$, then

$$y = 3(0) - 1 = -1.$$

Therefore $(0,-1)$ is a solution.
If $x = 1$, then

$$y = 3(1) - 1 = 2.$$

Therefore $(1,2)$ is a solution.
If $x = 2$, then

$$y = 3(2) - 1 = 5.$$

Therefore $(2,5)$ is a solution.
It is convenient to place these ordered pairs in a table.

x	-1	0	1	2
y	-4	-1	2	5

We now locate the points $(-1,-4)$, $(0,-1)$, $(1,2)$, and $(2,5)$ on the coordinate plane and connect them with a line.

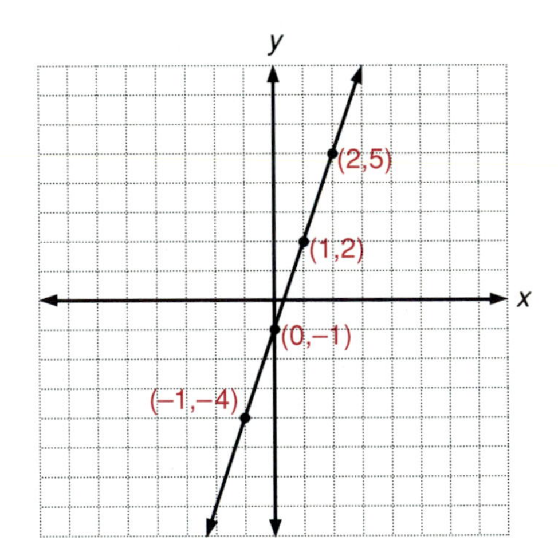

These points establish a pattern and we see that $y = 3x - 1$ is a straight line. It can be shown that the graph of any first-degree polynomial equation in two variables (a **linear equation**) will form a straight line.

Example 2 Sketch the graph of $y = x^2 - 3x - 4$.

Solution We again assign arbitrary values to x and find the corresponding values of y.

For instance, if $x = -2$, then

$$\begin{aligned} y &= (-2)^2 - 3(-2) - 4 \\ &= 4 + 6 - 4 \\ &= 6. \end{aligned}$$

Therefore $(-2,6)$ represents a point on our graph.
If $x = 0$, then

$$y = (0)^2 - 3(0) - 4 = -4.$$

Therefore $(0,-4)$ represents a point on our graph.

We continue this process until a sufficient number of points have been found to establish a pattern. The number of points needed will vary with the relation. However, we should always avoid taking values of x that are too near each other. This will insure that we obtain a total pattern rather than just a "local" pattern.

A possible table of values for $y = x^2 - 3x - 4$ is:

x	-2	-1	0	1	2	3	4	5
y	6	0	-4	-6	-6	-4	0	6

As we locate these points on the coordinate plane, it becomes clear that they are not on a straight line. Only first-degree polynomial equations graph as straight lines. Higher degree equations graph as curves.

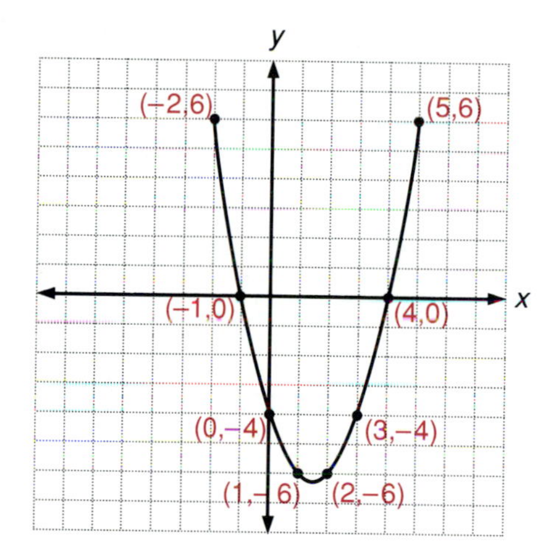

The curve in the preceding graph is a **parabola.** The curves formed by second-degree polynomial equations in two variables are the **circle, ellipse, parabola,** and **hyperbola.** These are studied in detail in a course in analytic geometry.

CAUTION When sketching the graph of a second-degree equation, it is necessary to find enough ordered pairs to establish a pattern. If too few points are used, the sketch will be inaccurate. When sketching the graph of a straight line (first-degree equation), only two points are necessary since two points determine a straight line. However, it is best to find at least three points, using any extra points as a check.

Sometimes the relation might be such that we wish to choose values of y and solve for values of x. Suppose we are asked to make a table of values for $2y + x = 1$. If we let $x = 2$, then solving for y, we obtain

$$2y + x = 1$$
$$2y + 2 = 1$$
$$2y = -1$$
$$y = -\frac{1}{2}.$$

Our ordered pair is $\left(2,-\frac{1}{2}\right)$, which contains a fraction, making the point more difficult to locate. If we assign values to y, we can avoid fractions. Let $y = 2$. Then

$$\begin{aligned} 2(2) + x &= 1 \\ x &= -3, \end{aligned}$$

giving the ordered pair $(-3,2)$.

Often there is no way to avoid fractions. However if we look ahead, we can choose values that will give whole numbers for solutions.

Example 3 Sketch $x = y^2 + 2y - 1$.

Solution First find a table of values by assigning values to y. Let $y = -1$. Then

$$x = (-1)^2 + 2(-1) - 1 = 1 - 2 - 1 = -2.$$

Therefore $(-2,-1)$ is a solution.

Let $y = 0$. Then

$$x = 0^2 + 2(0) - 1 = -1.$$

Therefore $(-1,0)$ is a solution.

Continuing to choose values for y, we might obtain the following table.

x	7	2	−1	−2	−1	2	7
y	−4	−3	−2	−1	0	1	2

Plotting these points gives us the following graph.

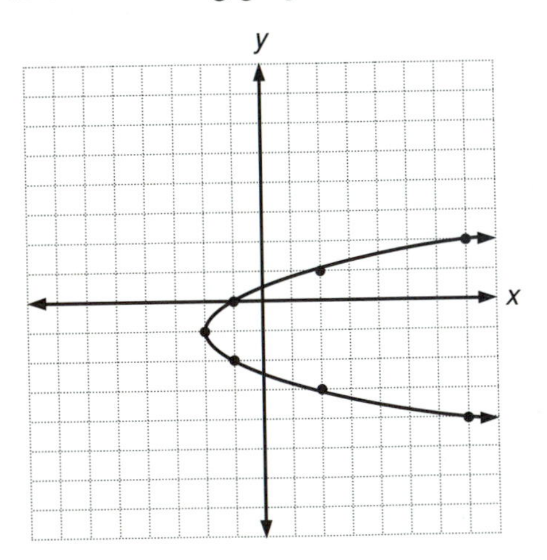

Graph of $x = y^2 + 2y - 1$
Is this relation a function of x?

EXERCISE 8–2–2

Sketch the graphs. In each case determine if y is a function of x.

1. $y = x$

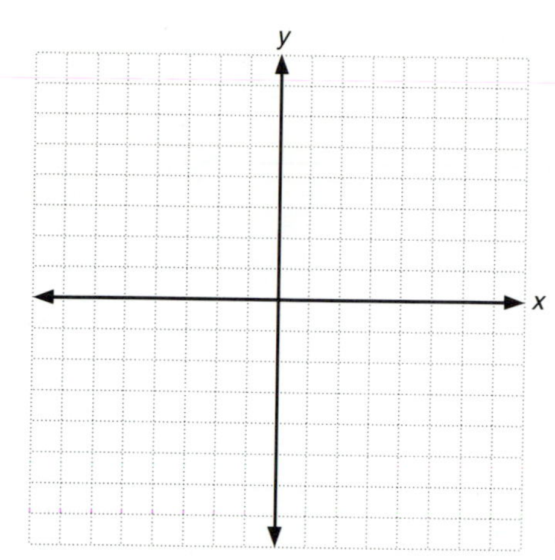

2. $y = x + 3$

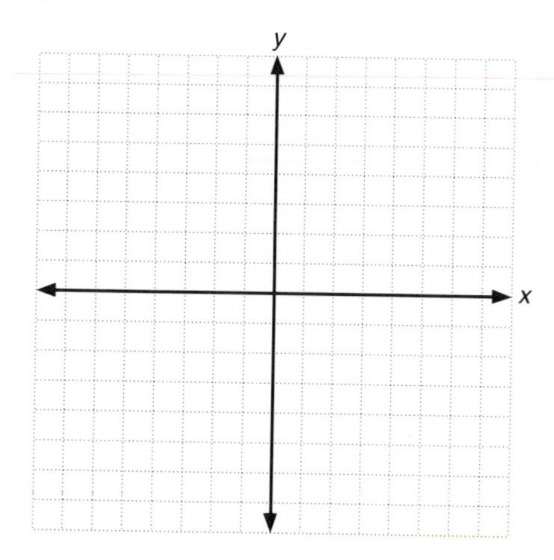

3. $y = x - 4$

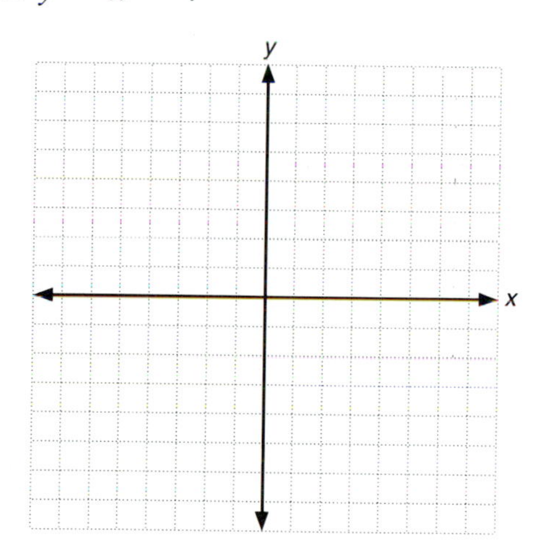

4. $y = 4 - x$

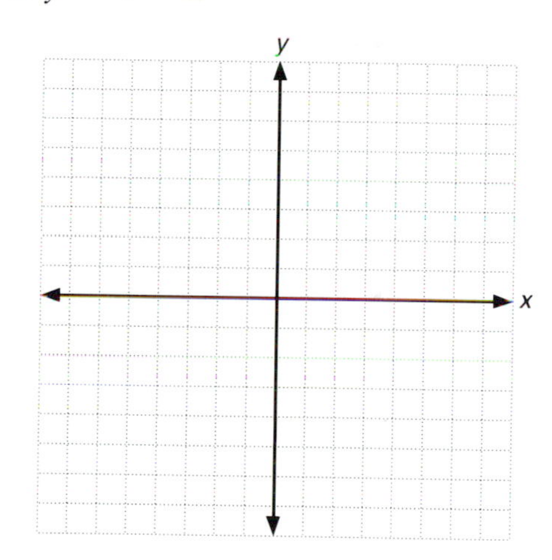

5. $y = 2x - 3$

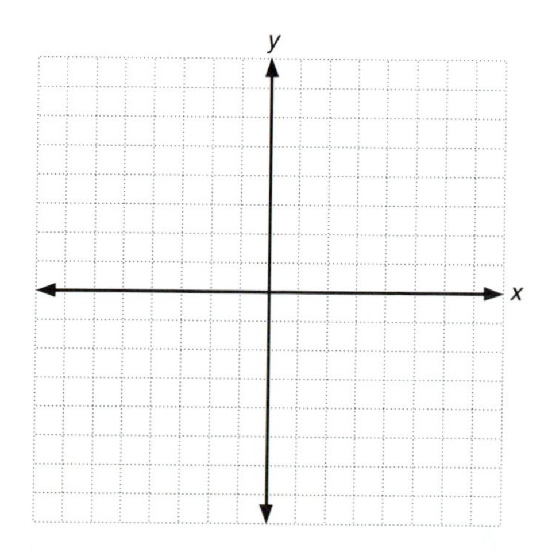

6. $y = 3 - 2x$

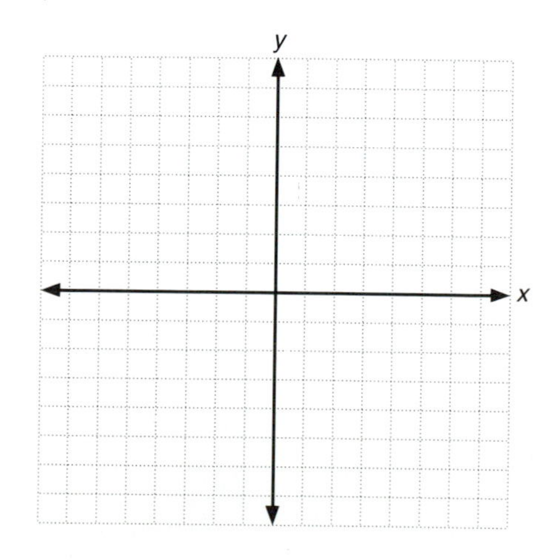

7. $x + 3y = 2$

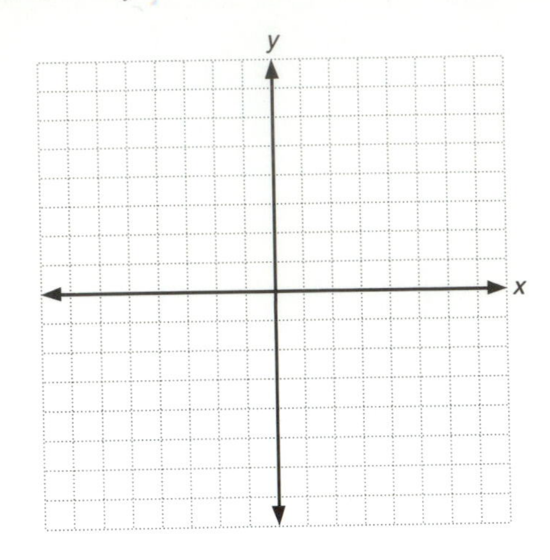

8. $3x - 2y = 7$

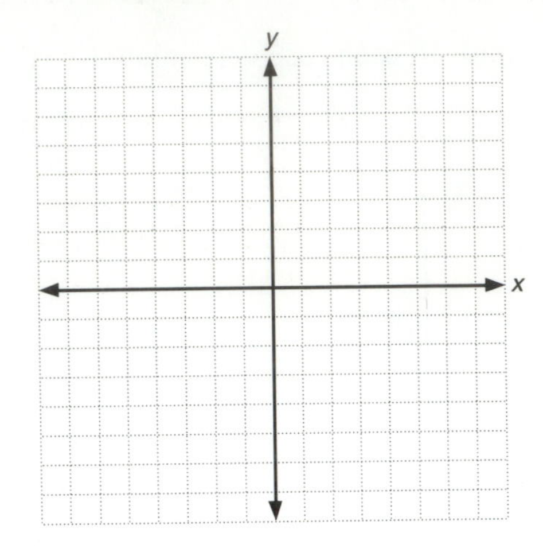

9. $y = x^2$

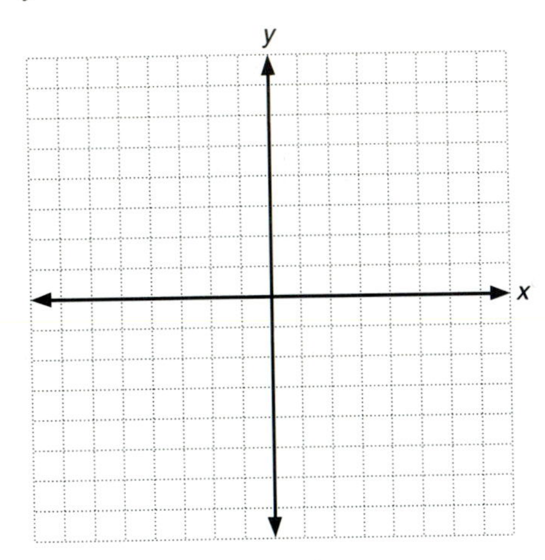

10. $y = x^2 + 2x - 3$

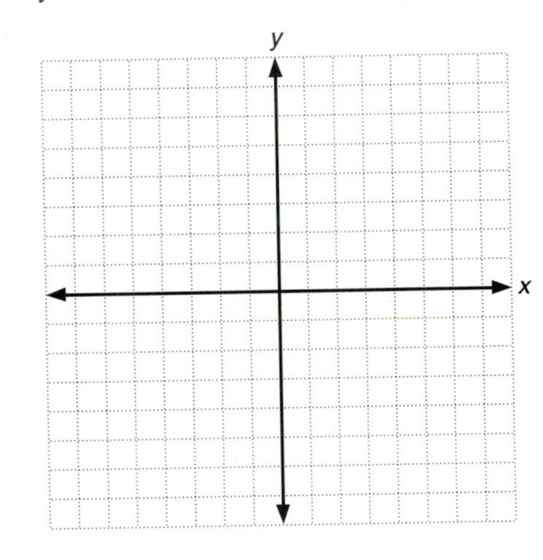

11. $y = x^2 - 4x + 3$

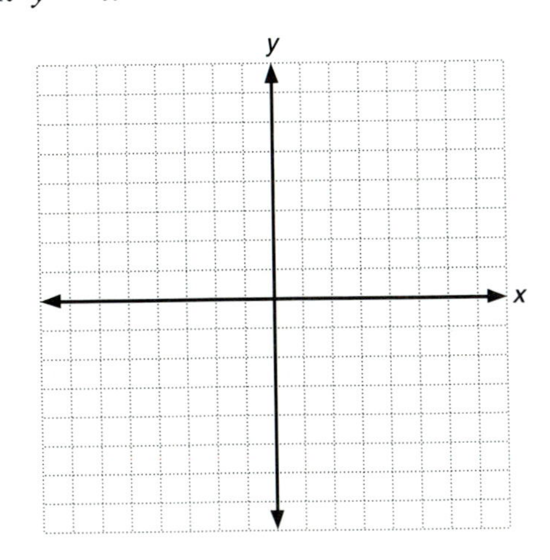

12. $y = -x^2 + 2x - 1$

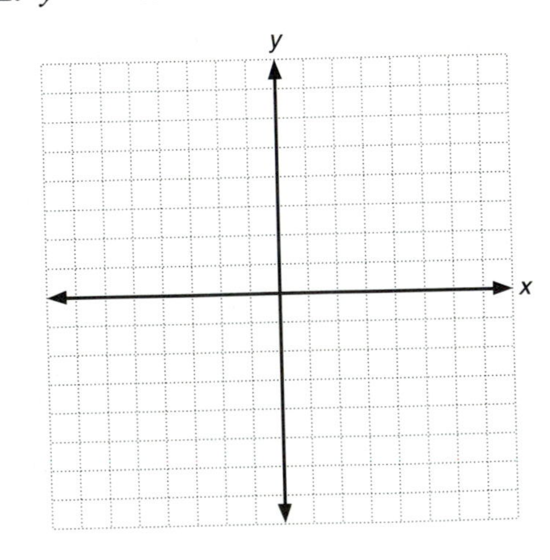

13. $x = y^2 - 4y + 2$

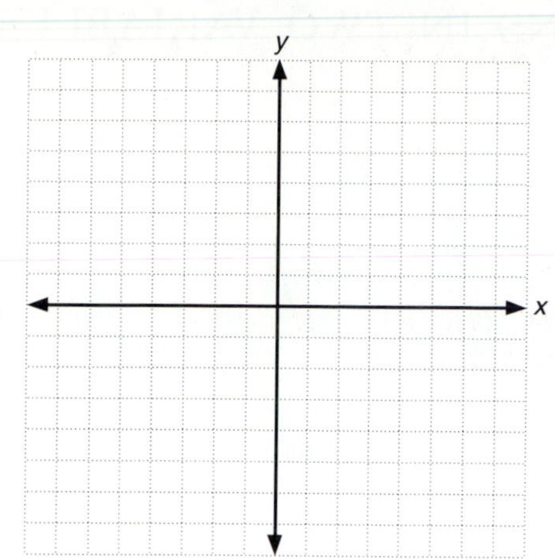

14. $x = -2y^2 + 8y - 5$

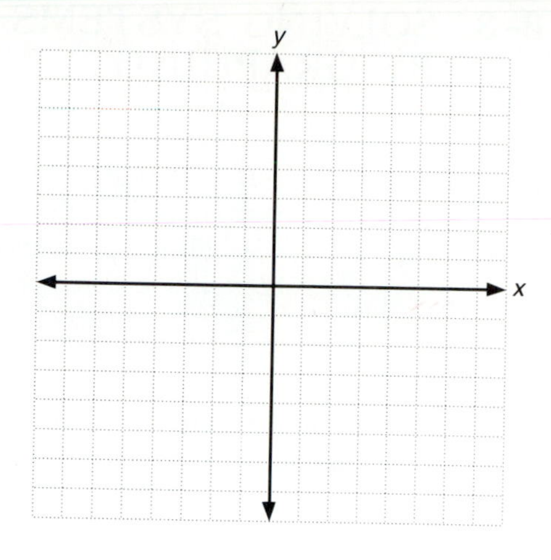

15. $y = \sqrt{x}$

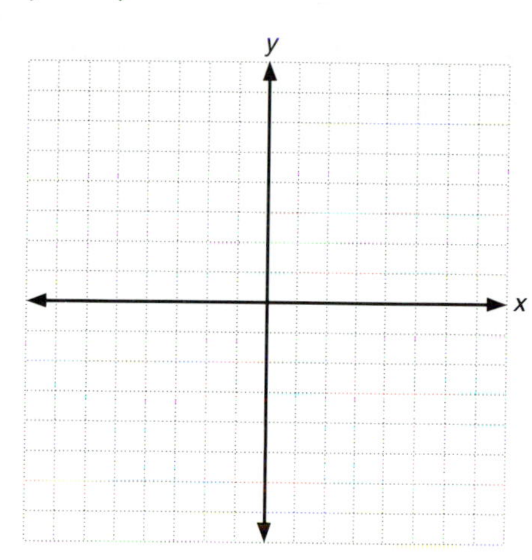

16. $y = x^3$

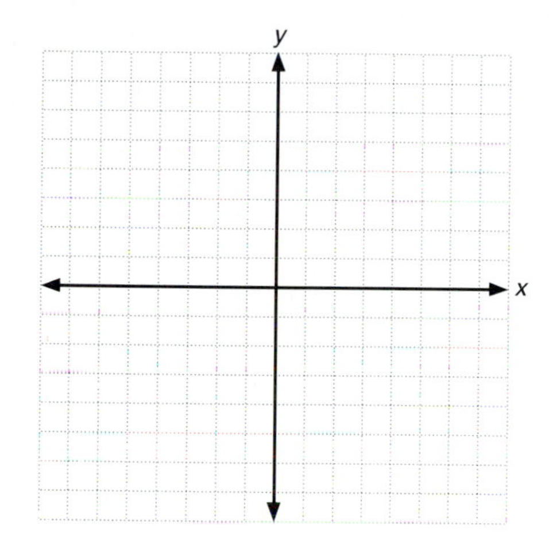

17. If the domain is taken as the x values and the range is taken as the y values, give a method of determining if a relation is a function by looking at its graph.

8–3 SOLVING SYSTEMS OF EQUATIONS IN TWO VARIABLES BY GRAPHING

OBJECTIVES

Upon completion of this section you should be able to:

1. Sketch the graphs of two equations on the same coordinate system.
2. Determine common solutions of the two graphs.

A **system of equations** in two variables is a set of two or more equations. We know that an equation such as $x + y = 5$ has infinitely many solutions, such as (3,2), (6,−1), and so on. The equation $2x + y = 8$ also has an infinite number of solutions such as (0,8), (4,0), and so on.

The solution to the system

$$\begin{cases} x + y = 5 \\ 2x + y = 8 \end{cases}$$

is the set of all ordered pairs that are solutions to both equations simultaneously. In other words, is there an ordered pair that makes the statements $x + y = 5$ and $2x + y = 8$ both true?

If we graph both equations on the same coordinate plane, the point or points of intersection will be the solution of the system. These graphs must be extremely accurate to arrive at the solution set.

Example 1 Solve by graphing: $\begin{cases} x + y = 5 \\ 2x + y = 8 \end{cases}$

Solution Set up a table of values for each equation and sketch their graphs. (We should recognize each of these equations as graphically representing a straight line.)

$x + y = 5$

x	−2	0	5
y	7	5	0

$2x + y = 8$

x	0	2	4
y	8	4	0

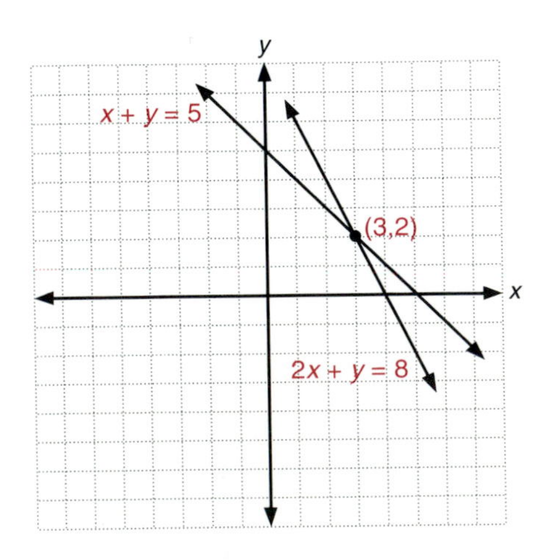

The point of intersection is (3,2) and should be labeled as in the illustration. We can check the correctness of our solution by substituting (3,2) into both of the equations to see that it is actually a solution.

Example 2 Solve by graphing: $\begin{cases} x - y = -1 \\ y = x^2 - 2x + 1 \end{cases}$

Solution As seen from the last section, these equations represent a straight line and a parabola, respectively. Set up tables of values and graph each equation.

$x - y = -1$

x	-3	1	5
y	-2	2	6

$y = x^2 - 2x + 1$

x	-2	-1	0	1	2	3	4
y	9	4	1	0	1	4	9

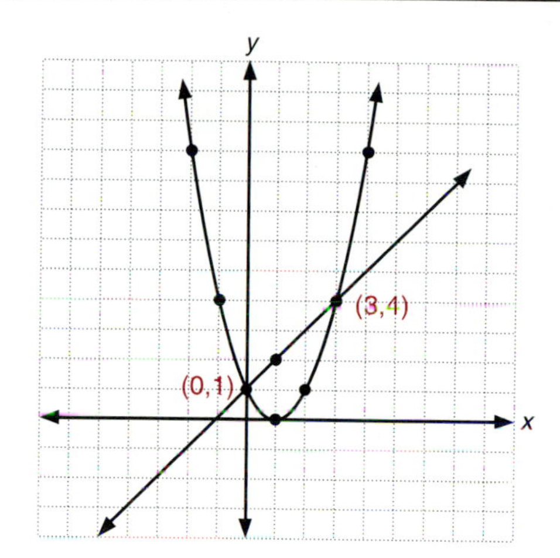

We see that the solutions are (0,1) and (3,4). These check in both equations.

EXERCISE 8–3–1

Solve the systems by graphing.

1. $\begin{cases} x + y = 2 \\ 2x - y = 1 \end{cases}$

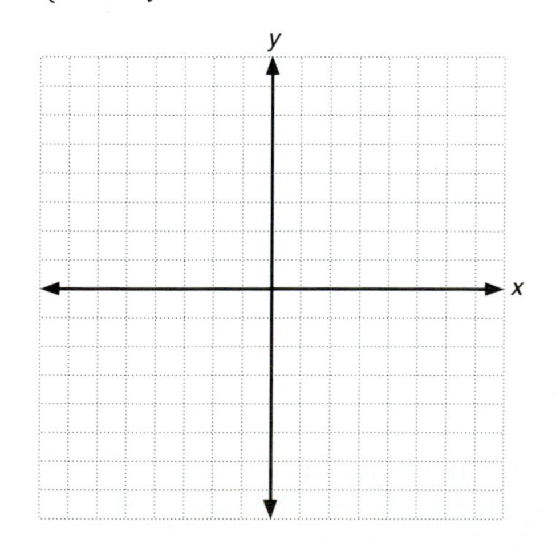

2. $\begin{cases} 3x + y = 0 \\ 2x - y = -5 \end{cases}$

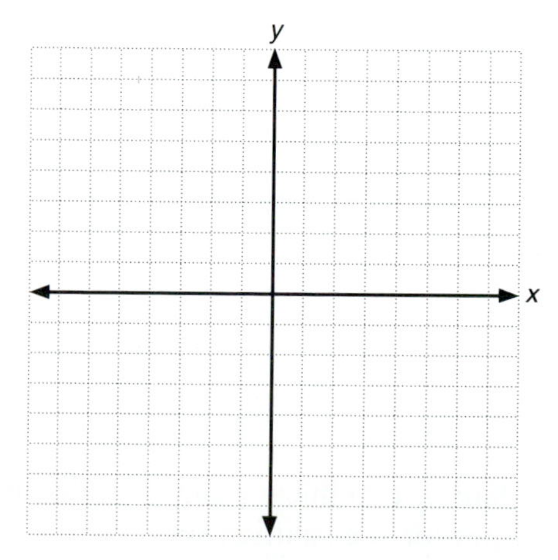

3. $\begin{cases} 3x + y = -9 \\ x - 2y = 4 \end{cases}$

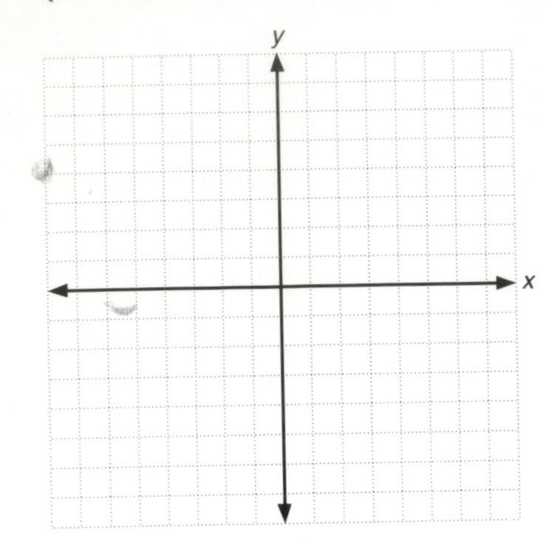

4. $\begin{cases} 3x + y = 8 \\ 2x - y = 7 \end{cases}$

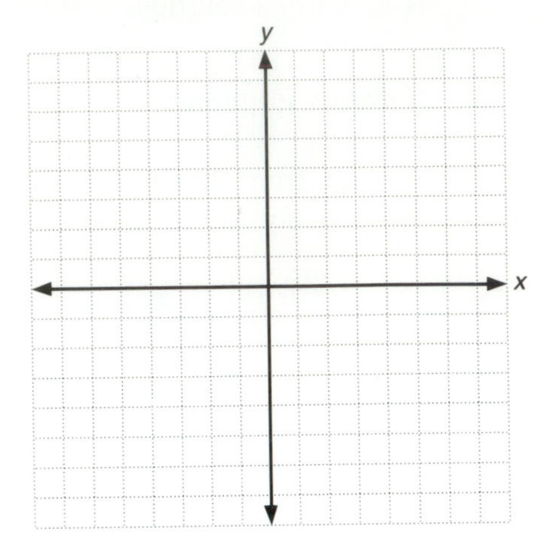

5. $\begin{cases} 3x + 2y = 7 \\ 2x - 3y = -4 \end{cases}$

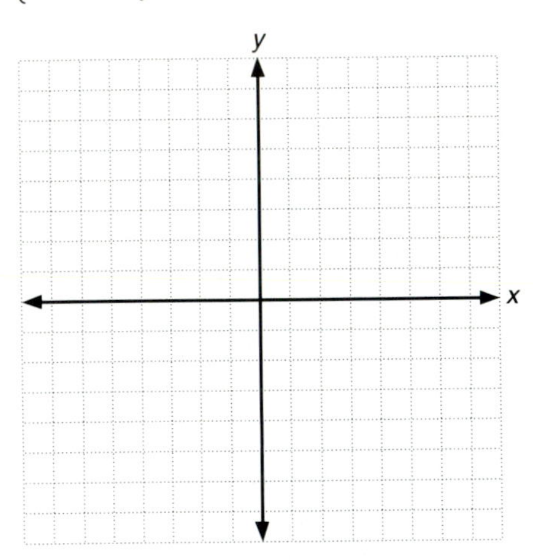

6. $\begin{cases} x - y = -2 \\ y = x^2 \end{cases}$

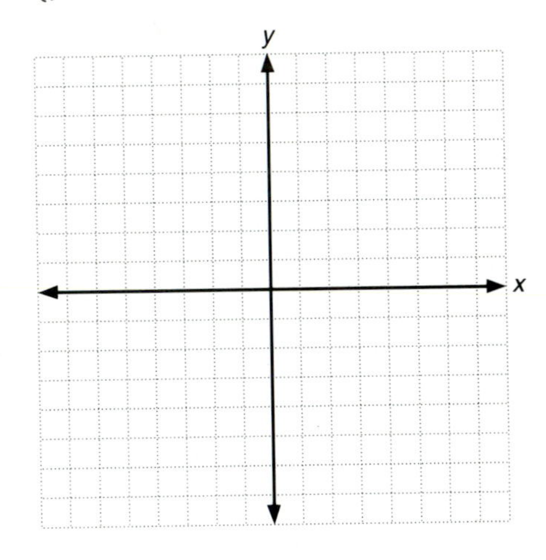

7. $\begin{cases} x + y = -1 \\ y = x^2 + x - 1 \end{cases}$

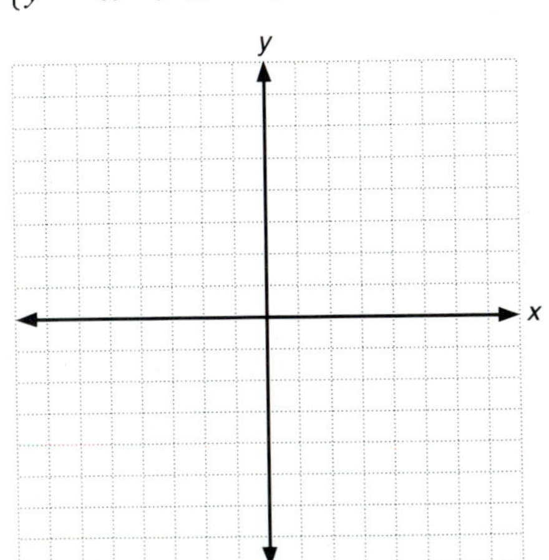

8. $\begin{cases} y = x^2 - 2x + 3 \\ x + y = 5 \end{cases}$

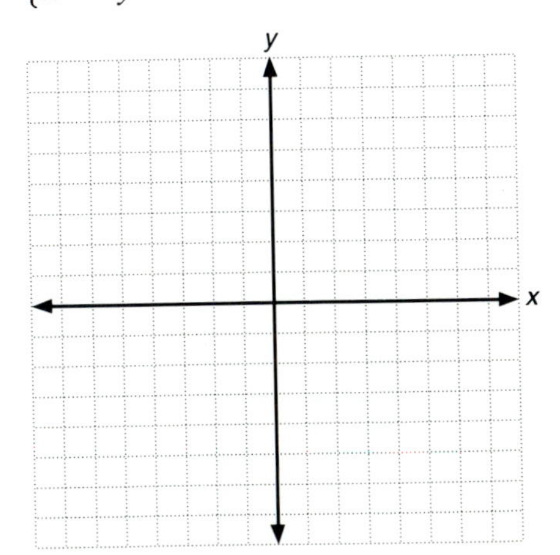

9. Given a system of two linear equations, discuss the possibilities for the solution set.

10. Given a system of one straight line and one parabola, discuss the possibilities for the solution set.

8–4 THE STRAIGHT LINE

OBJECTIVES

Upon completion of this section you should be able to:

1. Find the equation of a line given two points on the line.
2. Find the equation of a line given its slope and a point on the line.
3. Write the equation of a line in slope-intercept form.

In section 8–2 we introduced the equation of a straight line and learned how to plot its graph. We will now examine the straight line as a function.

The graph of a straight line parallel to the y-axis does not represent a function since each x value has many y values. All other straight lines represent functions. Remember that the graph of any first-degree equation in two variables will form a straight line.

We will call

$$ax + by = c,$$

where a, b, and c are integers and $a \geq 0$, the **standard form** of the equation of a straight line. An important concept, related to the equation of a straight line, is that of slope. Intuitively we think of the slope of a line as the "steepness" of the line. The following definition gives us a more precise meaning.

DEFINITION

The **slope** (m) of a line through the two distinct points (x_1,y_1) and (x_2,y_2) is given by the ratio

$$m = \frac{y_2 - y_1}{x_2 - x_1},\ x_1 \neq x_2.$$

Example 1 Find the slope of a line through the points $(1,-5)$ and $(4,0)$.

Solution If we let $(1,-5)$ be (x_1,y_1) and $(4,0)$ be (x_2,y_2), then applying the formula, we obtain

$$m = \frac{0 + 5}{4 - 1} = \frac{5}{3}.$$

The ratio $\dfrac{y_2 - y_1}{x_2 - x_1}$ is not dependent on the points chosen as long as they are two different points on the line. A given line has only one slope.

Notice that the definition of slope includes the statement $x_1 \neq x_2$. If $x_1 = x_2$, division by zero would result and this is not allowed. For instance, a straight line containing the points $(3,5)$ and $(3,7)$ is parallel to the y-axis and we see that the slope is

$$m = \frac{7 - 5}{3 - 3} = \frac{2}{0}.$$

Recall that division by zero is meaningless and therefore the slope of a line parallel to the y-axis (vertical line) is undefined.

If a straight line is parallel to the x-axis (horizontal line), the y value is a constant for each value of x. For instance, a straight line containing the points (3,5) and (7,5) is five units above and parallel to the x-axis. Computing the slope gives

$$m = \frac{5 - 5}{7 - 3} = \frac{0}{4} = 0.$$

Thus the slope of a line parallel to the x-axis (horizontal line) is zero.

The definition of slope and the fact that the slope of a given line is constant is the basis for the solution of many problems concerning linear functions.

For instance, if we are given that a line contains the points (x_1,y_1) and (x_2,y_2) and we are asked for the equation of the line, we proceed in the following manner.

First we choose any other point on the line and call it (x,y). The slope is now given by the ratio $\frac{y_2 - y_1}{x_2 - x_1}$ *or* by the ratio $\frac{y - y_1}{x - x_1}$ and since the slope is constant we can equate these two ratios giving

$$\frac{y - y_1}{x - x_1} = \frac{y_2 - y_1}{x_2 - x_1},$$

which becomes

$$y - y_1 = \left(\frac{y_2 - y_1}{x_2 - x_1}\right)(x - x_1), \qquad x_1 \neq x_2.$$

This is known as the **two-point form** of the equation of a line.

Example 2 Write the equation in standard form of the line through the points (3,5) and (−2,7).

Solution Using the two-point form, we have

$$y - 5 = \left(\frac{7 - 5}{-2 - 3}\right)(x - 3).$$

Simplifying, we obtain

$$y - 5 = -\frac{2}{5}(x - 3)$$

$$\text{or} \qquad 2x + 5y = 31.$$

Check to see that both points are on this line by substituting into the equation.

If we are given the slope m of a line and a point (x_1,y_1) and asked to write the equation, we proceed as follows.

We first choose some other point on the line and call it (x,y). The definition of the slope gives us the ratio $\frac{y - y_1}{x - x_1}$. But since we are given the slope to be m, we can write the equation

$$m = \frac{y - y_1}{x - x_1}$$

or

$$y - y_1 = m(x - x_1).$$

This is the **point-slope form** of the equation of a straight line.

Example 3 Write the equation in standard form of the line through the point (2,5) and having slope of $\frac{2}{5}$.

Solution Using the point-slope form gives us

$$y - 5 = \frac{2}{5}(x - 2),$$

which in standard form is

$$2x - 5y = -21.$$

DEFINITION
The ***y*-intercept** of a straight line is the ordinate of the point where the line intersects the *y*-axis. In other words, it is the value of y when $x = 0$. We will designate the *y*-intercept as b. The ordered pair representing the *y*-intercept is $(0,b)$.

If we are given the slope m of a line and the point $(0,b)$ on the line, then the point-slope form gives us

$$y - b = m(x - 0)$$

or

$$y = mx + b.$$

This very useful form of the equation of a straight line is called the **slope-intercept form.**

Example 4 If the slope of a line is $-\frac{5}{8}$ and the y-intercept is 6, write the equation in standard form.

Solution The slope-intercept form gives us

$$y = -\frac{5}{8}x + 6,$$

which in standard form is

$$5x + 8y = 48.$$

The slope-intercept form of the equation of a straight line is useful in graphing. We can use the equation from the last example to illustrate.

$$y = -\frac{5}{8}x + 6$$

We know that the y-intercept is 6. Thus the point whose coordinates are (0,6) is on the graph.

The slope is $-\frac{5}{8}$ and since it is a ratio it does not matter whether we place the negative sign in the numerator or denominator. If we put it in the numerator we obtain $\frac{-5}{8}$, which indicates that the change in y is -5 while the change in x is 8.

Using these values, we begin at (0,6) and move 8 places in the positive x direction. We then move 5 places in the negative y direction. The point we arrive at is also on the line. Since we now have two points, we may draw the graph of the line as shown.

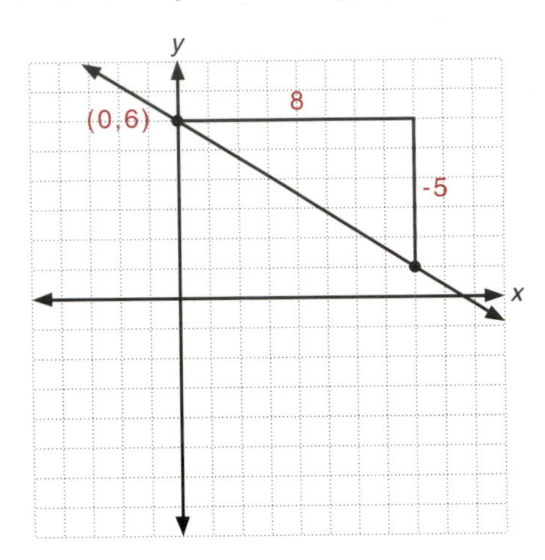

Example 5 Find the slope and the y-intercept of the line given by the equation $3x - 2y = 6$. Use this information to sketch the graph.

Solution Solving for y in terms of x gives us

$$y = \frac{3}{2}x - 3.$$

So the slope is $m = \frac{3}{2}$ and the y-intercept is $b = -3$. To sketch the graph we start at the y-intercept $(0, -3)$ and move two units in the positive x direction and then three units in the positive y direction. The point we arrive at is also on the line, so using this point and the y-intercept we draw the graph.

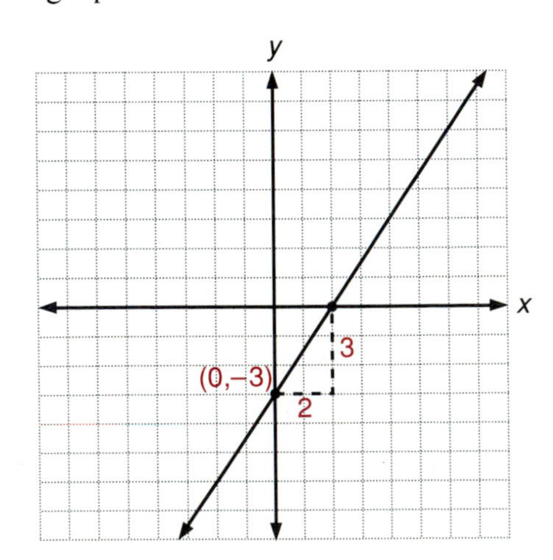

Special properties involving slopes of lines arise when considering equations of parallel and perpendicular lines.

Parallel lines are lines on the plane that do not intersect (that is, they do not meet).

> If two distinct lines have the same slope, they are **parallel.**

Example 6 The lines given by the equations

$$y = \frac{1}{3}x + 5 \text{ and } y = \frac{1}{3}x - 7$$

are parallel since their slopes are equal and their y-intercepts are different (that is, they are not the same line).

Perpendicular lines are lines that form right angles when they intersect.

> If the slope of one line is the negative reciprocal of the slope of another line, the lines are **perpendicular.**

Recall that one number is the negative reciprocal of another if their product is -1.

Example 7 The lines given by the equations

$$y = \frac{2}{3}x + 4 \text{ and } y = -\frac{3}{2}x + 1$$

are perpendicular since $\frac{2}{3}$ and $-\frac{3}{2}$ are negative reciprocals. That is

$$\left(\frac{2}{3}\right)\left(-\frac{3}{2}\right) = -1.$$

EXERCISE 8-4-1

Find the equation, in standard form, of the line through each of the pairs of points.

1. (2,1) and (-3,9)

2. (-6,2) and (4,-3)

3. (16,-5) and (4,0)

4. $(3,2)$ and $(-4,2)$

5. $(-8,-1)$ and $(3,-5)$

Use the information given to sketch the graph and then find the equation, in standard form, of the line represented.

6. $(1,7)$, $m = 3$

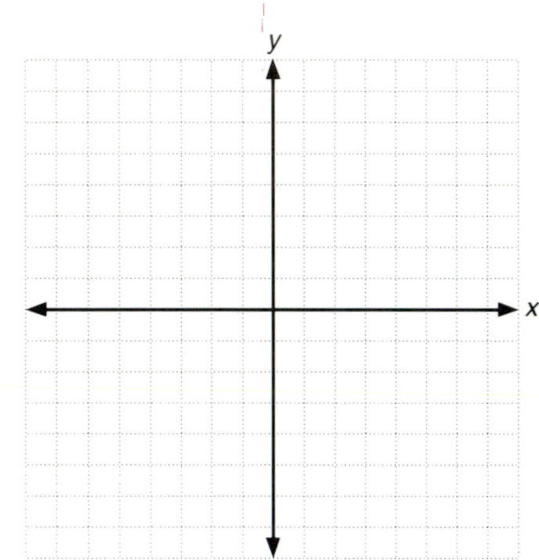

7. $(-2,0)$, $m = \frac{1}{2}$

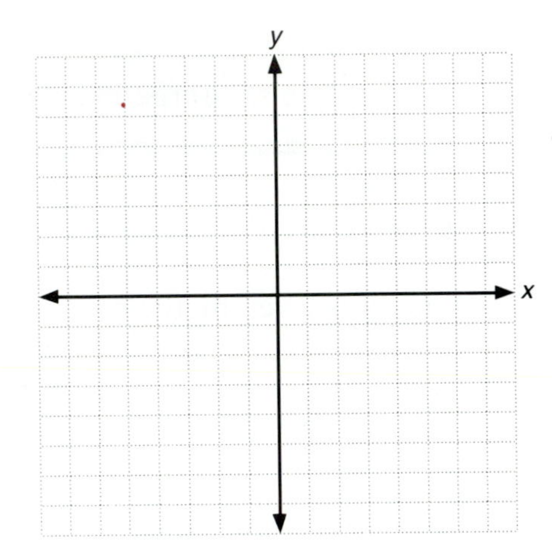

8. $(-5,9)$, $m = -8$

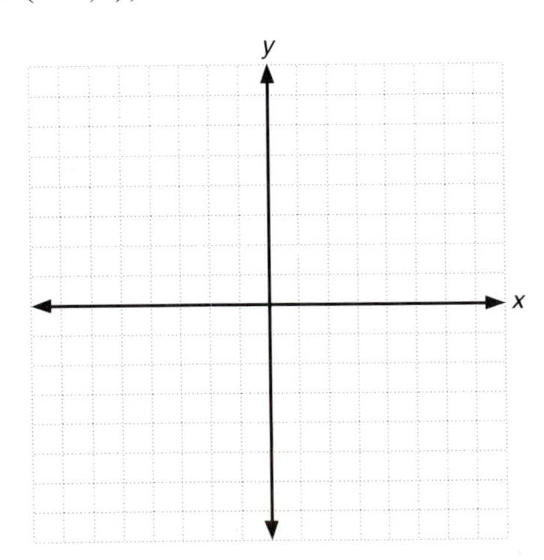

9. $(-2,3)$, parallel to the y-axis.

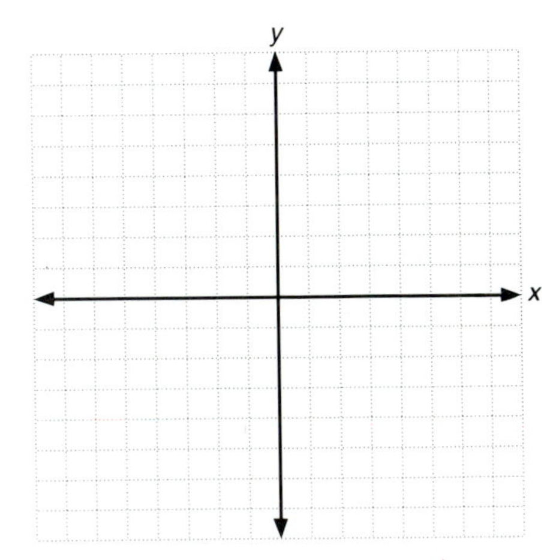

10. $(2,5)$, $m = 0$

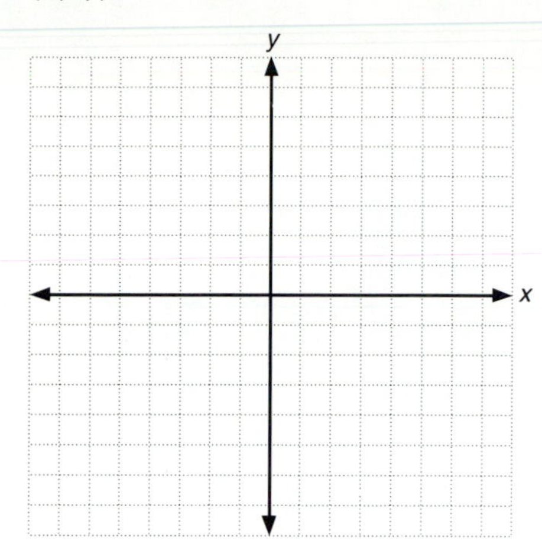

Write each of the equations in slope-intercept form. Specify the slope and y-intercept and use this information to sketch the graph.

11. $x - 4y = 20$

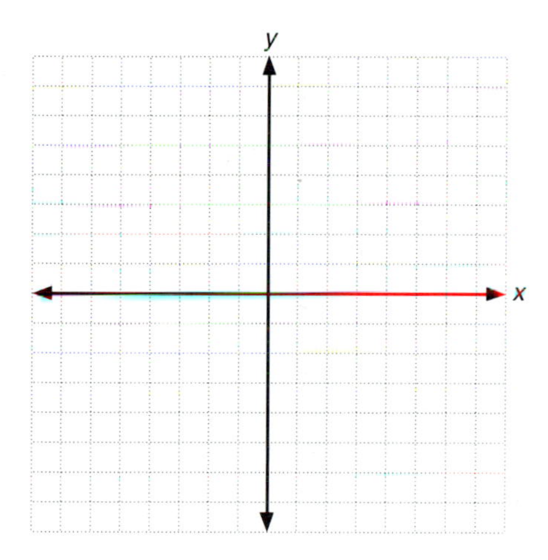

12. $3x + 4y = 8$

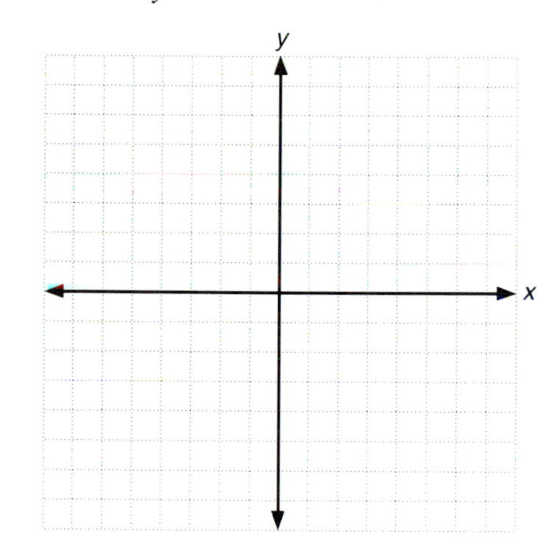

13. $2x - 5y = 15$

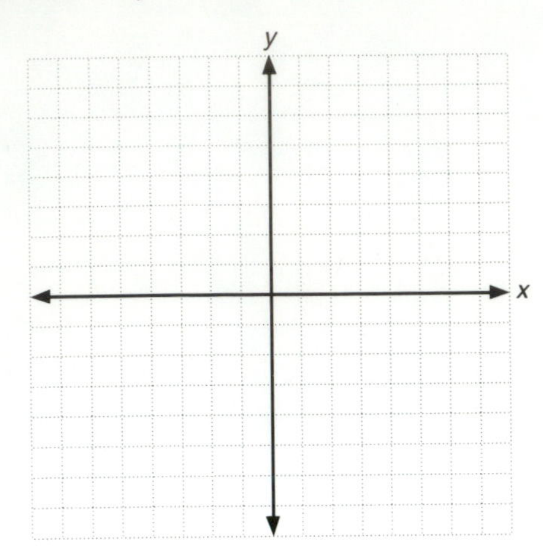

14. $5x - 2y = 0$

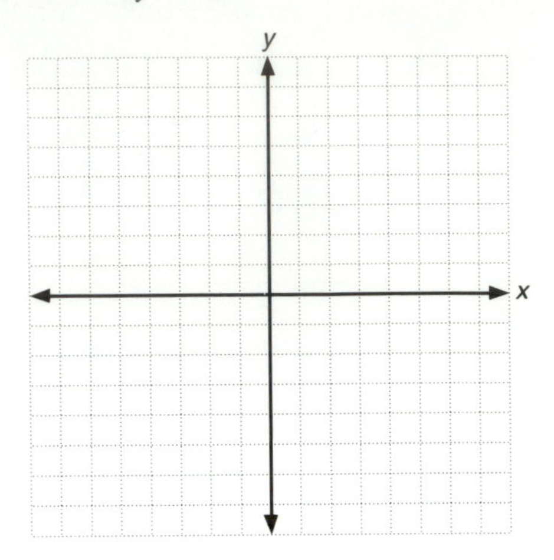

15. $2x + y = 5$

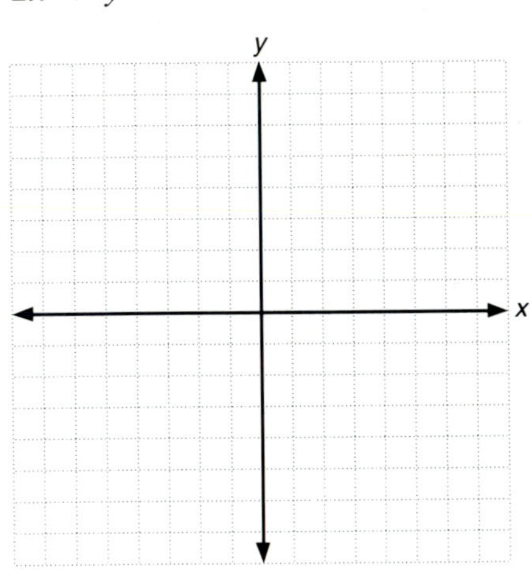

16. A line has an x-intercept of -2 and a y-intercept of 7. Find the equation of the line in standard form.

Write the equation, in standard form, of the line passing through the given point and parallel to the given line.

17. $(3,1)$, $y = 2x + 1$

18. $(5,-1)$, $2y - 5x = 1$

19. $(-8,1)$, $y = -3$

20. $(4,-9)$, $x = 2$

Write the equation, in standard form, of the line passing through the given point and perpendicular to the given line.

21. $(3,2)$, $y = \frac{2}{3}x + 1$

22. $(0,-5)$, $y = \frac{1}{2}x - 3$

23. $(-1,4)$, $y = -\frac{1}{5} + 1$

24. $(2,5)$, $y = -2x + 3$

8–5 GRAPHING LINEAR INEQUALITIES

OBJECTIVES

Upon completion of this section you should be able to graph linear inequalities.

In chapter 2 we constructed line graphs of inequalities such as

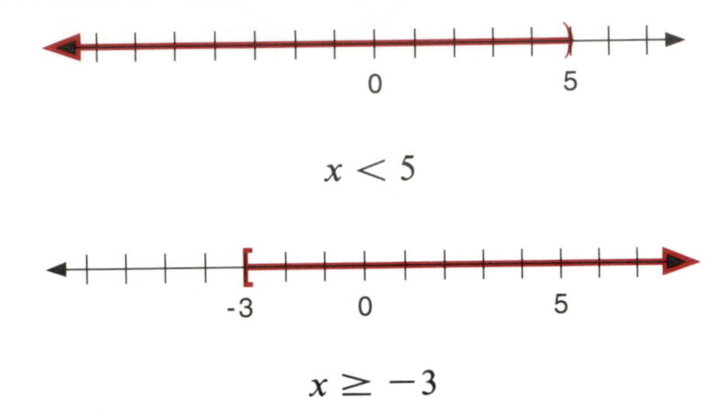

$x < 5$

$x \geq -3$

These were inequalities involving only one variable. We found that in all such cases the graph was some portion of the number line. Since an equation in two variables gives a graph on the plane, it seems reasonable to assume that an inequality in two variables would graph as some portion or region of the plane. This is in fact the case. The solution of the inequality $x + y < 5$ is the set of all ordered pairs of numbers (x,y) such that their sum is less than 5. ($x + y < 5$ is a linear inequality since $x + y = 5$ is a linear equation.)

Example 1 Is each of the following pairs of numbers in the solution set of $x + y < 5$? $(2,1),(3,-4),(5,6),(3,2),(0,0),(-1,4),(-3,8)$

Solution

$(2,1)$ $x + y < 5$
$2 + 1 < 5$
$3 < 5$ Yes

$(3,-4)$ $x + y < 5$
$3 + (-4) < 5$
$-1 < 5$ Yes

$(5,6)$ $x + y < 5$
$5 + 6 < 5$
$11 < 5$ No

$(3,2)$ $x + y < 5$
$3 + 2 < 5$
$5 < 5$ No

$(0,0)$ $x + y < 5$
$0 + 0 < 5$
$0 < 5$ Yes

$$\begin{array}{lrl} (-1,4) & x + y < 5 & \\ & (-1) + 4 < 5 & \\ & 3 < 5 & \text{Yes} \\ (-3,8) & x + y < 5 & \\ & (-3) + 8 < 5 & \\ & 5 < 5 & \text{No} \end{array}$$

Following is a graph of the line $x + y = 5$. The points from example 1 are indicated on the graph with answers to the question "Is $x + y < 5$?"

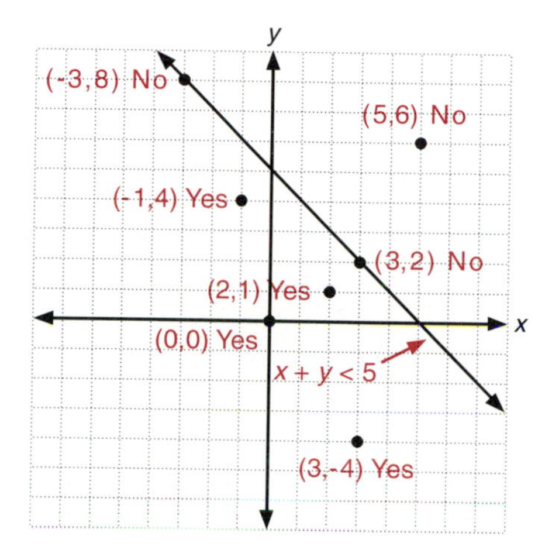

Observe that all "yes" answers lie on the same side of the line $x + y = 5$, and all "no" answers lie on the other side of the line or on the line itself.

The graph of the line $x + y = 5$ divides the plane into three parts: the line itself and the two sides of the line (called **half-planes**).

$x + y < 5$ is a *half-plane*

$x + y \leq 5$ is a *line* and a *half-plane.*

If one point of a half-plane is in the solution set of a linear inequality, then all points in that half-plane are in the solution set. This gives us a convenient method for graphing linear inequalities.

RULE

To graph a linear inequality:

Step 1 Replace the inequality symbol with an equal sign and graph the resulting line.

Step 2 Check *one* point that is obviously in a particular half-plane of that line to see if it is in the solution set of the inequality.

Step 3 If the point chosen *is* in the solution set, then that entire half-plane is the solution set. If the point chosen *is not* in the solution set, then the other half-plane is the solution set.

Example 2 Sketch the graph of $2x + 3y > 7$.

Solution First sketch the graph of the line $2x + 3y = 7$ using either a table of values or the slope-intercept form.

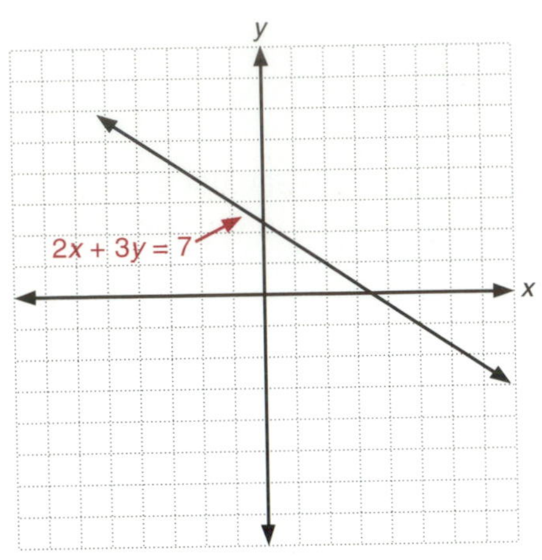

Next choose a point that is *not* on the line $2x + 3y = 7$. (If the line does not go through the origin, then the point (0,0) is always a good choice.) Now turn to the inequality $2x + 3y > 7$ to see if the chosen point is in the solution set.

$$
\begin{aligned}
(0,0) \quad 2x + 3y &> 7 \\
2(0) + 3(0) &> 7 \\
0 &> 7 \quad \text{No}
\end{aligned}
$$

The point (0,0) is *not* in the solution set, therefore the half-plane containing (0,0) is not the solution set. Hence the other half-plane determined by the line $2x + 3y = 7$ is the solution set.

Since the line itself is not a part of the solution, it is shown as a dashed line and the half-plane is shaded to show the solution set.

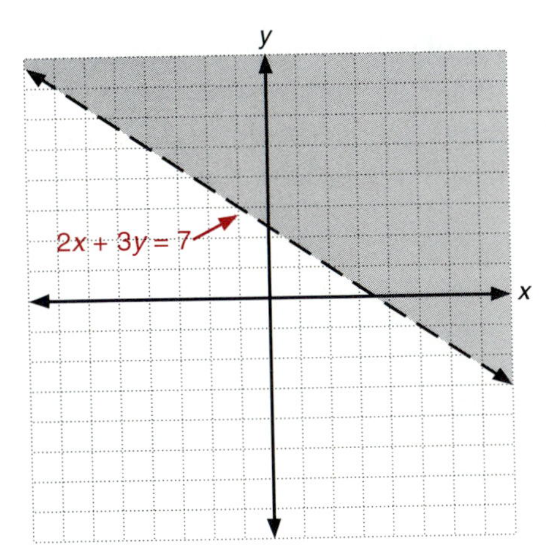

Example 3 Graph the solution for the linear inequality $2x - y \geq 4$.

Solution First graph $2x - y = 4$.

Since the line graph for $2x - y = 4$ does not go through the origin (0,0), check that point in the linear inequality.

$$\begin{aligned} (0,0) \quad 2x - y &\geq 4 \\ 2(0) - (0) &\geq 4 \\ 0 &\geq 4 \quad \text{No} \end{aligned}$$

Since the point (0,0) is not in the solution set, the half-plane containing (0,0) is not in the set. Hence the solution is the other half-plane. Notice, however, that the line $2x - y = 4$ is included in the solution set. Therefore draw a solid line to show that it is part of the graph.

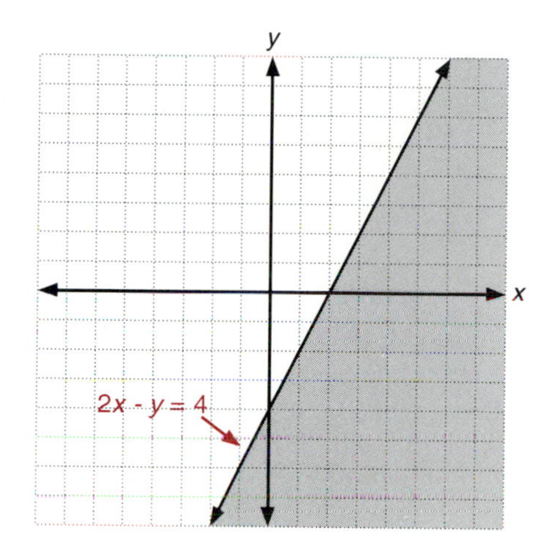

Example 4 Graph $x < y$.

Solution First graph $x = y$.

Next check a point not on the line. Notice that the graph of the line contains the point (0,0), so we cannot use it as a checkpoint. To determine which half-plane is the solution set use any point that is obviously not on the line $x = y$. The point $(-2,3)$ is such a point.

$$\begin{aligned} (-2,3) \quad x &< y \\ -2 &< 3 \quad \text{Yes} \end{aligned}$$

Using this information, graph $x < y$.

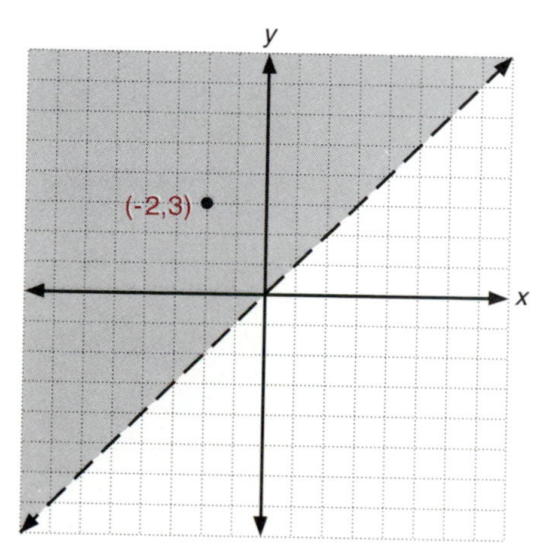

EXERCISE 8–5–1

Graph the linear inequalities:

1. $x + y > 3$

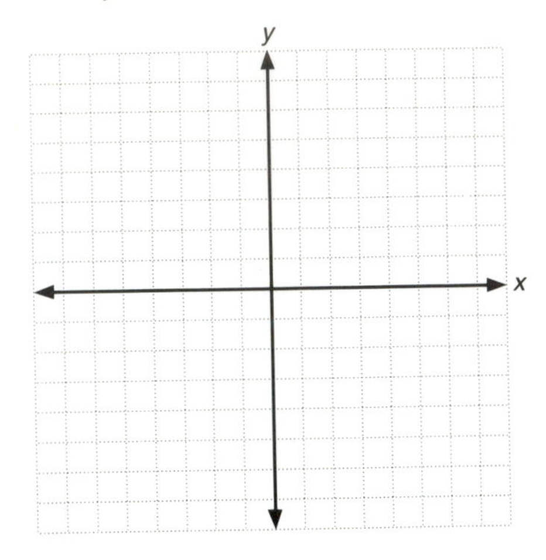

2. $x - y < 5$

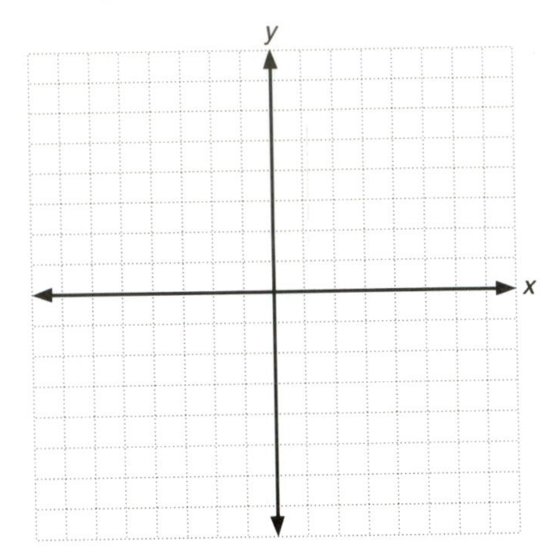

3. $x + 2y < 5$

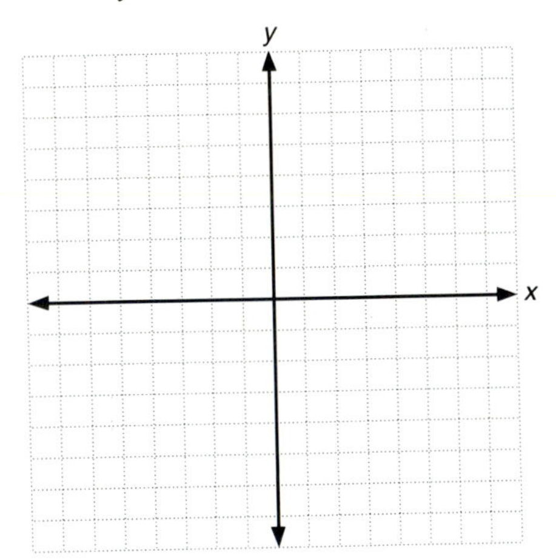

4. $x > y$

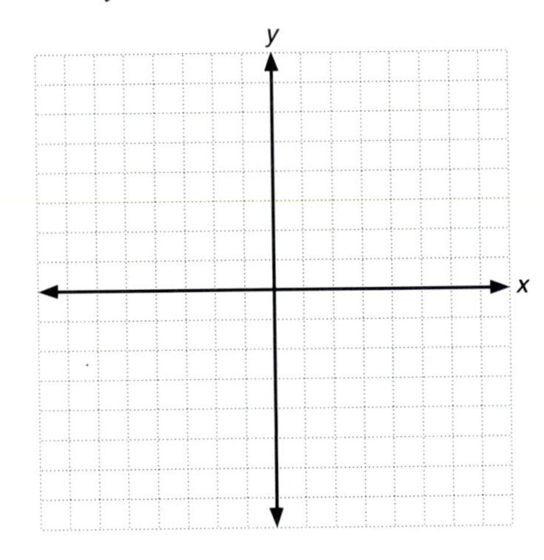

5. $x + 3y \leq 5$

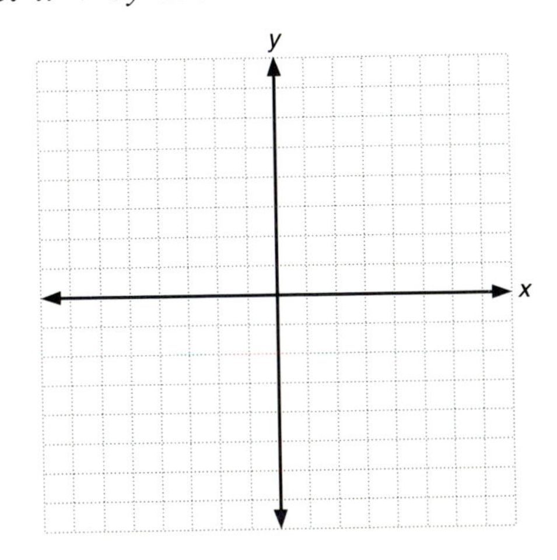

6. $x - 2y \leq 4$

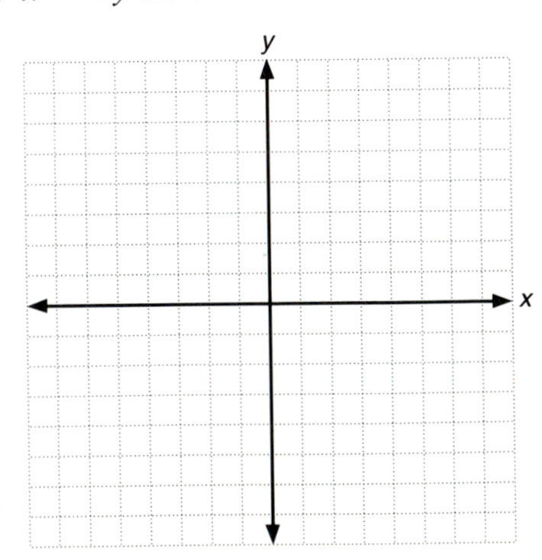

7. $3x - y > 0$

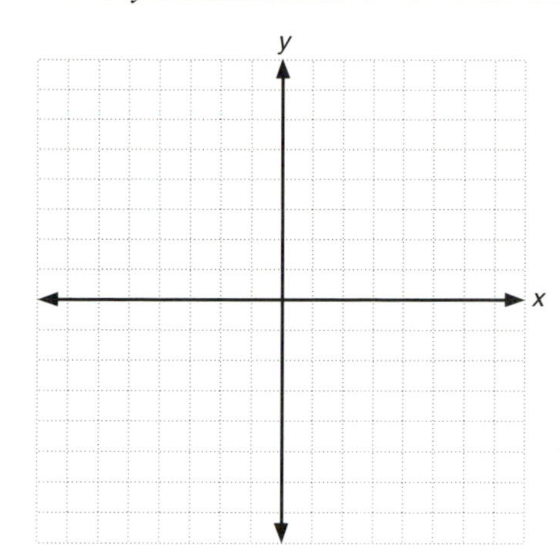

8. $x + 3y > 3$

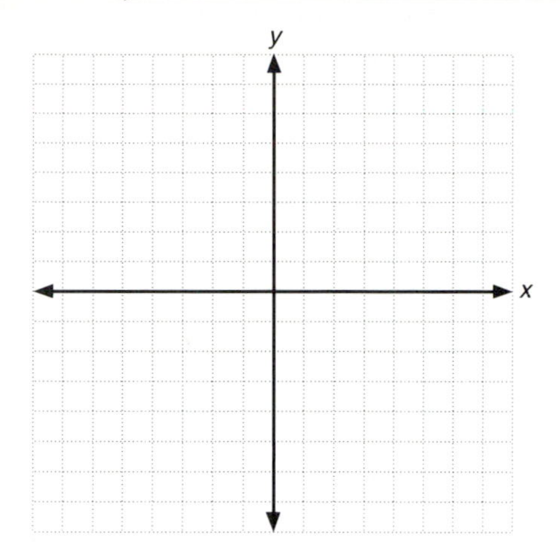

9. $2x + y \leq 3$

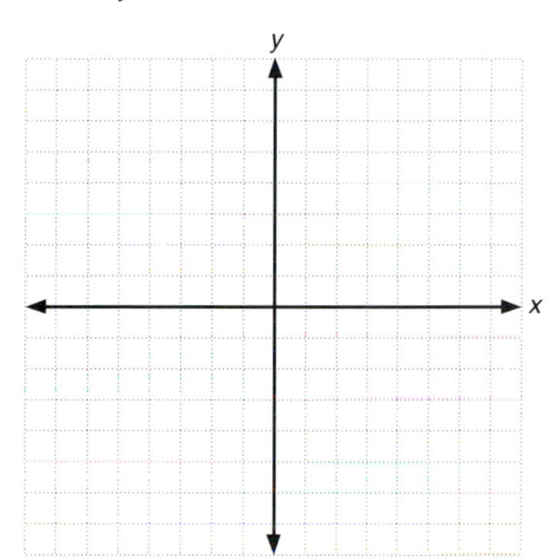

10. $3x - 4y \leq 2$

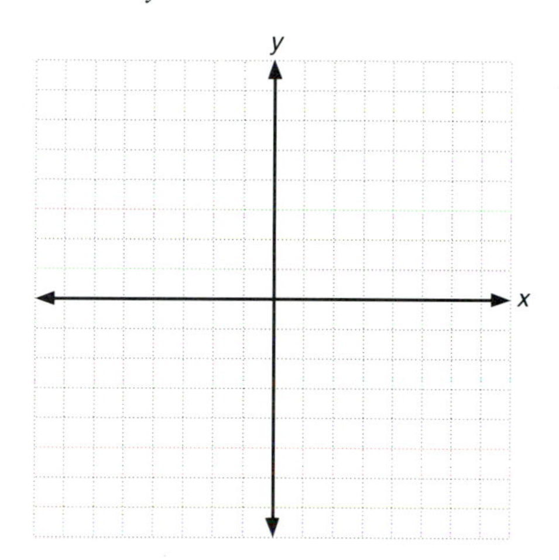

8-6 GRAPHICAL SOLUTION OF A SYSTEM OF LINEAR INEQUALITIES

OBJECTIVES

Upon completion of this section you should be able to:

1. Graph two linear inequalities on the same set of coordinate axes.
2. Determine the region of the plane that is the solution to the system.

You found in section 8–3 that the solution to a system of equations is the intersection of the solutions to each of the equations. In the same manner the solution to a system of linear inequalities is the intersection of the half-planes (and perhaps lines) that are solutions to each individual linear inequality.

In other words, $x + y > 5$ has a solution set and $2x - y < 4$ has a solution set. Therefore the system

$$\begin{cases} x + y > 5 \\ 2x - y < 4 \end{cases}$$

has as its solution set the region of the plane that is in the solution set of both inequalities.

To graph the solution to this system we graph each linear inequality on the same set of coordinate axes and indicate the intersection of the two solution sets.

Example 1 Graph the solution: $\begin{cases} x + y > 5 \\ 2x - y < 4 \end{cases}$

Solution First graph the lines $x + y = 5$ and $2x - y = 4$.

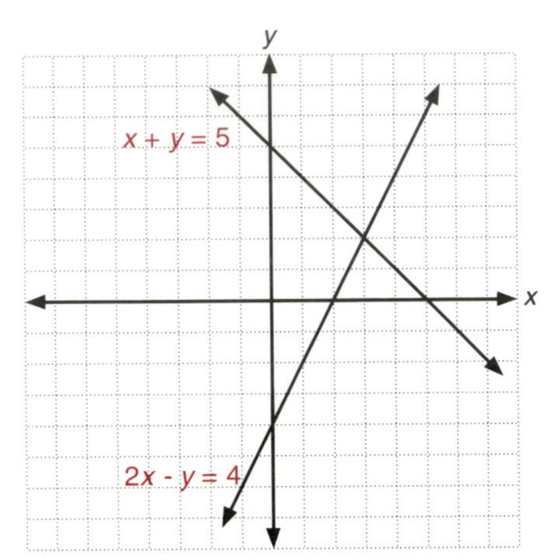

Checking the point (0,0) in the inequality $x + y > 5$ indicates that the point (0,0) is not in its solution set. We indicate the solution set of $x + y > 5$ with black shading.

Checking the point (0,0) in the inequality $2x - y < 4$ indicates that the point (0,0) is in its solution set. We indicate this solution set with color shading.

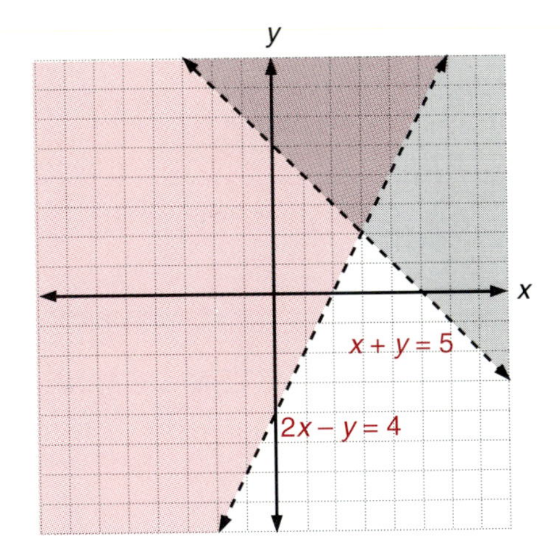

The intersection of the two solution sets is that region of the plane in which the two shadings intersect. This region is shown in the graph.

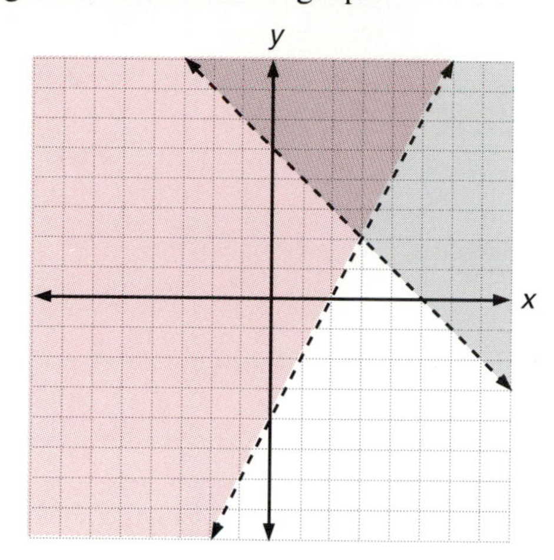

The results indicate that all points in the shaded section of the graph would be in the solution sets of $x + y > 5$ *and* $2x - y < 4$ at the same time. Note that the solution does not include the lines.

If, for example, we were asked to graph the solution of the system

$$\begin{cases} x + y \geq 5 \\ 2x - y < 4 \end{cases}$$

we would obtain

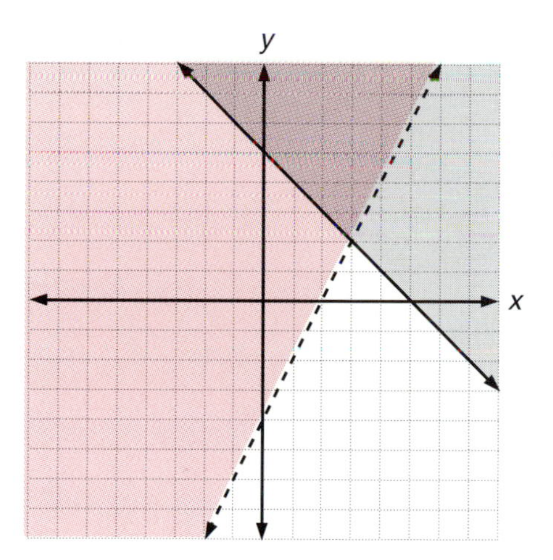

which indicates the solution includes points on the line $x + y = 5$.

EXERCISE 8–6–1

Solve the systems by graphing.

1. $\begin{cases} x + y > 2 \\ 2x - y < 1 \end{cases}$

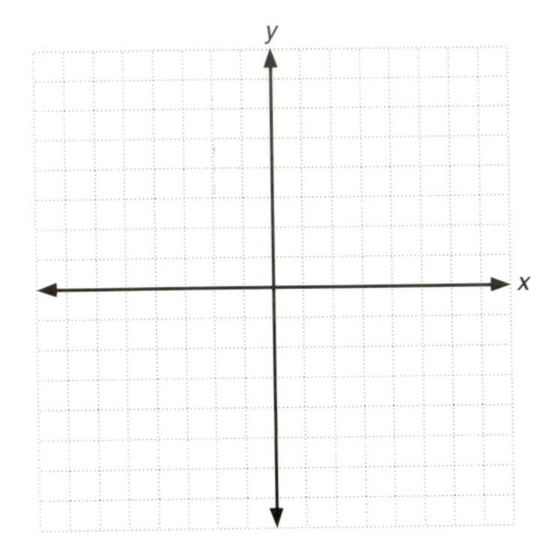

2. $\begin{cases} 3x + y > 6 \\ x - 2y < 4 \end{cases}$

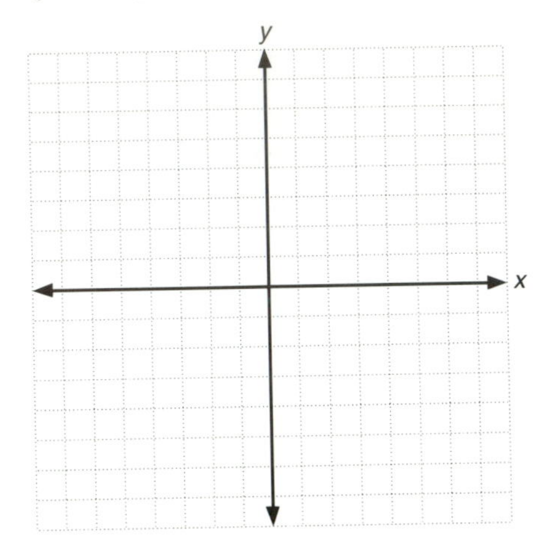

3. $\begin{cases} x + y > 0 \\ x - 3y > 3 \end{cases}$

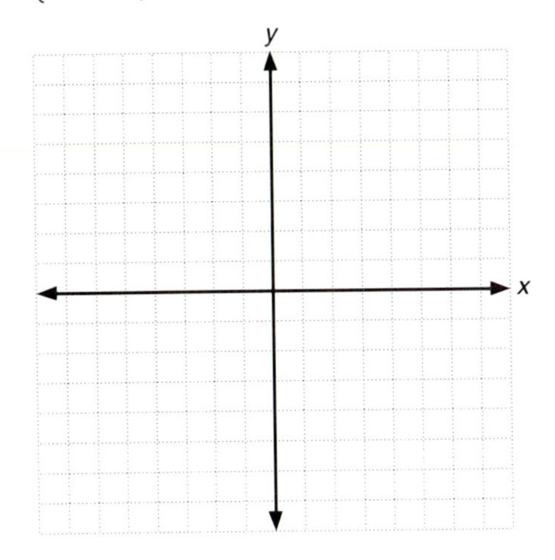

4. $\begin{cases} x + 2y \leq 4 \\ 2x - y \geq 6 \end{cases}$

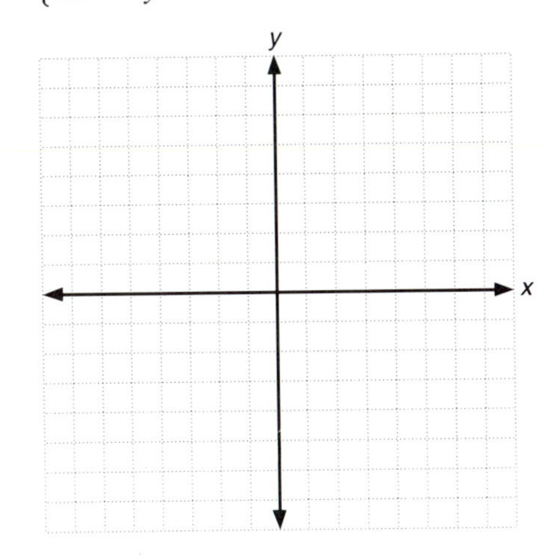

5. $\begin{cases} 4x - y < 4 \\ x + 2y \leq 2 \end{cases}$

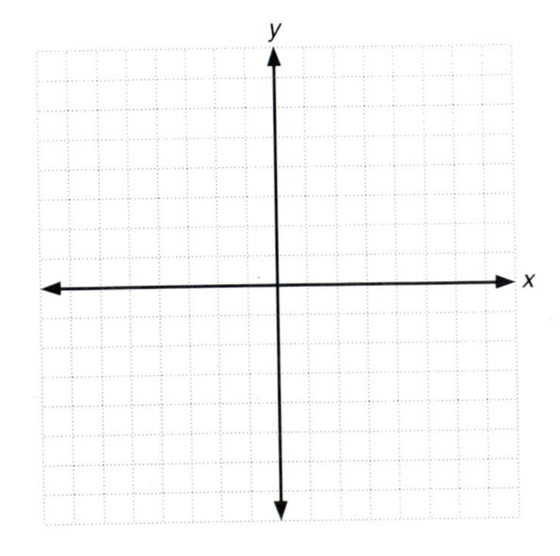

6. $\begin{cases} x - 2y < 3 \\ 2x + y > 8 \end{cases}$

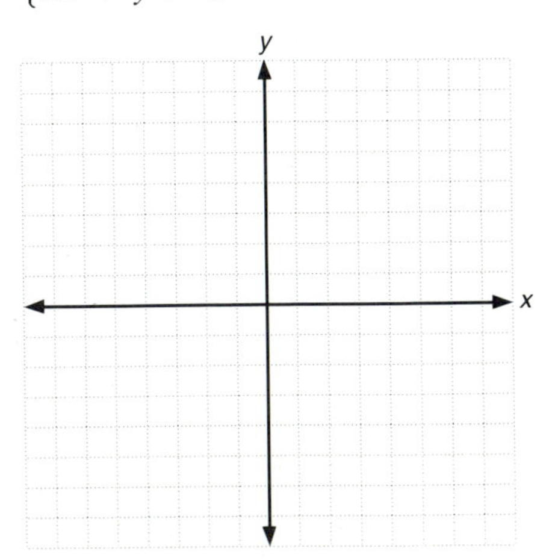

7. $\begin{cases} 4x - 3y \leq -12 \\ x + 4y > 6 \end{cases}$

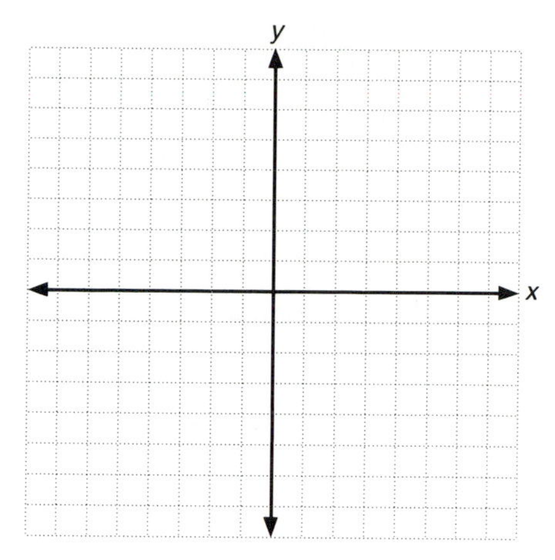

8. $\begin{cases} x - y \leq 0 \\ x + y \geq 0 \end{cases}$

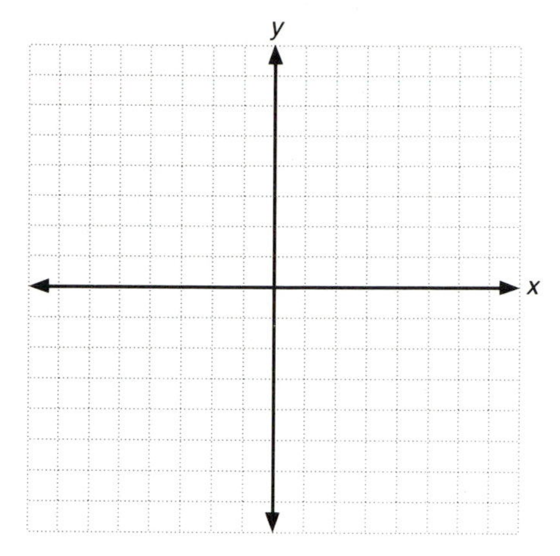

9. $\begin{cases} 3x + y \geq -9 \\ x - 2y \geq 4 \end{cases}$

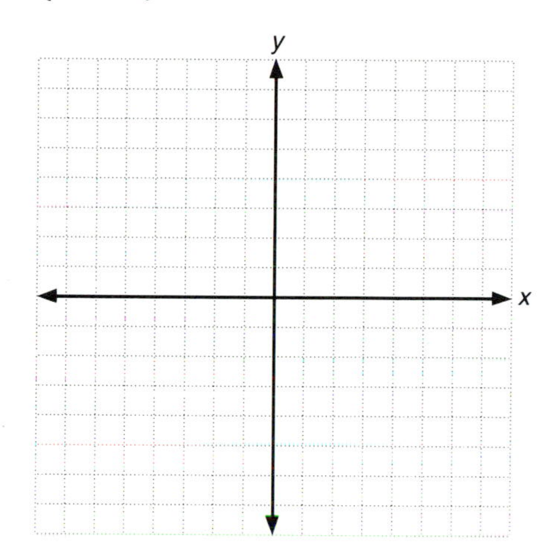

10. $\begin{cases} x - 3y < 5 \\ 2x + 3y \leq 8 \end{cases}$

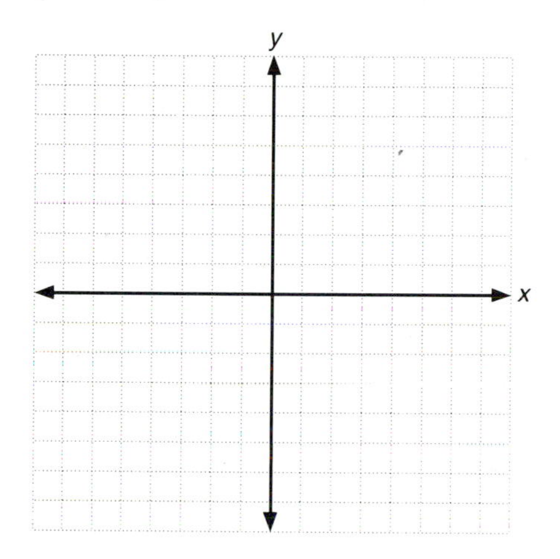

8–7 THE ALGEBRAIC SOLUTION OF A SYSTEM OF TWO LINEAR EQUATIONS

OBJECTIVES

Upon completion of this section you should be able to:

1. Solve a system of two linear equations by substitution.
2. Solve a system of two linear equations by addition.
3. Classify systems as independent, inconsistent, or dependent.
4. Solve word problems by using a system of two equations in two variables.

In this section we will discuss two algebraic methods of solving a system of two first-degree equations in two variables.

The first method to be discussed is the method of **substitution.** This method involves solving for one unknown (variable) in terms of the other in one of the two equations and then substituting this expression into the other equation.

Example 1 Solve by the substitution method: $\begin{cases} 2x + 3y = 1 \\ x - 2y = 4 \end{cases}$

Solution **Step 1** We must solve for one of the unknowns in one of the equations. We can choose either x or y in either the first or second equation. Our choice can be based on obtaining the simplest expression. In this case we will solve for x in the second equation, obtaining

$$x = 4 + 2y.$$

(Notice that any other choice would have resulted in a fraction.)

Step 2 Substitute this expression for x into the other equation. In this case the equation is

$$2x + 3y = 1.$$

Substituting $(4 + 2y)$ for x, we obtain

$$2(4 + 2y) + 3y = 1.$$

(Note that this equation has only one unknown.)

Step 3 Solve for the unknown.

$$\begin{aligned} 2(4 + 2y) + 3y &= 1 \\ 8 + 4y + 3y &= 1 \\ 8 + 7y &= 1 \\ 7y &= -7 \\ y &= -1 \end{aligned}$$

Step 4 Substitute $y = -1$ into either equation to find the corresponding value for x. Since we have already solved the second equation for x in terms of y, we may use it.

$$\begin{aligned} x &= 4 + 2y \\ &= 4 + 2(-1) \\ &= 4 - 2 \\ &= 2 \end{aligned}$$

Thus we have the solution $(2, -1)$.

Step 5 Check the solution in both equations. Remember that the solution for a system must be true for each equation in the system.

$$\begin{array}{r|l} 2x + 3y & = 1 \\ 2(2) + 3(-1) & 1 \\ 4 - 3 & \\ 1 & \end{array} \qquad \begin{array}{r|l} x - 2y & = 4 \\ 2 - 2(-1) & 4 \\ 2 + 2 & \\ 4 & \end{array}$$

We see that the solution $(2, -1)$ does check.

Example 2 Solve by substitution: $\begin{cases} 2x + 3y = 7 \\ 4x + 3y = 8 \end{cases}$

Solution **Step 1** We will obtain a fractional expression in any case, so one choice is as easy as another. We will solve for x in the first equation.

$$x = \frac{7 - 3y}{2}$$

Step 2 Substitute the expression $\left(\frac{7 - 3y}{2}\right)$ for x in the second equation.

$$\begin{aligned} 4x + 3y &= 8 \\ 4\left(\frac{7 - 3y}{2}\right) + 3y &= 8 \end{aligned}$$

Step 3 Solve the equation.

$$\begin{aligned} 4\left(\frac{7 - 3y}{2}\right) + 3y &= 8 \\ 2(7 - 3y) + 3y &= 8 \\ 14 - 6y + 3y &= 8 \\ -3y &= -6 \\ y &= 2 \end{aligned}$$

Step 4 Substitute $y = 2$ in either of the equations. If we use the first equation, we obtain

$$\begin{aligned} 2x + 3y &= 7 \\ 2x + 3(2) &= 7 \\ 2x + 6 &= 7 \\ 2x &= 1 \\ x &= \frac{1}{2}. \end{aligned}$$

Step 5 Checking, we find that the ordered pair $\left(\frac{1}{2}, 2\right)$ satisfies both equations and is thus the solution to the system.

EXERCISE 8-7-1

Solve by the substitution method.

1. $\begin{cases} x + y = 3 \\ 2x + y = 5 \end{cases}$

2. $\begin{cases} x - 2y = 5 \\ 2x + y = 10 \end{cases}$

3. $\begin{cases} 3x - y = 4 \\ x + 5y = -4 \end{cases}$

4. $\begin{cases} x + 5y = 2 \\ 2x + 3y = -3 \end{cases}$

5. $\begin{cases} 2x + y = 5 \\ 6x + 2y = 11 \end{cases}$

6. $\begin{cases} 2x - y = 1 \\ 5x + 2y = -11 \end{cases}$

7. $\begin{cases} x - 2y = 1 \\ x + 4y = 22 \end{cases}$

8. $\begin{cases} 2x - y = 16 \\ 3x + 2y = 3 \end{cases}$

9. $\begin{cases} 5x + y = 5 \\ 3x - 2y = 29 \end{cases}$

10. $\begin{cases} x - 2y = 5 \\ 2x - 3y = 1 \end{cases}$

11. The total number of students in a class is 48. If there are six more men than women, find the number of men and women in the class.

12. The perimeter of a rectangular lot is 450 feet. If the length is twice the width, find the length and width.

The next method we will discuss is the **addition** method. It is based on two facts that we have used previously.

First we know that the solutions to an equation do not change if every term of that equation is multiplied by a nonzero number. Second we know that if we add the same or equal quantities to both sides of an equation the results are still equal.

Example 3 Solve by addition: $\begin{cases} 2x + y = 5 \\ 3x + 2y = 6 \end{cases}$

Solution **Step 1** Our purpose is to add the two equations and eliminate one of the unknowns so that we can solve the resulting equation in one unknown. If we add the equations as they are, we will not eliminate an unknown. This means we must first multiply each side of one or both of the equations by a number or numbers that will lead to the elimination of one of the unknowns when the equations are added.

After carefully looking at the problem, we note that the easiest unknown to eliminate is y. This is done by multiplying each side of the first equation by -2.

$$\begin{cases} 2x + y = 5 \leftarrow \text{multiply this equation by } (-2) \\ 3x + 2y = 6 \end{cases}$$

We obtain the equivalent system

$$\begin{cases} -4x - 2y = -10 \\ 3x + 2y = 6. \end{cases}$$

Step 2 Add the equations.

$$\begin{array}{r} -4x - 2y = -10 \\ 3x + 2y = 6 \\ \hline -x = -4 \end{array}$$

Step 3 Solve the resulting equation.

$$\begin{aligned} -x &= -4 \\ x &= 4 \end{aligned}$$

Step 4 Find the value of the other unknown by substituting this value into one of the original equations. We will choose the first equation.

$$\begin{aligned} 2x + y &= 5 \\ 2(4) + y &= 5 \end{aligned}$$

value of x

$$\begin{aligned} 8 + y &= 5 \\ y &= -3 \end{aligned}$$

Step 5 If we check the ordered pair $(4,-3)$ in both equations, we see that it is a solution of the system.

$$\begin{array}{r|l} 2x + y = & 5 \\ 2(4) + (-3) & 5 \\ 8 - 3 & \\ 5 & \end{array} \qquad \begin{array}{r|l} 3x + 2y = & 6 \\ 3(4) + 2(-3) & 6 \\ 12 - 6 & \\ 6 & \end{array}$$

Example 4 Solve by addition: $\begin{cases} 2x + 3y = 7 \\ 3x + 2y = 3 \end{cases}$

Solution **Step 1** We observe that both equations will have to be changed to eliminate one of the unknowns. Neither equation will be easier than the other, so we may choose to eliminate either x or y.

If we choose to eliminate x, we can multiply each side of the first equation by 3 and the second equation by -2.

$$\begin{cases} 2x + 3y = 7 \leftarrow \text{multiply this equation by } 3 \\ 3x + 2y = 3 \leftarrow \text{multiply this equation by } -2 \end{cases}$$

We obtain the equivalent system

$$\begin{cases} 6x + 9y = 21 \\ -6x - 4y = -6. \end{cases}$$

Step 2 Adding the equations, we obtain

$$\begin{array}{r} 6x + 9y = 21 \\ -6x - 4y = -6 \\ \hline 5y = 15. \end{array}$$

Step 3 Solving for y yields

$$y = 3.$$

Step 4 Using the first equation in our original system to find the value of the other unknown gives

$$\begin{aligned} 2x + 3y &= 7 \\ 2x + 3(3) &= 7 \\ 2x + 9 &= 7 \\ 2x &= -2 \\ x &= -1. \end{aligned}$$

Step 5 Check to see that the ordered pair $(-1,3)$ is a solution of the system.

Example 5 It takes a boat 3 hours to travel 120 kilometers downstream and 5 hours to make the return trip upstream. Find the average speeds of the boat and the stream.

Solution We must use a formula from physics to solve this problem. The formula states that

$$\text{rate} \times \text{time} = \text{distance}.$$

If we first consider the trip downstream, we recognize that the rate of travel will be the sum of the speed of the boat and the speed of the stream. If we let

$$b = \text{speed of the boat}$$

and

$$s = \text{speed of the stream},$$

then

$$(b + s)(3) = 120.$$

The rate of travel for the trip upstream will be the speed of the boat decreased by the speed of the stream. Thus

$$(b - s)(5) = 120.$$

We now have the system

$$\begin{cases} 3b + 3s = 120 \\ 5b - 5s = 120. \end{cases}$$

If we divide the top equation by 3 and the bottom equation by 5 we obtain the system

$$\begin{cases} b + s = 40 \\ b - s = 24. \end{cases}$$

Adding the equations yields

$$\begin{aligned} 2b &= 64 \\ b &= 32. \end{aligned}$$

If we now substitute this value of b in the equation $3b + 3s = 120$, we obtain

$$\begin{aligned} 3(32) + 3s &= 120 \\ 96 + 3s &= 120 \\ 3s &= 24 \\ s &= 8. \end{aligned}$$

Checking in both original equations, we have

$$\begin{array}{r|l} 3b + 3s = & 120 \\ 3(32) + 3(8) & 120 \\ 96 + 24 & \\ 120 & \end{array} \qquad \begin{array}{r|l} 5b - 5s = & 120 \\ 5(32) - 5(8) & 120 \\ 160 - 40 & \\ 120 & \end{array}$$

Thus the speed of the boat is 32 km/hr and the speed of the stream is 8 km/hr.

EXERCISE 8-7-2

Solve by the addition method.

1. $\begin{cases} x + y = 2 \\ 4x - y = 13 \end{cases}$

2. $\begin{cases} 2x - y = 0 \\ x + y = 6 \end{cases}$

3. $\begin{cases} 3x - y = -7 \\ x + 2y = 7 \end{cases}$

4. $\begin{cases} 5x - 2y = 54 \\ 2x + 3y = 14 \end{cases}$

5. $\begin{cases} 8x - 3y = 13 \\ 5x - 4y = 6 \end{cases}$

6. $\begin{cases} 2x + y = -3 \\ 7x + 2y = -18 \end{cases}$

7. $\begin{cases} 2x - 5y = 44 \\ 10x + 3y = -4 \end{cases}$

8. $\begin{cases} 2x + 5y = 21 \\ 4x - 3y = -10 \end{cases}$

9. $\begin{cases} 4x + 7y = 27 \\ 3x - 2y = -16 \end{cases}$

10. $\begin{cases} 8x + 3y = -5 \\ 9x + 7y = -2 \end{cases}$

11. The sum of the lengths of the Golden Gate Bridge and the Brooklyn Bridge is 5,795 feet. The Golden Gate Bridge is 1,010 feet longer than twice the length of the Brooklyn Bridge. Find the length of each bridge.

12. The length of a rectangular lot is eight feet less than three times the width. Find the length and width if the perimeter of the lot is 600 feet.

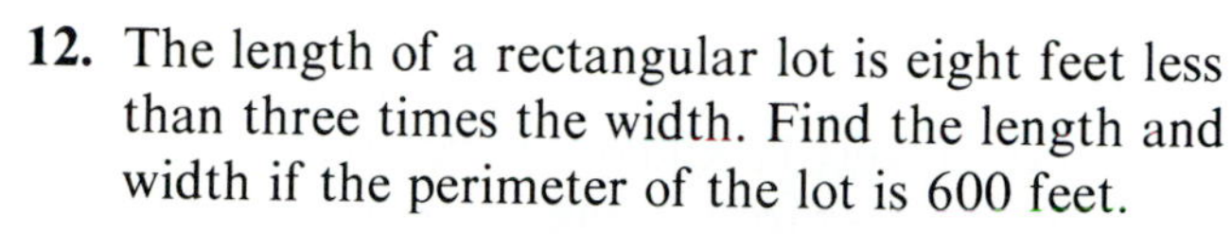

13. The sum of two numbers is 147 and their difference is 19. Find the numbers.

14. An airliner took 6 hours to travel 3,864 kilometers from New York to Las Vegas. The return trip took 4 hours. Find the speeds of the plane and the wind.

15. A professor has 85 students enrolled in two classes. 49 of these students are freshmen. If two-thirds of the first class and one-half of the second class are freshmen, how many students are in each class?

16. A total of 5,000 tickets were sold for a football game. The price was $5.00 per adult and $2.50 per child. If the total receipts were $21,915.00, find the number of each type of ticket sold.

17. A boat travels 12 kilometers upstream in two hours. The return trip takes one hour. Find the rate of the boat in still water and the rate of the current.

18. A plane travels 1,600 kilometers with the wind in two hours. The return trip against the wind takes one-half hour longer. Find the speed of the plane (airspeed) and the speed of the wind.

19. Mike has two more dimes than nickels. The total value of the coins is $2.60. How many of each type of coin does he have?

20. A 40% copper alloy is to be used with a 70% copper alloy to produce 180 kilograms of a 60% alloy. How much of each alloy must be used?

A system of two linear equations does not always have a unique solution and we should be aware of the other possibilities. (See problem 9 in exercise 8–3–1.)

Since we are dealing with linear equations, we may think of the lines represented by the equations. Two linear equations will represent one of the following possible situations.

1. **Independent equations** The two lines intersect in a single point. In this case we have a unique solution.

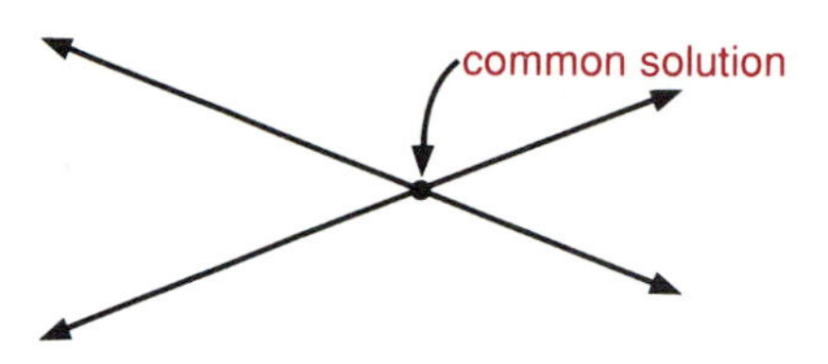

2. **Inconsistent equations** The two lines are parallel. In this case we have no solution.

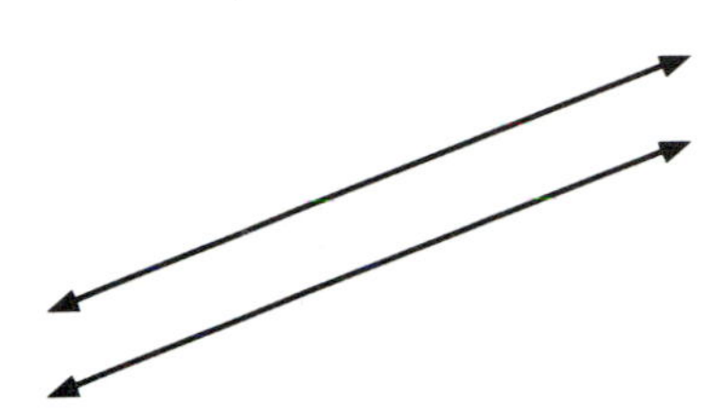

3. **Dependent equations** The two equations give the same line. In this case any solution of one equation is a solution of the other.

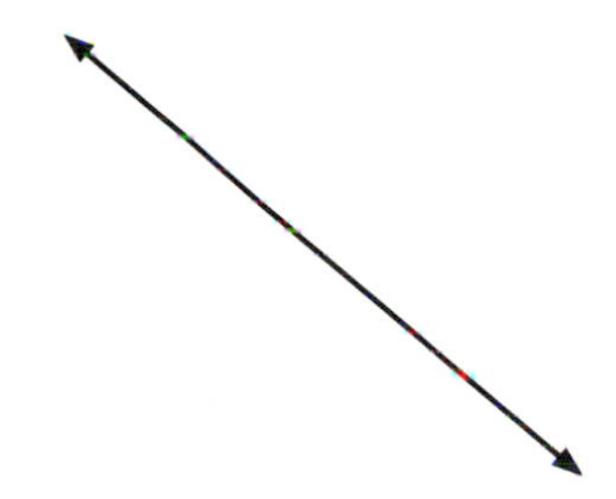

We can recognize any of these three situations by proceeding to work the problem either by the substitution or addition method.

If the equations are **independent,** we will obtain a **unique solution.**
If the equations are **inconsistent,** we will obtain a **contradiction.**
If the equations are **dependent,** we will obtain an **identity.**

Example 6 Solve: $\begin{cases} x + y = 6 \\ x + y = 8 \end{cases}$

Solution If we multiply both sides of the second equation by -1 and add, we obtain

$$0 = -2.$$

Since this is a *contradiction,* the equations are *inconsistent* and their graphs would be parallel lines. This system has no solution.

Example 7 Solve: $\begin{cases} x + y = 6 \\ 2x + 2y = 12 \end{cases}$

Solution Multiplying both sides of the first equation by -2 and adding yields

$$0 = 0.$$

Since this is an *identity,* the equations are *dependent* and represent the same line on the coordinate plane. Therefore, any ordered pair that satisfies one equation will also satisfy the other. There are infinitely many solutions.

EXERCISE 8-7-3

Classify each system as independent, inconsistent, or dependent. If the system is independent, find its solution.

1. $\begin{cases} 2x + y = 1 \\ 3x - y = 9 \end{cases}$

2. $\begin{cases} x + y = 4 \\ x - y = 6 \end{cases}$

3. $\begin{cases} x + 2y = 9 \\ 3x - y = -1 \end{cases}$

4. $\begin{cases} 2x - y = 1 \\ 6x - 3y = 3 \end{cases}$

5. $\begin{cases} 3x - y = 2 \\ 2x + 3y = -6 \end{cases}$

6. $\begin{cases} 3x - 2y = 3 \\ 6x - 4y = 1 \end{cases}$

7. $\begin{cases} 6x - 4y = 2 \\ 3x - 2y = 1 \end{cases}$

8. $\begin{cases} x - 2y = 15 \\ 3x + y = 3 \end{cases}$

9. $\begin{cases} 3x + 2y = 5 \\ 2x + y = 1 \end{cases}$

10. $\begin{cases} 2x + y = 5 \\ 10x + 5y = 10 \end{cases}$

11. $\begin{cases} 2x + 3y = 10 \\ 6x - 2y = -3 \end{cases}$

12. $\begin{cases} x + 3y = 1 \\ 2x - 9y = -8 \end{cases}$

13. $\begin{cases} 2x - 4y = 6 \\ 3x - 6y = 9 \end{cases}$

14. $\begin{cases} 6x + 3y = 3 \\ 10x + 5y = 15 \end{cases}$

15. $\begin{cases} 5x + 2y = -1 \\ 4x + 3y = -12 \end{cases}$

8–8 THE ALGEBRAIC SOLUTION OF THREE EQUATIONS WITH THREE VARIABLES

OBJECTIVES

Upon completion of this section you should be able to:

1. Solve a system of three equations in three variables.
2. Solve word problems using a system of three equations in three variables.

A first-degree polynomial equation in three variables is the equation of a plane in three-dimensional space. Three planes in space might intersect in a single point. Of course, there are other possibilities (such as all three being parallel, and so forth) and you may wish to list them. In this section we are interested only in those that intersect in a single point (that is, those that have a unique solution), and in an algebraic method of finding the solution.

A point in three-dimensional space is represented by an **ordered triple** of numbers. We use x, y, and z as the three unknowns and the ordered triple is (x,y,z). The ordered triple $(2,3,-1)$ represents a point such that $x = 2$, $y = 3$, and $z = -1$.

The method we will use reduces a system of three equations in three unknowns to two equations in two unknowns by addition. This method is best illustrated by example.

Example 1 Solve the system:

$$\begin{cases} 2x + 3y - z = 11 & (1) \\ x + 2y + z = 12 & (2) \\ 3x - y + 2z = 5 & (3) \end{cases}$$

Solution **Step 1** Choose any two equations and eliminate any one of the three unknowns by addition.

Note that this gives a wide range of choices. We can choose on the basis of the unknown that is easiest to eliminate. In this case we will choose to eliminate z by adding equations (1) and (2).

Equation (1) added to equation (2) yields

$$3x + 5y = 23.$$

Step 2 Choose another pair of equations and eliminate the *same* unknown by addition.

The choices here are not as broad because we must eliminate the same unknown as in step 1 but cannot use the same two equations as before. In this example we must eliminate z and can use equations (1) and (3) or equations (2) and (3). We will choose equations (1) and (3) and eliminate z by multiplying both sides of equation (1) by 2 and adding the result to equation (3).

Twice equation (1) added to equation (3) yields

$$7x + 5y = 27.$$

Step 3 Solve the system of two equations with two unknowns that results from steps 1 and 2.

$$\begin{cases} 3x + 5y = 23 \\ 7x + 5y = 27 \end{cases}$$

Using either substitution or addition to solve this system, we find

$$x = 1$$
$$y = 4.$$

Step 4 Use the values from step 3 in any one of the original equations to find the other unknown.

Using equation (2) and $x = 1$, $y = 4$, we have

$$\begin{aligned} x + 2y + z &= 12 \\ 1 + 2(4) + z &= 12 \\ z &= 3. \end{aligned}$$

Step 5 Check the solution in all three of the original equations to see that each equation is satisfied.

In our example the solution (1,4,3) checks in all three equations.

Example 2 Solve the system:
$$\begin{cases} x - y + 2z = 6 & (1) \\ 2x + 3y - z = -3 & (2) \\ 3x + 2y + 2z = 5 & (3) \end{cases}$$

Solution **Step 1** Eliminate y by multiplying both sides of equation (1) by 3 and adding the result to equation (2).

Three times equation (1) added to equation (2) gives

$$5x + 5z = 5.$$

Step 2 Eliminate y by multiplying both sides of equation (1) by 2 and adding the result to equation (3). This yields

$$5x + 6z = 17.$$

Step 3 Solve the system:
$$\begin{cases} 5x + 5z = 15 \\ 5x + 6z = 17 \end{cases}$$

We find the solution to be

$$\begin{aligned} x &= 1 \\ z &= 2. \end{aligned}$$

Step 4 Substitute $x = 1$ and $z = 2$ in equation (1) to solve for y. We find the solution to be $y = -1$.

Step 5 The ordered triple $(1,-1,2)$ checks in all three equations.

If one of the unknowns is missing from one or more of the equations, finding the solution becomes easier.

Example 3 Solve the system:
$$\begin{cases} x + z = 3 & (1) \\ 2x + y + z = 3 & (2) \\ 3x - y + 2z = 8 & (3) \end{cases}$$

Solution Note that equation (1) contains only two of the unknowns. We can eliminate y using equations (2) and (3) and then we will have two equations in two unknowns (x and z).

Equation (2) added to equation (3) gives

$$5x + 3z = 11.$$

Now we solve the system

$$\begin{cases} x + z = 3 \\ 5x + 3z = 11. \end{cases}$$

We obtain as our solution

$$x = 1$$
$$z = 2.$$

Substituting these values into equation (2) yields $y = -1$.

The solution $(1, -1, 2)$ checks in all three equations.

Example 4 A girl has \$5.10 in nickels, dimes, and quarters. She has a total of 37 coins. The number of nickels and quarters combined is three more than the number of dimes. Find the number of each kind of coin she has.

Solution We set up the following key:

n = number of nickels
d = number of dimes
q = number of quarters

Since the value of all the coins is \$5.10, we may write one equation as

$$.05n + .10d + .25q = 5.10$$
$$\text{or} \quad 5n + 10d + 25q = 510.$$

This reduces to

$$n + 2d + 5q = 102.$$

The next equation derives from the fact that there are a total of 37 coins.

$$n + d + q = 37$$

The third equation is obtained from the statement that the number of nickels and quarters combined is three more than the number of dimes.

$$n + q = d + 3$$
$$\text{or} \quad n - d + q = 3$$

We now have the system

$$\begin{cases} n + 2d + 5q = 102 \\ n + d + q = 37 \\ n - d + q = 3. \end{cases}$$

Solving, we obtain

$$n = 8$$
$$d = 17$$
$$q = 12.$$

EXERCISE 8–8–1

Solve.

1. $\begin{cases} x + y + z = 6 \\ 2x - y + z = 3 \\ x - y + 2z = 5 \end{cases}$

2. $\begin{cases} x + 2y + z = 0 \\ x - 3y - z = -2 \\ x + y - z = -2 \end{cases}$

3. $\begin{cases} 2x + y + z = 0 \\ 3x - 2y - z = -11 \\ x - y + 2z = 3 \end{cases}$

4. $\begin{cases} x - y + z = 8 \\ 5x + 4y - z = 7 \\ 2x + y - 3z = -7 \end{cases}$

5. $\begin{cases} x + 5y - 2z = 13 \\ 6x + y + 3z = 4 \\ x - y + 2z = -5 \end{cases}$

6. $\begin{cases} x + y = 6 \\ 2x - y + z = 7 \\ x + y - 3z = 12 \end{cases}$

7. $\begin{cases} 3x + 4y + z = -2 \\ y + z = 1 \\ 2x - y - z = -5 \end{cases}$

8. $\begin{cases} y - z = -3 \\ x + y = 1 \\ 2x + 3y + z = 1 \end{cases}$

9. A man has \$4.50 in nickels, dimes, and quarters. He has a total of 28 coins. The number of dimes is twice the number of nickels. How many of each type of coin does he have?

10. The sum of three numbers is 58. Twice the first number added to the sum of the second and third numbers is 71. If the first number is added to four times the second number and the sum is decreased by three times the third number, the result is 18. Find the numbers.

11. A chemist wishes to make 9 liters of a 30% acid solution by mixing three solutions of 5%, 20%, and 50%. How much of each solution must the chemist use if twice as much 50% solution is used as 5% solution?

12. Karen buys 11 rolls of three different kinds of wallpaper, some at \$8.00 a roll, some at \$7.00 a roll, and some at \$5.00 a roll. She has twice as many rolls of \$5.00 paper as she does of \$7.00 paper. If the total bill for the wallpaper is \$67.00, how many rolls of \$5.00 paper did she buy?

The number in brackets refers to the section of the chapter that discusses that concept.

Terminology

- A **relation** is a set of ordered pairs. [8–1]
- The **domain** is the set of all first numbers in the ordered pairs of a relation. [8–1]
- The **range** is the set of all second numbers in the ordered pairs of a relation. [8–1]
- A **function** is a relation such that each element of the domain is related to exactly one element in the range. [8–1]
- The **Cartesian coordinate system** is a method of uniquely naming each point in the plane. [8–2]
- A **linear equation** is a first-degree polynomial equation in two variables. [8–2]
- A **system of equations** is a set of two or more equations. [8–3]
- The **solution** of a system of equations is the set of all ordered pairs that are solutions to each equation of the system. [8–3]
- The **standard form** of the equation of a line is $ax + by + c = 0$, where a, b, and c are integers and $a \geq 0$. [8–4]
- The **slope** (m) of a line is given by $m = \dfrac{y_2 - y_1}{x_2 - x_1}$, $x_1 \neq x_2$. [8–4]
- The **two-point form** of the equation of a line is

$$y - y_1 = \left(\frac{y_2 - y_1}{x_2 - x_1}\right)(x - x_1),\ x_1 \neq x_2. \text{ [8–4]}$$

- The **point-slope form** of the equation of a line is $y - y_1 = m(x - x_1)$. [8–4]
- The ***y*-intercept** of a straight line is the value of y when $x = 0$. [8–4]
- The **slope-intercept form** of the equation of a line is $y = mx + b$, where m is the slope of the line and b is the y-intercept. [8–4]
- **Parallel lines** have the same slope. [8–4]
- A **half-plane** is that portion of the plane that lies on one side of a line. [8–5]
- A **linear inequality** is a first-degree polynomial inequality in two variables. [8–5]
- A **system of inequalities** is a set of two or more inequalities. [8–6]
- **Independent equations** have unique solutions. [8–7]
- **Inconsistent equations** have no solution. [8–7]
- **Dependent equations** have infinitely many solutions. [8–7]

Rules and Procedures

Graphing

- To graph an equation in two variables on the Cartesian coordinate system find ordered pairs that are solutions to the equation and represent these as points on the coordinate system. Connect the points thus represented as a sketch of the graph of the equation. [8–2]
- To sketch the graph of a straight line using its slope: [8–4]
  - *STEP 1* Write the equation of the line in the form $y = mx + b$.
  - *STEP 2* Locate the y-intercept $(0,b)$.
  - *STEP 3* Starting at $(0,b)$, use the slope m to locate a second point.
  - *STEP 4* Connect the two points with a straight line.
- To graph a linear inequality: [8–5]
  - *STEP 1* Replace the inequality symbol with an equal sign and graph the resulting line.
  - *STEP 2* Check *one* point that is obviously in a particular half-plane of that line to see if it is in the solution set of the inequality.
  - *STEP 3* If the point chosen *is* in the solution set, then that entire half-plane is the solution set. If the point chosen *is not* in the solution set, then the other half-plane is the solution set.

Solving Systems of Equations and Inequalities

- To solve a system of equations in two variables by graphing graph the equations carefully on the same coordinate system. Their points of intersection will be the solutions of the system. [8–3]
- To solve a system of linear inequalities by graphing determine the region of the plane that satisfies both inequality statements. [8–6]

- A system of two linear equations is solved algebraically by using addition or substitution to eliminate one of the variables. The value of the remaining variable is then substituted into one of the equations to obtain the value of the other variable. The ordered pair thus obtained must be a solution of each equation to be a solution of the system. [8–7]
- The algebraic solution of three equations with three variables involves choosing any two of the three equations and then eliminating one of the variables by addition. Next choose any other two of the three equations and eliminate the same variable. The two equations with two variables thus obtained are solved as a system of two linear equations and the solution is substituted into any one of the three original equations to obtain the value of the third variable. The numbers obtained must be a solution of each of the three equations to be a solution of the system. [8–8]

CHAPTER 8 REVIEW

For each of the following: (a) is y a function of x and (b) if so, what is the domain?

1. $y = x^2 + 1$

2. $y = \dfrac{1}{x^2 - 1}$

3. $x = y^2 - 2$

4. $y = \dfrac{1}{x + 7}$

5. $y = \dfrac{1}{3 - \sqrt{x}}$

Sketch the graphs.

6. $y = x - 3$

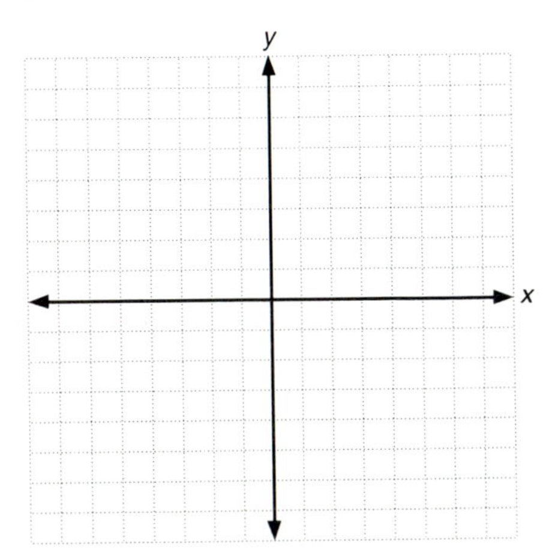

7. $y = 2x - 1$

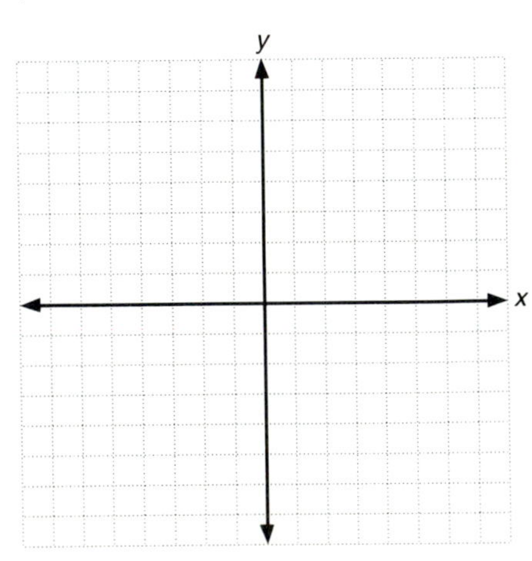

8. $y = 2 - 3x$

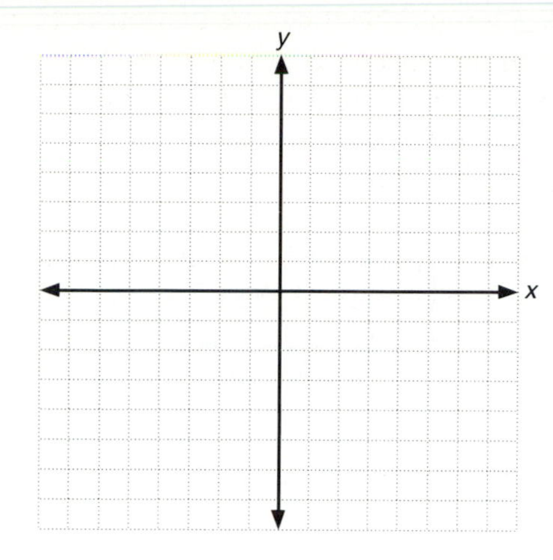

9. $y = x^2 - 1$

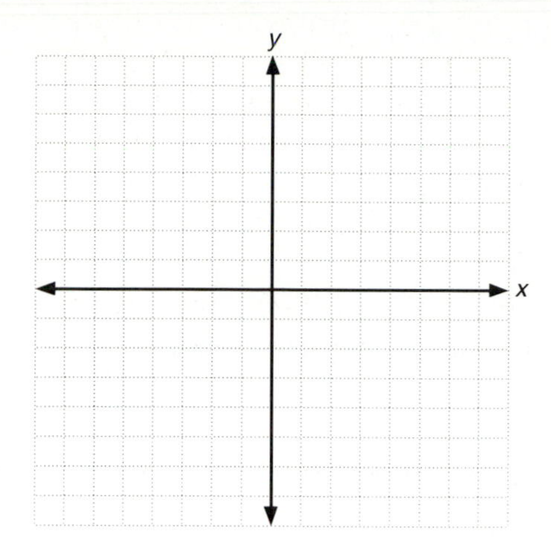

10. $y = x^2 - x + 1$

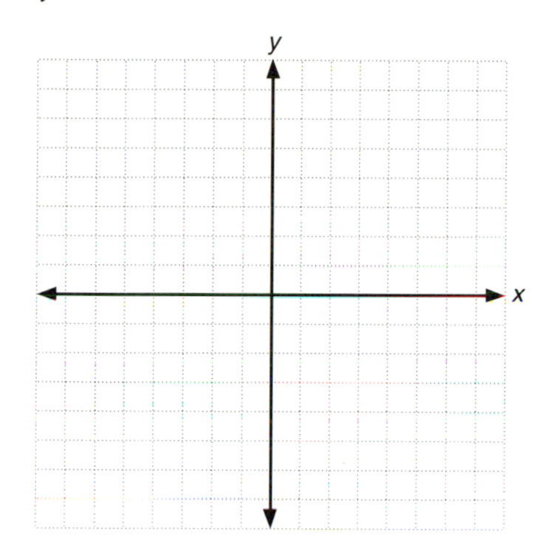

Solve the systems by graphing.

11. $\begin{cases} x + y = -1 \\ 2x - y = 4 \end{cases}$

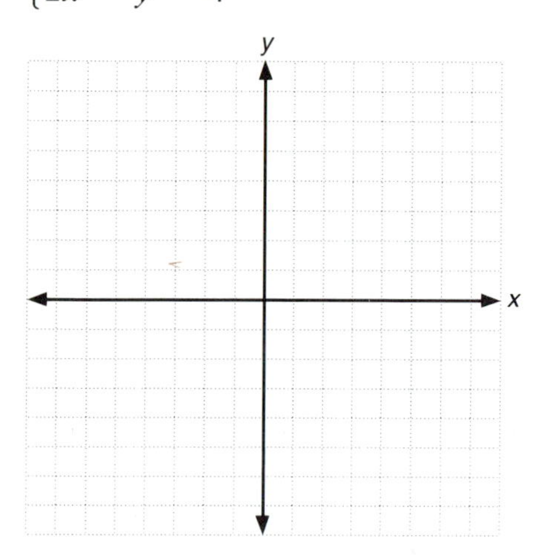

12. $\begin{cases} x - y = 3 \\ x + 2y = -6 \end{cases}$

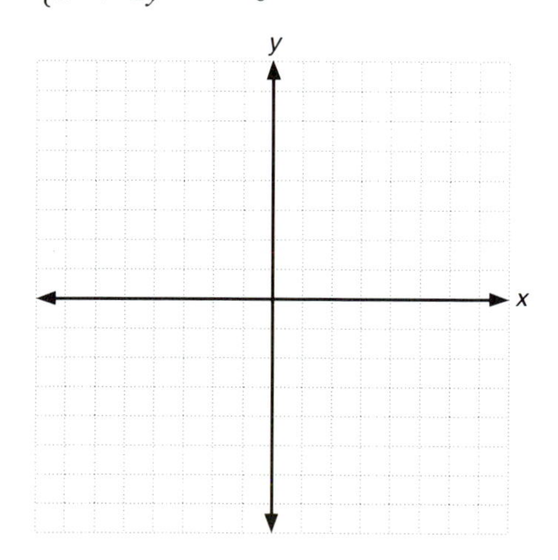

13. $\begin{cases} 2x + y = -3 \\ x - 2y = -4 \end{cases}$

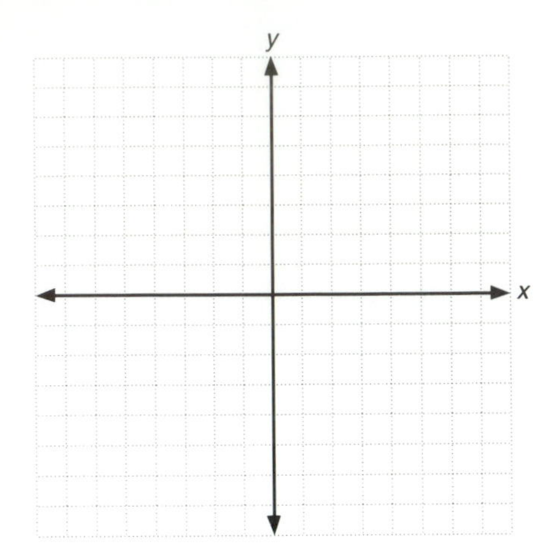

14. $\begin{cases} 3x - y = 7 \\ 2x + y = 8 \end{cases}$

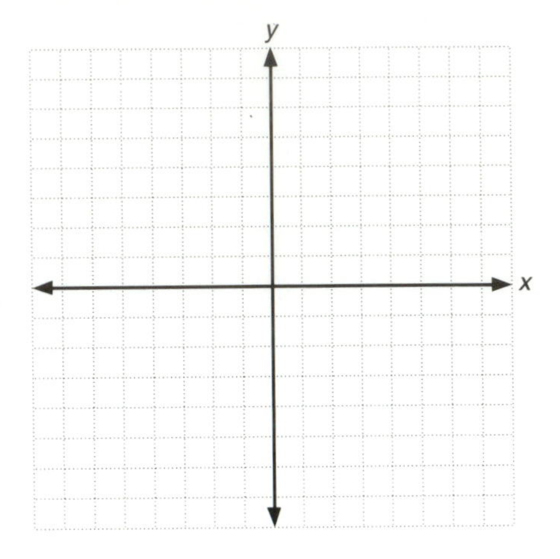

15. $\begin{cases} x + y = -1 \\ y = x^2 - 1 \end{cases}$

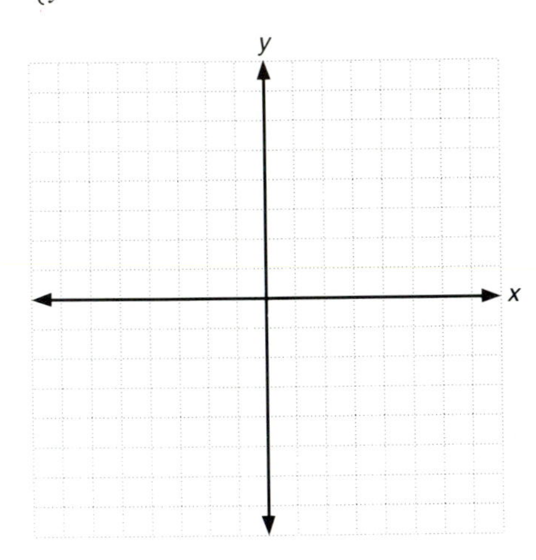

16. Find the equation, in standard form, of the line through the points $(4,-6)$ and $(-3,5)$.

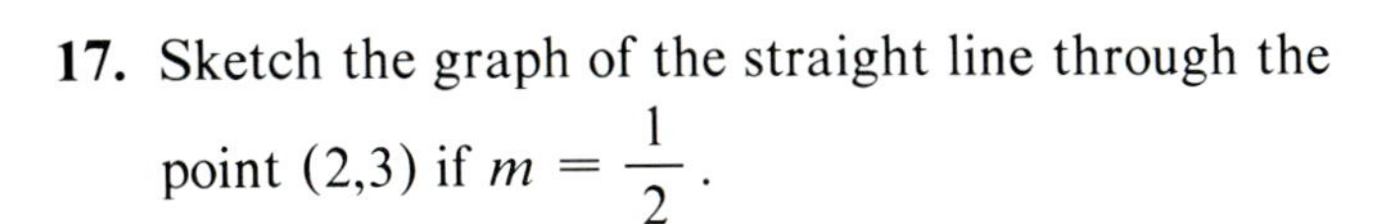

17. Sketch the graph of the straight line through the point $(2,3)$ if $m = \dfrac{1}{2}$.

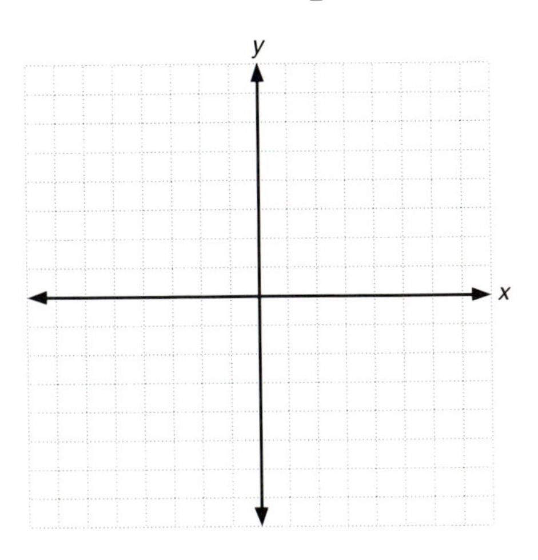

18. Write the equation $2x - 5y = 10$ in slope-intercept form.

19. Find the equation of a line, in standard form, that has a slope of $\frac{1}{2}$ and a y-intercept of 3.

20. Write the equation, in standard form, of the line passing through the point $(-2,7)$ that is parallel to the line $3x + 8y = 1$.

Graph the linear inequalities.

21. $x + y < 0$

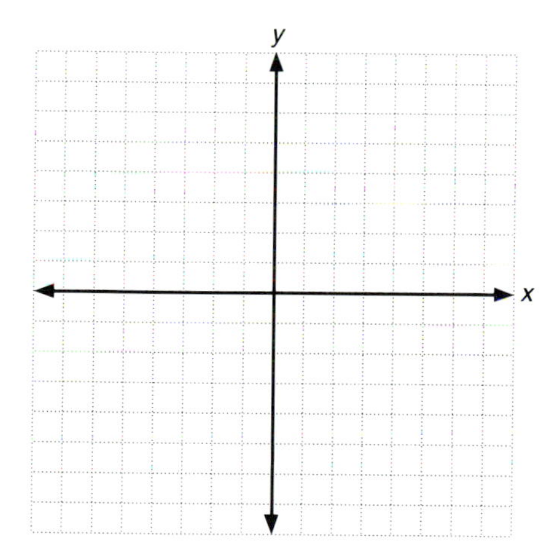

22. $2x + 3y > 1$

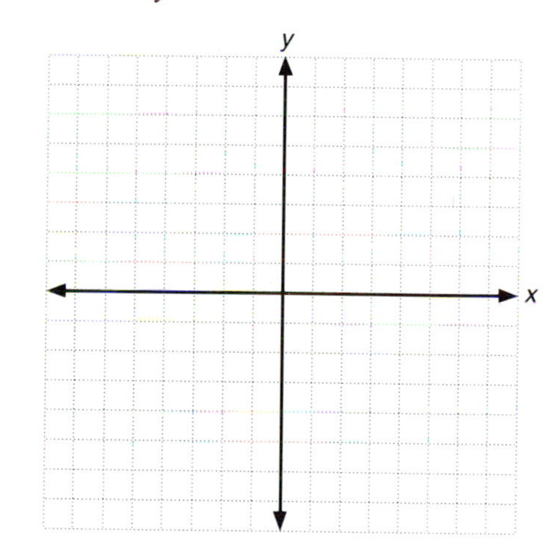

23. $x + 2y \geq 3$

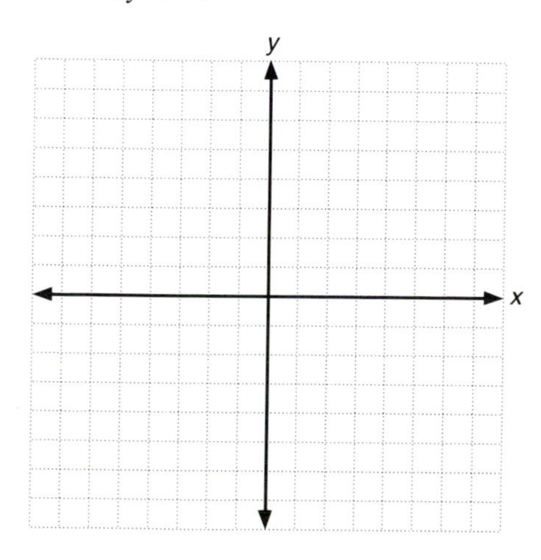

24. $x - 3y \leq 0$

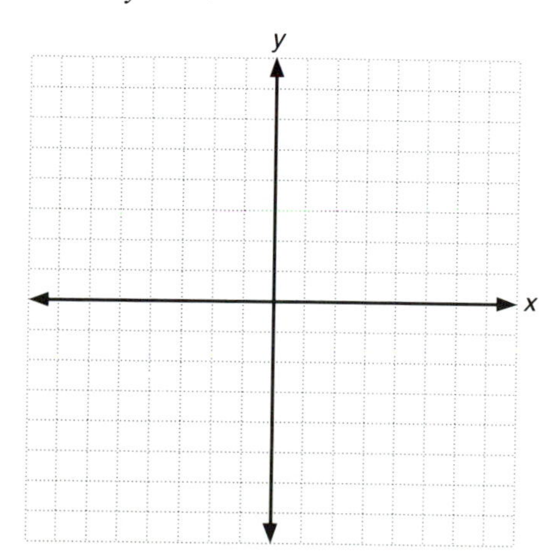

25. $2x + 3y > 8$

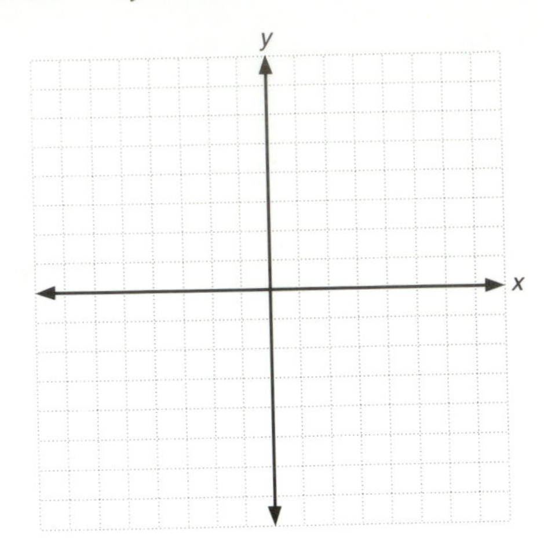

Solve the systems by graphing.

26. $\begin{cases} 3x + y > 0 \\ 2x - y < -4 \end{cases}$

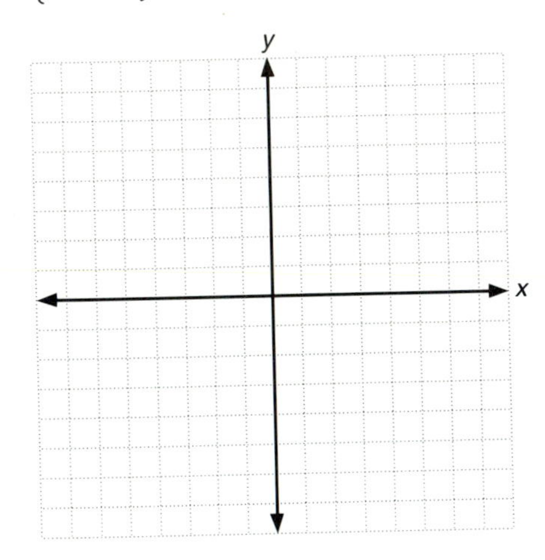

27. $\begin{cases} 2x - y \geq 2 \\ x + y \leq 3 \end{cases}$

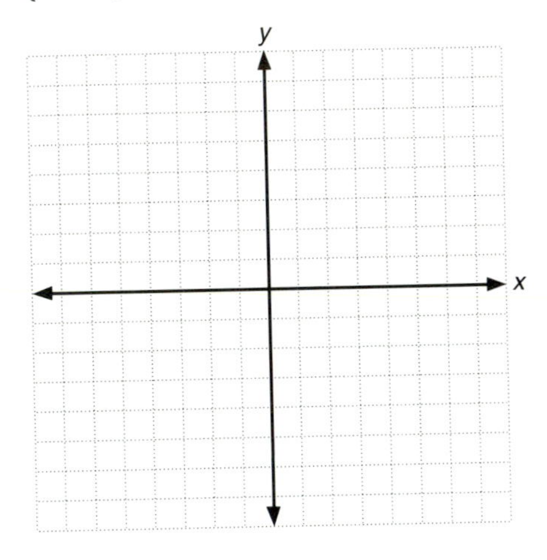

28. $\begin{cases} 5x - 2y \leq 10 \\ x + 3y > 9 \end{cases}$

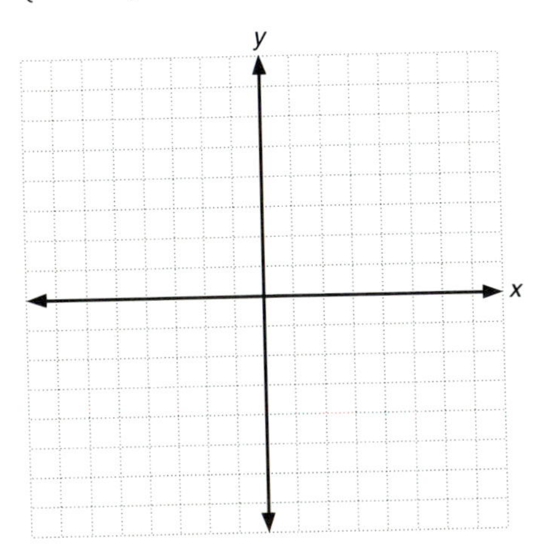

29. $\begin{cases} 5x - 3y > 15 \\ 2x + 3y \leq -6 \end{cases}$

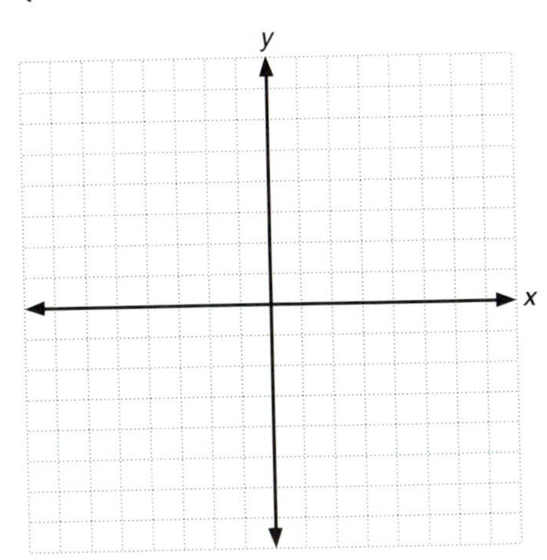

30. $\begin{cases} 2x - 3y \leq -6 \\ x + 4y \leq 4 \end{cases}$

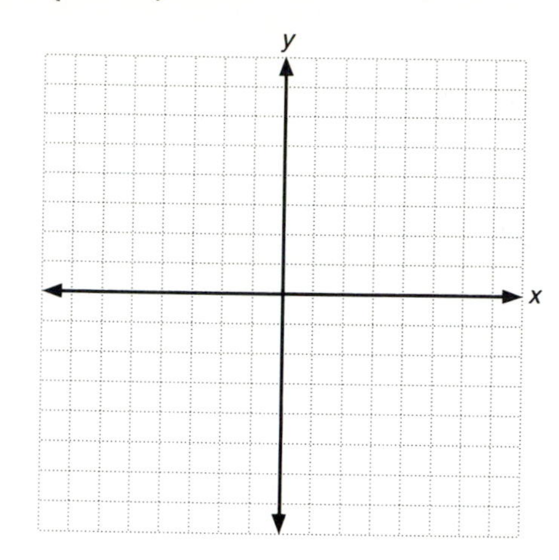

Classify each of the following systems as independent, inconsistent, or dependent. If the system is independent, find its solution.

31. $\begin{cases} x + y = 1 \\ 2x - y = 5 \end{cases}$

32. $\begin{cases} 2x - y = 4 \\ 6x - 3y = 8 \end{cases}$

33. $\begin{cases} 2x + y = -5 \\ 3x + 2y = -10 \end{cases}$

34. $\begin{cases} 3x + 6y = 15 \\ 2x + 4y = 10 \end{cases}$

35. A boat travels 23 kilometers per hour downstream and 15 kilometers per hour upstream. Find the speed of the current and the speed of the boat.

Solve the systems.

36. $\begin{cases} x + y - z = 4 \\ 2x + y + z = 3 \\ 2x + 2y + z = 5 \end{cases}$

37. $\begin{cases} x + y - z = -3 \\ x + y + z = 3 \\ 3x - y + z = 7 \end{cases}$

38. $\begin{cases} 2x - y + z = 5 \\ x + 2y - z = -2 \\ x + y - 2z = -5 \end{cases}$

39. $\begin{cases} 2x - 3y + z = 11 \\ x + y + 2z = 8 \\ x + 3y - z = -11 \end{cases}$

40. A man buys 11 liters of three different kinds of paint, some at $5.00 a liter, some at $4.00 a liter, and some at $3.00 a liter. He has twice as many liters of $3.00 paint as he does of $4.00 paint. If his total bill for the paint is $40.00, how many liters of $3.00 paint did he buy?

CHAPTER 8

PRACTICE TEST

1. In the equation $y = \frac{1}{\sqrt{x-1}}$, is y a function of x? If so, what is the domain?

1. ____________________

2. Given the function $y = x^2 - 1$:

a. Complete the accompanying table.

2. a. ____________________

x	-2	-1	0	1	2
y					

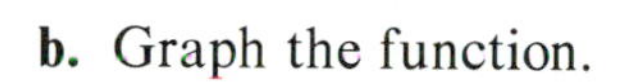

b. Graph the function.

2. b. ____________________

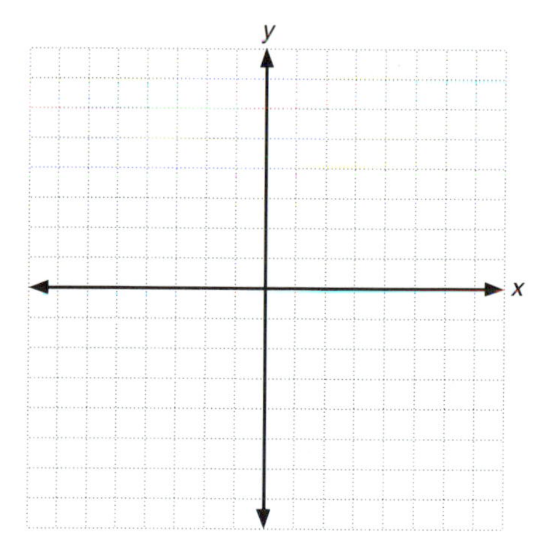

3. Solve the system by graphing: 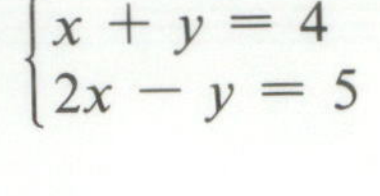

$$\begin{cases} x + y = 4 \\ 2x - y = 5 \end{cases}$$

3. ______________________

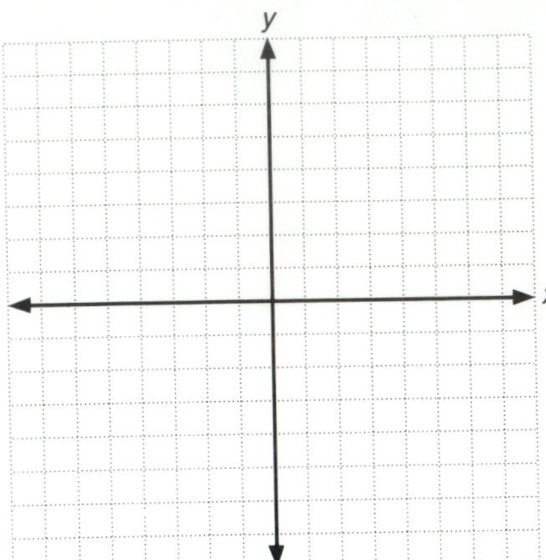

4. ______________________

4. Write the equation, in standard form, of the straight line passing through the points $(0,-2)$ and $(5,1)$.

5. Sketch the graph of the linear inequality:
$x + y < 6$

5. ______________________

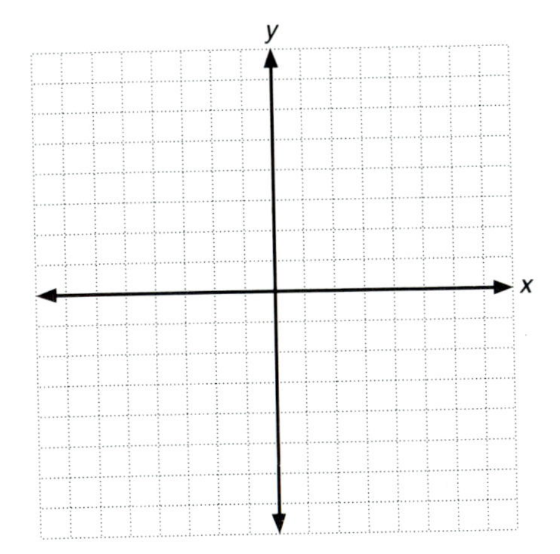

6. Solve by graphing:
$$\begin{cases} x + y \leq -4 \\ 5x - 3y \geq 10 \end{cases}$$

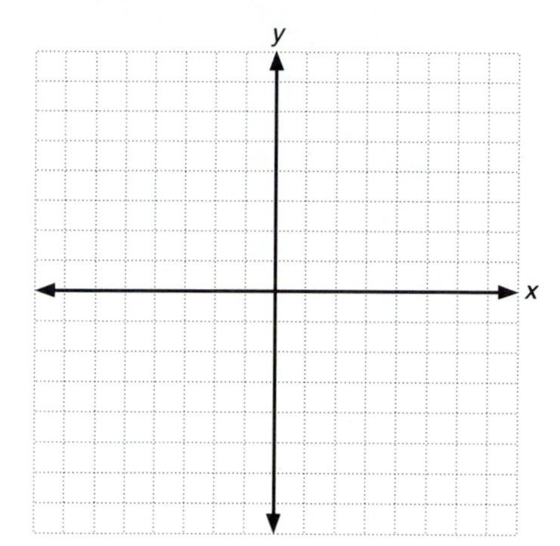

6. ______________________

7. Solve by algebraic methods:
$$\begin{cases} 2x + y = 2 \\ x - 3y = 15 \end{cases}$$

7. ______________________

8. Solve:
$$\begin{cases} 2x + 3y = -2 \\ 3x + 2y = 7 \end{cases}$$

8. ______________________

9. Solve:

$$\begin{cases} x + 2y + z = 1 \\ 2x + 3y - z = 0 \\ x - 2y + 3z = 7 \end{cases}$$

9. ______________________

10. The sum of three numbers is zero. Twice the first number added to the second is 11 less than the third. The third number is 13 more than the second. Find the numbers.

10. ______________________

END-OF-BOOK TEST

1. Combine: $-4 - (+5) - (-7)$

1. ______

2. Divide: $\frac{+24a^4}{-3a^2}$

2. ______

3. Multiply: $2a^2b^3(2a^3b - 3b)$

3. ______

4. Simplify: $5x^2 - [7x - 3x(4x + 1)]$

4. ______

Solve for x:

5. $7(x - 2) = 3x + 6$

5. ______

6. $3(x + 2) - 3 = 5(x - 1)$

6. ______

7. $\frac{3x}{2} + \frac{x}{4} = 7$

7. ______

8. $\frac{1}{3}(2x - 3a) = 2(x + a)$

8. ________________

9. $-2x > -6$

9. ________________

10. $|x - 3| < 7$

10. ________________

Factor completely:

11. $9a^2b^2 - 6ab^3 - 12a^3b^2$

11. ________________

12. $7x^2 - 7$

12. ________________

13. $4x^2 - 28x + 49$

13. ________________

14. $8x^2 + 20x - 12$

14. ______________________

Simplify:

15. $\dfrac{x^2 + 5x - 14}{x^2 - 3x + 2} \cdot \dfrac{x^2 - 6x + 5}{x^2 - 25}$

15. ______________________

16. $\dfrac{6x + 3}{x^2 - 4x} \div \dfrac{2x^2 - 5x - 3}{x^2 - 3x}$

16. ______________________

17. $\dfrac{5}{x} + \dfrac{4}{x + 3}$

17. ______________________

18. $\dfrac{2x}{x - 1} - \dfrac{5x + 2}{x^2 + 4x - 5}$

18. ______________________

19. $\dfrac{1 - \dfrac{3}{x+3}}{\dfrac{1}{x^2 - 9}}$

19. ______________________________

20. ______________________________

20. Solve for x:

$$\frac{5}{x^2 + 4x - 21} - \frac{2}{x+7} = \frac{3}{x-3}$$

21. ______________________________

21. Evaluate: $16^{-3/2}$

Simplify and eliminate negative exponents:

22. ______________________________

22. $(-3xy^4)^3$

23. $(3x^2y)^2(-2xy^3)^3$

23. ______________________________

24. $\left(\dfrac{3x^{-4}}{5y^{-2}}\right)^{-2}$

24. ______________________

25. Simplify: $\dfrac{1}{\sqrt{x} - 2}$

25. ______________________

26. Simplify: $\dfrac{1}{\sqrt{x} + 2}$

26. ______________________

27. Solve for real or complex roots:
$3x^2 - 4 = 0$

27. ______________________

28. $6x^2 - 27x - 15 = 0$

28. ______________________

29. $\frac{2x}{7} - 2 + \frac{7}{2x} = 0$

29. ____________________

30. $\sqrt{x + 2} = x$

30. ____________________

31. Solve by algebraic methods:

$$\begin{cases} 2x + 3y = 6 \\ 3x - 2y = -17 \end{cases}$$

31. ____________________

32. Solve by algebraic methods:

$$\begin{cases} x = \frac{2y - 3}{3} \\ 6x - y = 12 \end{cases}$$

32. ____________________

33. Write the equation, in standard form, of the line passing through the points (7,0) and (3,−5).

33. ____________________

34. Solve by graphing:

$$\begin{cases} 3x + y = 8 \\ 2x - y = 7 \end{cases}$$

34. ____________________

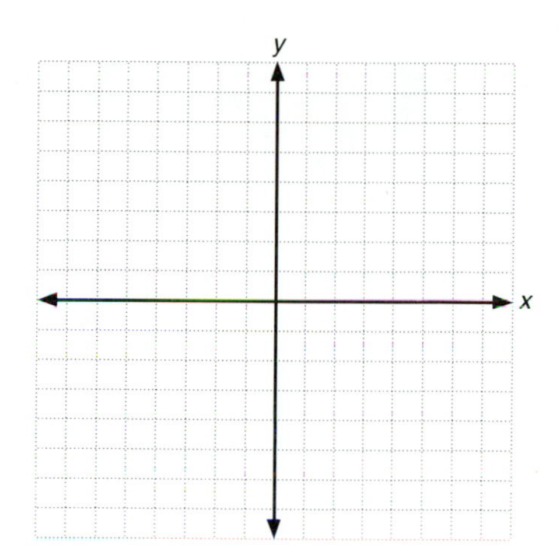

35. Solve by graphing:

$$\begin{cases} x + y \le 0 \\ 2x - y > 4 \end{cases}$$

35. ____________________

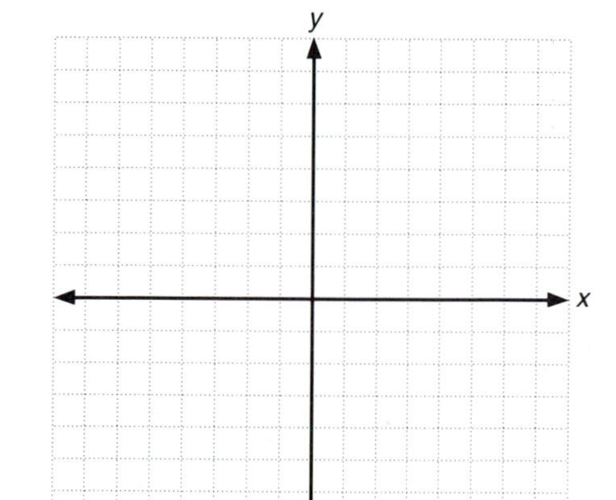

36. Solve:

$$\begin{cases} 4x - 2y + z = 4 \\ 2x - y + 3z = -3 \\ x + y - 2z = 7 \end{cases}$$

36. ____________________

37. During a certain month of sales, salesperson A earned $23 more commission than salesperson B. Salesperson C earned twice as much as salesperson B. Find the commission for each salesperson if their total combined commission was $3,555.

37. ____________________

38. A spacecraft leaves the Moon for Earth (a distance of 238,250 miles) traveling at 6,500 miles per hour. Three hours later, a spacecraft leaves Earth for the Moon traveling at 6,000 miles per hour. In how many hours after the Moonbound spacecraft departs will they pass each other?

38. ____________________

39. A merchant wishes to mix a brand of coffee that sells for $2.50 per pound with a second brand that sells for $2.00 per pound to obtain a 100-pound mixture that will sell for $2.20 per pound. How much of each brand should be used?

39. ____________________

40. A farmer has 200 meters of fence on hand and wishes to enclose a rectangular field so that it will contain 2,400 square meters in area. What should the dimensions of the field be?

40. ____________________

Answers

Chapter 1 Survey

The number in brackets following each answer indicates the section that contains that type of problem.
1. $+31$ [1–1] **2.** -47 [1–1] **3.** -6 [1–2] **4.** -31 [1–2] **5.** 42 [1–3] **6.** -6 [1–3] **7.** -30 [1–3] **8.** x^7 [1–4]
9. $-35a^5b^5$ [1–4] **10.** $-\frac{9x^6}{y^2}$ [1–4] **11.** $x^2y - 5xy + 7xy^2$ [1–5] **12.** $-3a^3 + 6a^2 - 3a$ [1–6] **13.** 10 [1–7]
14. $7x^2 + 4xy$ [1–7] **15.** 22 [1–8]

1–1–1

1. -5 **3.** $-\frac{3}{4}$ **5.** $-x$ **7.** $+(7 + 2)$ or $+9$ **9.** 0 (since $-0 = 0$)

1–1–2

1. $+17$ **3.** -15 **5.** $+22$ **7.** -4 **9.** -26 **11.** $+124$ **13.** -9 **15.** $+2$ **17.** $-\frac{2}{7}$ **19.** 0 **21.** $+6°$ **23.** $+4$
25. -4 yd

1–2–1

1. $+6$ **3.** -8 **5.** $+10$ **7.** $+17$ **9.** -23 **11.** 15 **13.** -11 **15.** -27 **17.** 37 **19.** 0

1–2–2

1. $+15$ **3.** -25 **5.** $+3$ **7.** -15 **9.** 0 **11.** -3 **13.** 14 **15.** -9 **17.** 8 **19.** 40 **21.** 56 degrees **23.** -8 lb

1–3–1

1. 12 **3.** -10 **5.** 12 **7.** 1 **9.** $-\frac{14}{15}$ **11.** 5 **13.** 4 **15.** -3 **17.** -4 **19.** $-\frac{24}{5}$

1–3–2

1. 70 **3.** -400 **5.** 48 **7.** 108 **9.** -1 **11.** 150 **13.** 30 **15.** -300 **17.** -80 **19.** -37

1–4–1

1. 9 **3.** 8 **5.** -8 **7.** x^8 **9.** 3^7 **11.** x^9

1–4–2

1. $10x^5$ **3.** $-21x^5$ **5.** $32a^2b^3$ **7.** $-6x^3y^4$ **9.** $-30x^{11}$ **11.** $-48x^3y^6$ **13.** $105x^3y^9z^3$ **15.** $60a^2b^3c^5d^2$

1–4–3

1. x **3.** $\frac{1}{x^5}$ **5.** $\frac{1}{5}$ **7.** $-2x^2$ **9.** $\frac{x^3}{4}$ **11.** x^5y^3 **13.** $-\frac{4y^4}{x^3}$ **15.** 1 **17.** $-\frac{5x}{7z}$ **19.** $-\frac{3wx^2}{4}$

1–5–1

1. $12a$ **3.** $-7a$ **5.** $-4xy$ **7.** Cannot be combined **9.** Cannot be combined **11.** x **13.** $-15a^2b + 5ab^2$
15. Cannot be combined **17.** $8a - 3xy$ **19.** $4xyz + 15xy + 7xy^2$

1-6-1

1. $3x + 6y$ **3.** $x^2 + xy$ **5.** $-2x + 7y$ **7.** $10x + 5y - 15z$ **9.** $-2x^2 + 4xy$ **11.** $-3x - 5y + z$ **13.** $-7xy + 14y^2 + 21yz$ **15.** $6x + 15x^2 - 12x^3$ **17.** $6x^3y^2 - 4x^2y^3 + 2x^2y^2$ **19.** $2x^3y^2z - 10xy^3z^2 + 18x^2yz^2$

1-7-1

1. 18 **3.** 4 **5.** 4 **7.** 4 **9.** -3 **11.** 20 **13.** 10 **15.** 21 **17.** -17 **19.** -30 **21.** 3 **23.** 26 **25.** 18 **27.** -18 **29.** 16

1-7-2

1. $3x - 1$ **3.** $2x + 1$ **5.** 1 **7.** $11x + 9$ **9.** $2x - 3$ **11.** $2a - 4$ **13.** $12x - 18$ **15.** $11a - 5$ **17.** $5x + 4$ **19.** $13 - 13a$ **21.** $16a - 9b$ **23.** $28x - 61y$ **25.** $a + 3b$ **27.** $6x + 2$ **29.** $5x^2 - 30x$

1-8-1

1. -16 **3.** 25 **5.** 25 **7.** 400 **9.** 14 **11.** 64 **13.** 38 cm **15.** 32 m **17.** 42.9 ft **19.** 220 mi **21.** $1,920 **23.** 154 in.^{2} **25.** 86 **27.** 42.875 m^3 **29.** 90 in.^{2} **31.** 400 ft **33.** 790 cm **35.** 7 **37.** 16 **39.** 72,000 ft lb **41.** 21 amps

Chapter 1 Review

The number in brackets following each answer indicates the section that contains that type of problem.

1. -2 [1–1] **3.** $\frac{2}{9}$ [1–1] **5.** -5 [1–1] **7.** 9 [1–2] **9.** 6 [1–2] **11.** 280 [1–3] **13.** $-\frac{2}{3}$ [1–3] **15.** $+180$ [1–3] **17.** $24x^7$ [1–4] **19.** $-3x^5y^2$ [1–4] **21.** $6a$ [1–5] **23.** $2x^2 + 2x$ [1–5] **25.** $-20ab + 5a + 2$ [1–5] **27.** $-3x^2 - 2x + 5$ [1–6] **29.** $6x^3y^2 - 15x^2y^4 + 6xy$ [1–6] **31.** 30 [1–7] **33.** $-8x^3 + 17x^2 + 10x - 3$ [1–7] **35.** $-33x^2 + 184x$ [1–7] **37.** 8 [1–8] **39.** -51 [1–8]

Chapter 1 Practice Test

The number in brackets following each answer indicates the section that contains that type of problem.

1. $\frac{3}{5}$ [1–1] **2.** -40 [1–1] **3.** $-\frac{2}{11}$ [1–1] **4.** 41 [1–2] **5.** 41 [1–2] **6.** -10 [1–2] **7.** -28 [1–3] **8.** $\frac{2}{5}$ [1–3] **9.** -44 [1–3] **10.** -243 [1–4] **11.** $-24x^6y^7$ [1–4] **12.** $3a^2$ [1–4] **13.** $\frac{-4x^4}{y}$ [1–4] **14.** $13x - 13y$ [1–5] **15.** $a^2b + 6ab^2 + 7ab$ [1–5] **16.** $20a^3b^2 - 8a^4b^3 - 20a^2b^4$ [1–6] **17.** $21x - 3y + 2$ [1–7] **18.** $7x^2 - 17x$ [1–7] **19.** 46 [1–8] **20.** 36 [1–8]

Chapter 2 Survey

The number in brackets following each answer indicates the section that contains that type of problem.

1. Yes [2–1] **2.** $x = -2$ [2–2] **3.** $x = -4$ [2–2] **4.** $x = 12$ [2–2] **5.** $x = \frac{9}{7}$ [2–2] **6.** $x = -4$ [2–3] **7.** $x = \frac{11}{3}$ [2–3] **8.** $x = \frac{2a}{17}$ [2–4] **9.** $x \leq 10$ [2–5] **10.** $x < -2$ or $x > 8$ [2–6]

2-1-1

1. True **3.** Conditional **5.** Conditional **7.** 3 is a solution **9.** 3 is a solution

2-1-2

1. Equivalent **3.** Not equivalent **5.** Not equivalent **7.** Not equivalent **9.** Equivalent

2-2-1

1. 2 **3.** 10 **5.** 2 **7.** -14 **9.** 13

2–2–2

1. 12 **3.** 3 **5.** $\frac{1}{2}$ **7.** $\frac{1}{3}$ **9.** 5

2–2–3

1. 12 **3.** -6 **5.** 125 **7.** 64 **9.** -5

2–2–4

1. $\frac{7}{2}$ **3.** $\frac{21}{16}$ **5.** $\frac{5}{6}$ **7.** 1 **9.** $\frac{16}{3}$ **11.** 25

2–3–1

1. 1 **3.** -9 **5.** -4 **7.** 2 **9.** $\frac{10}{3}$ **11.** $\frac{7}{3}$ **13.** $\frac{14}{5}$ **15.** 5 **17.** $\frac{2}{15}$ **19.** $\frac{45}{4}$ **21.** -5 **23.** $\frac{46}{45}$ **25.** $\frac{5}{9}$
27. $-\frac{40}{7}$ **29.** $-\frac{1}{5}$ **31.** \$5,700 **33.** \$84.00 **35.** 75 ft **37.** \$67.50 **39.** $a = 18$ in., $b = 9$ in., $c = 12$ in.

2–4–1

1. $2y$ **3.** $2y$ **5.** $-6a$ **7.** $2y$ **9.** $2a$ **11.** $\frac{22bc}{a}$ **13.** $\frac{5a + 3y}{3ab}$ **15.** $\frac{10b - 3a}{9}$ **17.** $\frac{a - 75}{29}$ **19.** $\frac{2a + 3b}{2}$
21. $\frac{2A}{b}$ **23.** $\frac{2s}{t^2}$ **25.** $\frac{5F - 160}{9}$ **27.** $\frac{Inr + IR}{n}$ **29.** $\frac{365I}{rD}$

2–5–1

1. $6 < 10$ **3.** $-3 < 3$ **5.** $4 > 1$ **7.** $-2 > -3$ **9.** $0 < 7$

2–5–2

1. 

3. 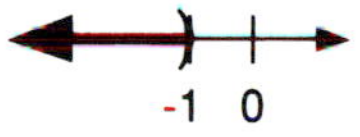

5. 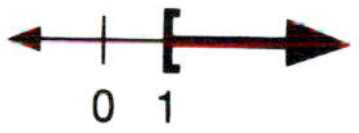

7. 

9. 

11. $x > 2$ **13.** $x < 0$ **15.** $x \geq 1$ **17.** $x < 3$ **19.** $-5 < x \leq 2$

2–5–3

1. $x < 4$ 

3. $x < -5$ 

5. $x \leq 3$ 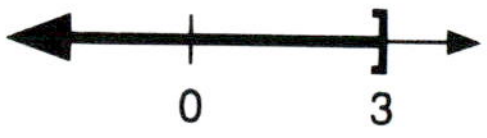

7. $x > -17$ 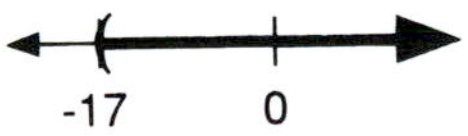

9. $x \geq -5$ 

2–5–4

1. $x < 2$ 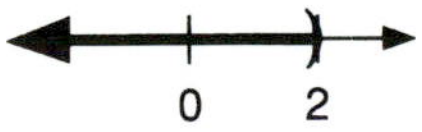

3. $x \leq -4$ 

5. $x > 2$ 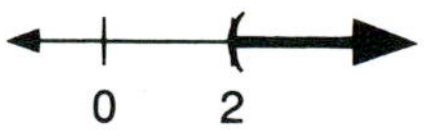

7. $x < 4$ 

9. $x > 6$ 

11. $x \leq 3$ 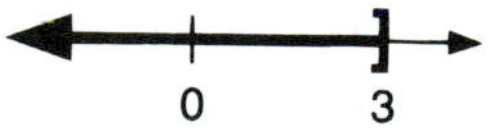

13. $x \geq -2$ 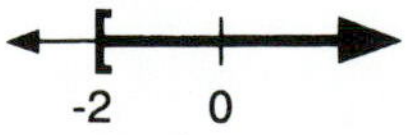

15. $x > 3$ 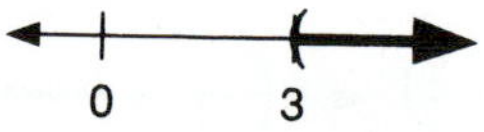

17. $x \geq 2$ 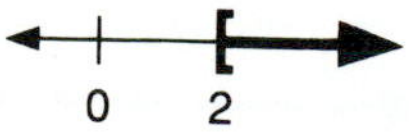

19. $x \le -20$ **21.** $x \le \frac{7}{3}$

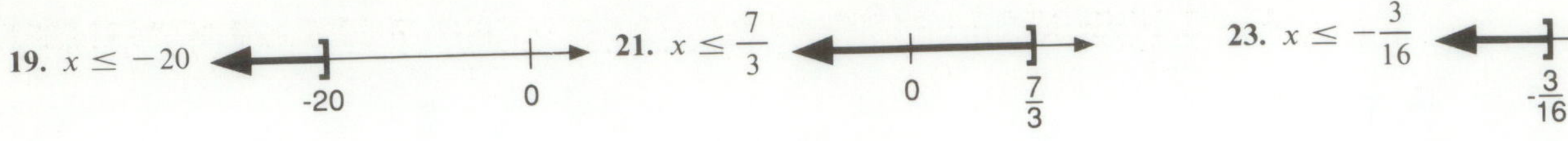

23. $x \le -\frac{3}{16}$

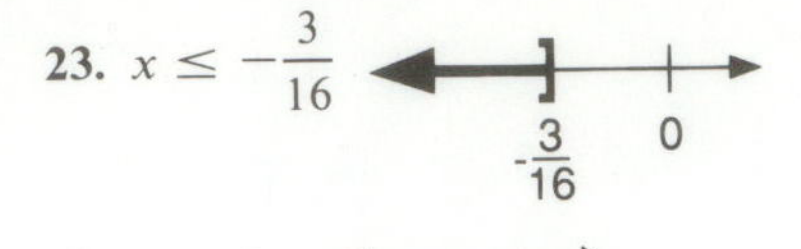

25. $x > 3$ **27.** $x < 5$ **29.** $x < 0$

2-6-1

1. $-3, 3$ **3.** $-13, 3$ **5.** $1, 11$ **7.** $-3, 3$ **9.** $3, -\frac{5}{3}$

2-6-2

1.

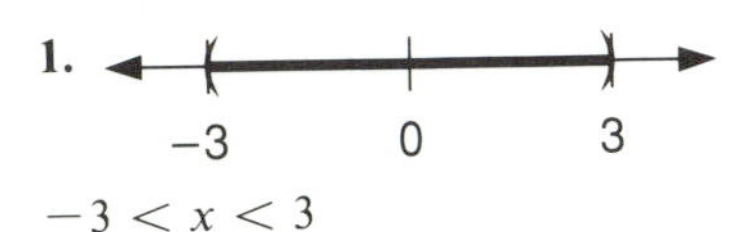

$-3 < x < 3$

3. $-13 < x < 3$

5.

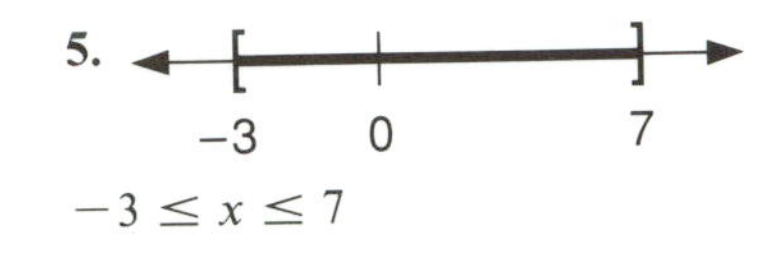

$-3 \le x \le 7$

7. $-2 < x < 1$

9. $x < -\frac{3}{4}$ or $x > -\frac{3}{4}$

11. $x \le -\frac{4}{3}$ or $x \ge 2$

Chapter 2 Review

The number in brackets following each answer indicates the section that contains that type of problem.

1. Equivalent [2–1] **3.** Not equivalent [2–1] **5.** Equivalent [2–1] **7.** -10 [2–2] **9.** -75 [2–2] **11.** -3 [2–3] **13.** 6 [2–3] **15.** $\frac{35}{6}$ [2–3] **17.** $-\frac{8a}{7}$ [2–4] **19.** $\frac{P - 2y}{2}$ [2–4]

21. **23.**

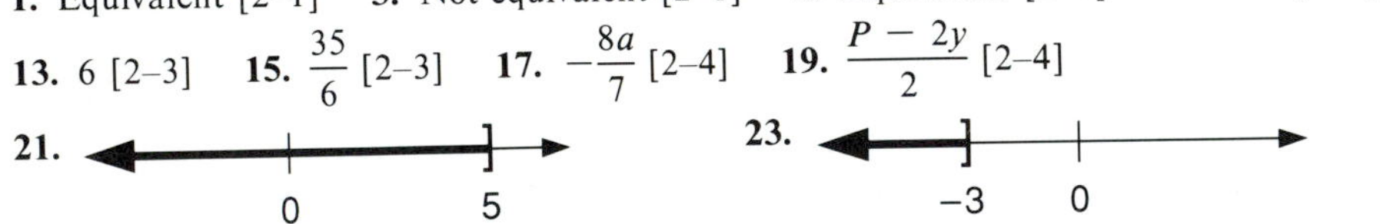

$x \le 5$ [2–5]

$x \le -3$ [2–5]

25.

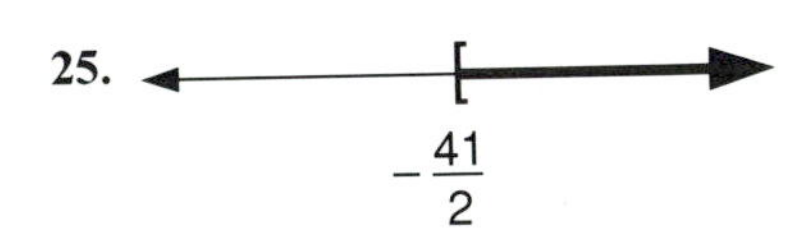

$x \ge -\frac{41}{2}$ [2–5]

27. $x = -\frac{5}{3}$ or $x = 1$ [2–6]

29. $x \ge 4$ or $x \le -1$ [2–6]

31. \$5.80 [2–3] **33.** 74.3° [2–3]

Chapter 2 Practice Test

The number in brackets following each answer indicates the section that contains that type of problem.

1. d [2–1] **2.** -12 [2–2] **3.** 6 [2–3] **4.** -5 [2–3] **5.** 9 [2–3] **6.** -5 [2–3] **7.** 12 [2–3] **8.** $\frac{6 - 5y}{3}$ [2–4]

9. $\frac{5a}{9}$ [2–4] **10.** $-\frac{11}{3}, 3$ [2–6]

11. [2–5] **12.** [2–5]

13. 

[2–5] **14.** $x > 0$ [2–5] **15.** $x \geq \frac{29}{2}$ [2–5]

16. $-\frac{8}{5} < x < 6$ [2–6] **17.** 8 [2–2] **18.** $w = 5$, $\ell = 12\frac{1}{2}$ [2–3] **19.** $b = \frac{2A - ha}{h}$ [2–4]

20. $x \leq \frac{14}{3}$ 

[2–5]

Chapter 3 Survey

The number in brackets following each answer indicates the section that contains that type of problem.

1. a. $2x - 9$ [3–1] **b.** $\frac{y}{3} + 1$ [3–1] **c.** $.173x$ [3–1] **d.** $10d$ [3–1] **2.** 6, 12, 8 [3–2] **3.** 10 students [3–3] **4.** 4 hr [3–4] **5.** 1.4 liters [3–5]

3–1–1

1. $x + 5$ **3.** $x - 9$ **5.** $2x$ **7.** $6x - 4$ **9.** $x - 3$ **11.** $\frac{5x}{2}$ **13.** $3(x - 9)$ **15.** $.1x$ **17.** $10d$ **19.** $12y$

3–1–2

The following represent one way of answering each problem.

1. $x + 8$ = first number
x = second number

3. x = units produced first week
$x + 612$ = units produced second week

5. x = population of Long Beach
$2x$ = population of San Francisco

7. x = width
$2x + 3$ = length

9. x = first number
$x + 18$ = second number

11. x = enrollment last year
$x - .07x$ = enrollment this year

13. x = first number
$2x$ = second number
$\frac{x}{3}$ = third number

15. x = cost in 1970
$\frac{3}{2}x$ = cost in 1976
$3x$ = cost now

17. $x + 3$ = mileage of Buick
x = mileage of Chevrolet
$x + 9$ = mileage of Toyota

19. x = length of first piece
$100 - x$ = length of second piece

3–2–1

1. First number = 15, second number = 7 **3.** First week = 8,902; second week = 9,514 **5.** 716,000 **7.** 65 cm **9.** First number = 32, second number = 50 **11.** First number = 26, second number = −11 **13.** Coach ticket = $109, first class ticket = $218, economy ticket = $79 **15.** First side = 32 cm, second side = 16 cm, third side = 36 cm **17.** $39,250 **19.** Population of Ohio = 11,000,000; population of California = 22,000,000; population of Indiana = 5,500,000

3–3–1

1. 39, 44 **3.** 12, 63 **5.** Length = 66 cm, width = 46 cm **7.** Price of notebook = $4.50, price of calculator = $9.95 **9.** 28, 30, 32 **11.** Population of town A = 25,795; population of town B = 23,450 **13.** Car A = $6,421; car B = $7,832; car C = $5,910 **15.** A copies = 20, B copies = 12 **17.** Length = 10, width = 5 **19.** 8

3–4–1

1. 264 mi **3.** 19 mph **5.** 238,080 mi **7.** 2 hr **9.** 2 hr **11.** 7 hr **13.** Walker = 6 mph, cyclist = 15 mph **15.** Car A = 77 km/hr, car B = 85 km/hr **17.** 390 km/hr **19.** 40 km/hr

3-5-1

1. 10 lb caramels, 20 lb creams **3.** 18 nickels, 57 quarters **5.** 25 bushels oranges, 10 bushels grapefruit **7.** 63 nickels, 21 quarters **9.** 50 children, 160 adults **11.** 10 nickels, 25 dimes, 18 quarters **13.** 100 gal skim milk, 125 gal 3.6% milk **15.** 24 first class, 120 coach, 32 super saver **17.** 2.5 liters **19.** $6\frac{2}{3}$ liters of 10% solution, $3\frac{1}{3}$ liters of 25% solution

Chapter 3 Review

The number in brackets following each answer indicates the section that contains that type of problem.

1. $2x - 7$ [3–1] **3.** $x + 5$ [3–1] **5.** $\frac{y}{100}$ [3–1] **7.** x, $84 - x$ [3–1] **9.** $2x$, $181 - 3x$, x [3–1] **11.** $x + (x + 5) = 23$ [3–2] **13.** $3x - (x - 7) = 13$ [3–2] **15.** $.09x + x = 1.26$ [3–2] **17.** 1, 8 [3–2] **19.** 25.7 kg [3–5] **21.** 12, 24, 28 [3–3] **23.** 4 nickels, 8 dimes, 10 quarters [3–5] **25.** A: 15 kg, B: 20 kg, C: 15 kg [3–5] **27.** 126 kg [3–4] **29.** 80 km/hr [3–4] **31.** 4.5 m, 13.5 m, 22 m [3–3] **33.** 30 nickels, 24 dimes, 53 quarters [3–5] **35.** 1.2 liters [3–5] **37.** $\frac{1}{2}$ hr [3–4]

Chapter 3 Practice Test

The number in brackets following each answer indicates the section that contains that type of problem.

1. a. $3x + 6$ **b.** $\frac{y}{2} - 7$ **c.** $.094x$ **d.** $\frac{d}{7}$ [3–1] **2.** 5 nickels, 8 dimes, 16 quarters [3–2] **3.** 12 m, 24 m, 27 m [3–3] **4.** 3 km/hr [3–4] **5.** 5 liters [3–5]

Chapter 4 Survey

The number in brackets following each answer indicates the section that contains that type of problem.

1. $2x^2 + 9x - 5$ [4–2] **2.** $x^3 + x^2 - 11x + 10$ [4–2] **3.** $7x(5x - 4)$ [4–1] **4.** $2x^2y^2(3xy - 4x + 5y)$ [4–1] **5.** $(x - y)(a - 1)$ [4–3] **6.** $(a - b)(x + 3)$ [4–3] **7.** $(x - 5)(x - 8)$ [4–4] **8.** $(x + 4)(3x + 5)$ [4–4] **9.** $(2x - 5)(2x + 5)$ [4–5] **10.** $(x + 10)^2$ [4–5] **11.** $3(x - 5)(2x + 1)$ [4–6] **12.** $4x(2x - 3)^2$ [4–6] **13.** $12x$ [4–5] **14.** 6 [4–8] **15.** Yes [4–8]

4-1-1

1. $2(6a + 5b)$ **3.** $a(a + 5)$ **5.** $2a(a + 3)$ **7.** $x(x - 1)$ **9.** $4xy(2x + 3)$ **11.** $2ab(5a - b + 3)$ **13.** Prime **15.** $7a(2ab - 5b - 9)$ **17.** $11ab^3(a^2b + 4 - 3ab)$ **19.** Prime

4-2-1

1. $a^2 + 5a + 6$ **3.** $x^2 + 7x + 10$ **5.** $a^2 - 5a + 6$ **7.** $a^2 - 8a + 15$ **9.** $x^2 + 4x - 5$ **11.** $a^2 + 4a - 45$ **13.** $6x^2 + 7x + 2$ **15.** $6a^2 + 5a - 4$ **17.** $6x^2 - 19x + 15$ **19.** $10a^2 - 5ab + 6a - 3b$ **21.** $4a^2 - 20a + ac - 5c$ **23.** $x^2 + 10x + 25$ **25.** $x^2 - 14x + 49$ **27.** $x^2 - 9$ **29.** $4x^2 - 49$

4-2-2

1. $x^3 + 8x^2 + 16x + 5$ **3.** $15a^3 + 8a^2b - 19ab^2 - 4b^3$ **5.** $2x^3 - 13x^2 + 7x - 6$ **7.** $x^3 - 125$ **9.** $x^3 + 5x^2 + x - x^2y^2 + 4xy + 5 + x^2y - x^3y$

4-3-1

1. $(x + 4)(2x + 3)$ **3.** $(a - 1)(a + 5)$ **5.** $(y + 4)(6x - 7)$ **7.** $(x - 2)(3x + 1)$ **9.** $(a - 4)(7a - 15)$ **11.** $(x + 3)(x - 1)$ **13.** $x(x + 1)(3x + 2)$ **15.** $a(a + b)(4 + a)$ **17.** $(x - 1)(x - 1)$ **19.** $2a(a + 4)(4a + 1)$

4-3-2

1. $(x + y)(a + 3)$ **3.** $(b + c)(a - 2)$ **5.** $(2x + y)(a + 3)$ **7.** $(x - y)(a + 2)$ **9.** $(x + 2y)(5 + a)$ **11.** $(x - y)(3a + 2)$ **13.** $(x + y)(2a - 1)$ **15.** $(3x + y)(2a - 1)$ **17.** $(3 + y)(x - 2)$ **19.** $(x + 5)(y - 2)$

4–4–1

1. $(x + 40)(x + 1)$ **3.** $(x + 30)(x - 1)$ **5.** $(x + 8)(x - 6)$ **7.** $(x + 9)(x - 2)$ **9.** $(x - 6)(x - 2)$
11. $(x + 35)(x - 1)$ **13.** $(x + 12)(x - 3)$ **15.** $(x - 8)(x - 3)$ **17.** $(x + 16)(x - 3)$ **19.** $(x + 10)(x - 4)$
21. $(x + 10)(x + 5)$ **23.** $(x + 50)(x - 1)$ **25.** $(x + 18)(x + 4)$ **27.** $(x + 3)(x - 8)$ **29.** $(x + 5)(x - 3)$
31. $(x + 8)(x + 6)$ **33.** $(x + 28)(x + 3)$ **35.** $(x + 84)(x - 1)$ **37.** $(x - 8)(x + 5)$ **39.** $(x + 22)(x - 1)$
41. $(x + 14)(x - 6)$ **43.** $(x - 14)(x + 2)$ **45.** $(x + 26)(x + 5)$ **47.** $(x + 13)(x - 10)$ **49.** $(x - 48)(x - 2)$

4–4–2

1. $(x + 2)(2x + 1)$ **3.** $(2x + 3)(x + 1)$ **5.** $(x - 3)(2x - 1)$ **7.** $(3x + 2)(2x + 1)$ **9.** $(2x - 5)(x + 4)$
11. $(3x + 5)(x - 2)$ **13.** $(4x - 3)(2x + 1)$ **15.** $(6x + 1)(x + 5)$ **17.** $(5x + 3)(x + 1)$ **19.** $(8x + 1)(2x - 1)$
21. $(x + 2)(4x - 9)$ **23.** $(2x + 1)(2x + 1)$ **25.** $(2x - 15)(3x - 2)$ **27.** $(3x - 2)(3x - 2)$ **29.** $(3x + 1)(2x - 1)$
31. $(2x + 3)(3x + 2)$ **33.** $(5x + 2)(3x - 5)$ **35.** $(6x - 5)(2x - 1)$ **37.** $(2x - 3)(3x + 5)$ **39.** $(3x + 4)(5x - 6)$

4–5–1

1. $(x - 2)(x + 2)$ **3.** $(a - 9)(a + 9)$ **5.** $(y - 11)(y + 11)$ **7.** $(4x - 1)(4x + 1)$ **9.** $(2x - 5y)(2x + 5y)$
11. $(x + y - 8)(x + y + 8)$ **13.** $(x^3 - 6)(x^3 + 6)$ **15.** $[(x + y)^2 - z][(x + y)^2 + z]$

4–5–2

1. $x^2 + 2xy + y^2$ **3.** $x^2 + 12x + 36$ **5.** $a^2 - 4a + 4$ **7.** $4x^2 + 28x + 49$ **9.** $9a^2 + 6a + 1$ **11.** $4x$ **13.** $20x$ **15.** $2x$
17. $4xy$ **19.** $14xy$ **21.** $(x + 3)^2$ **23.** $(x - 7)^2$ **25.** $(2x + 3)^2$ **27.** $(3x + 5)^2$ **29.** $(6x - 5y)^2$

4–5–3

1. $(x + y)(x^2 - xy + y^2)$ **3.** $(a - 2)(a^2 + 2a + 4)$ **5.** $(x - 1)(x^2 + x + 1)$ **7.** $(3a + 1)(9a^2 - 3a + 1)$
9. $(3a + 2b)(9a^2 - 6ab + 4b^2)$ **11.** $(4x - 3y)(16x^2 + 12xy + 9y^2)$

4–6–1

1. $2(2x + 1)(x + 3)$ **3.** $3(x + 8)(x - 6)$ **5.** $3(x - 2)(x + 2)$ **7.** Prime **9.** $7(3x + 1)(2x - 1)$ **11.** $x(x - 6)(x - 2)$
13. $(a + 3)(x + 1)(x - 1)$ **15.** $2(2x - 1)^2$ **17.** $(x - 2)(x + 7)$ **19.** $(2x - 3)(2x + 1)$ **21.** $3x(3x - 5)(3x + 5)$
23. $(x - 5)(x + 6)$ **25.** Prime **27.** $5(x - 1)(x^2 + x + 1)$

4–7–1

1. 6 **3.** -40 **5.** -16 **7.** -21 **9.** 24

4–7–2

1. $(x + 2)(2x + 3)$ **3.** $(3x + 2)(x + 4)$ **5.** $(x - 5)(2x + 1)$ **7.** $(x - 5)(3x - 2)$ **9.** $(2x + 1)(3x + 2)$
11. $(4x + 3)(x - 9)$ **13.** $(x - 6)(6x - 5)$ **15.** $(3x + 2)(6x - 5)$

4–7–3

1. $(x + 4)(x + 5)$ **3.** $(x - 6)(x - 2)$ **5.** $(x + 1)(3x + 2)$ **7.** $(5x - 1)(x + 2)$ **9.** $(x - 4)(3x - 2)$
11. $(x + 3)(4x - 3)$ **13.** $(2x + 1)(2x + 3)$ **15.** $(4x - 3)(4x + 1)$

4–8–1

1. Q: $(x - 2)$, R: 0, yes **3.** Q: $(2x^2 + 4x + 7)$, R: 27, no **5.** Q: $(x^2 - 2x + 8)$, R: -64, no **7.** Q: $(x^2 + 11x + 120)$, R: 1,300, no **9.** Q: $(2x^2 + 3x - 2)$, R: 0, yes

Chapter 4 Review

The number in brackets following each answer indicates the section that contains that type of problem.
1. $2x^2 - x - 6$ [4–2] **3.** $9x^2 - 64$ [4–2] **5.** $x^2 - y^2 + 2x - 4y - 3$ [4–2] **7.** $a(a - 1)$ [4–1]
9. $3ab(2a + 3b - 2)$ [4–1] **11.** $(5x - 7)(5x + 7)$ [4–5] **13.** $(3x - 5)^2$ [4–5] **15.** $3(2x - 1)(4x^2 + 2x + 1)$ [4–6]
17. $(x + 7)(x - 2)$ [4–4] **19.** $(x + 5)(x - 3)$ [4–4] **21.** $(2x + 1)(x + 2)$ [4–4] **23.** $(x + 5)(3x - 2)$ [4–4]

25. $(3x - 2)(2x - 1)$ [4–4] **27.** $3(2x - 1)(3x + 2)$ [4–6] **29.** $3(x^2 + 3x + 5)$ [4–6] **31.** $(x + y)(c + d)$ [4–3] **33.** $(x - 5)(a - 4)$ [4–3] **35.** $(2x + 3)(a - 1)(a + 1)$ [4–6] **37.** $x^2 - 7x + 15$, R: 0, yes [4–8] **39.** $x^2 + x - 1$, R: 0, yes [4–8]

Chapter 4 Practice Test

The number in brackets following each answer indicates the section that contains that type of problem.
1. $6x^2 + 7x - 3$ [4–2] **2.** $2x^3 - 5x^2 + 8x - 3$ [4–2] **3.** $11a(2a + 3)$ [4–1] **4.** $2xy(5x^2 - 3xy + y^2)$ [4–1] **5.** $(a - 1)(2x - 3)$ [4–3] **6.** $(x + 9)(x - 9)$ [4–5] **7.** $(x - 11)^2$ [4–5] **8.** $(a + b)(x + 3)$ [4–3] **9.** $(x - 7)^2$ [4–5] **10.** $(x - 4)(4x - 3)$ [4–4] **11.** $(2x + 3)(x + 5)$ [4–4] **12.** $(x + 2)(x + 9)$ [4–4] **13.** $(5x - 12)(5x + 12)$ [4–5] **14.** $(x - 9)(x + 3)$ [4–4] **15.** $(x - 3)(x + 3)(a + 3)$ [4–3, 4–5] **16.** $2(3x + 1)(x - 2)$ [4–6] **17.** $8x(2x - 3)(2x + 3)$ [4–6] **18.** $(x - 7)(x + 9)$ [4–4] **19.** $3x(x + 2)(2x - 3)$ [4–6] **20.** $3x(3x + 2)^2$ [4–6] **21.** $(3x - 5)(2x + 3)$ [4–4] **22.** $4(2x + 1)(3x + 5)$ [4–6] **23.** $3x^2 + 2x + 2$, R: 20, no [4–8] **24.** $2x^2 + 3x + 2$, R: 0, yes [4–8]

Chapters 1–4 Cumulative Test

The number in brackets refers to the chapter and section that discusses that type of problem.
1. $+2$ [1–1] **2.** $8x^6$ [1–4] **3.** $15x^3y^4 - 10x^4y$ [1–6] **4.** $20x^2 - 3x$ [1–7] **5.** 9 [1–8] **6.** $x = -6$ [2–3]
7. $x = 12$ [2–3] **8.** $x = \frac{9a}{5}$ [2–4]

9. $x < -4$ [2–5]

−4 0

10. $-1 \le x \le 9$ [2–6]

−1 0 9

11. $7a^2b^2(2b - a^2 + 3ab)$ [4–1] **12.** $2(3x + 2)(3x - 2)$ [4–6] **13.** $(2x - 5)^2$ [4–5] **14.** $7(2x - 1)(x + 2)$ [4–6]
15. $x^2 + 5x - 2$, R: 0, yes [4–8] **16.** $\frac{m}{60}$ [3–1] **17.** 14, 26 [3–2] **18.** \$3,500 @ 9.5%; \$7,000 @ 8.6% [3–2]
19. 35 mph [3–4] **20.** 10 liters [3–5]

Chapter 5 Survey

The number in brackets following each answer indicates the section that contains that type of problem.
1. $\frac{x + 3}{x(2x + 5)}$ [5–1] **2.** 1 [5–2] **3.** $\frac{x - 4}{x + 5}$ [5–2] **4.** $(x + 4)(x + 6)(3x - 2)$ [5–3] **5.** $2x^2 - 13x + 21$ [5–3]
6. $\frac{x^2 + 12x + 7}{(x - 2)(x + 4)(x - 1)}$ [5–4] **7.** $\frac{x + 2}{(x + 1)(x - 1)}$ [5–5] **8.** $-\frac{1}{x + 3}$ [5–6] **9.** $x = -11$ [5–7] **10.** 2 hr [5–7]

5–1–1

1. $\frac{x}{x + 1}$ **3.** $\frac{x + 1}{x + 2}$ **5.** $\frac{x + 1}{x - 5}$ **7.** $\frac{x - 1}{x - 3}$ **9.** $\frac{x - 4}{x}$ **11.** $\frac{x}{x + 3}$ **13.** $\frac{a + 5}{a - 3}$ **15.** $\frac{x - 3}{x - 4}$ **17.** $\frac{3x + 4}{x - 4}$
19. $\frac{3a + 2}{2a - 3}$

5–2–1

1. $\frac{x + 1}{x - 3}$ **3.** $\frac{1}{x + 4}$ **5.** $\frac{x + 5}{x + 7}$ **7.** $\frac{x + 2}{x - 6}$ **9.** $\frac{x - 1}{(x + 5)(x + 1)}$ **11.** 1 **13.** $-\frac{1}{(x - 2)(x + 1)}$ **15.** $\frac{x + 7}{(x + 9)(x - 4)}$
17. $-\frac{3x + 2}{x - 6}$ **19.** $-\frac{3x + 4}{2x - 1}$ **21.** 1 **23.** $\frac{x - 9}{(2x + 5)^2}$ **25.** $\frac{x + 5}{x - 6}$ **27.** $\frac{1 - 2x}{x + 4}$ **29.** 1

5-3-1

1. $\frac{5}{7}$ **3.** $\frac{8}{x}$ **5.** $\frac{1-a}{y}$ **7.** $\frac{a-1}{x}$ **9.** $\frac{a+b+c}{x}$ **11.** $\frac{4-x}{x}$

5-3-2

1. xyz **3.** $x(x+2)$ **5.** $(a-3)(a+3)$ **7.** $(a-2)(a+2)(a-5)$ **9.** $(2x+3)(2x-3)(x+4)$
11. $(x+3)(x+2)(x-6)(x+6)$

5-3-3

1. $x-3$ **3.** $8x+12$ **5.** $2x^2-13x-6$ **7.** $x^2-3x-10$ **9.** $2x^2+28x+98$ **11.** $6x^2-5x-6$

5-4-1

1. $\frac{x+y}{xy}$ **3.** $\frac{x+11}{3(x+5)}$ **5.** $\frac{x-1}{(x+2)(x-3)}$ **7.** $\frac{3x-8}{(x+2)(x-5)}$ **9.** $\frac{5x+3}{(x+3)^2(x-3)}$ **11.** $\frac{3x^2-7}{(x+2)(x+1)(x-3)}$
13. $\frac{10x^2+3x-43}{(x+1)(x-3)(x+4)}$ **15.** $\frac{x^2+x-7}{(x+5)(x-2)}$ **17.** $\frac{3x^2-4x-3}{(x+3)^2(x-3)}$ **19.** $\frac{4}{x-2}$

5-5-1

1. $\frac{3x-20}{5x}$ **3.** $\frac{x-2}{3(x+1)}$ **5.** $\frac{12-2x}{x(x+3)}$ **7.** $\frac{2x^2+6x+12}{(x-3)(x+5)}$ **9.** $\frac{5x-18}{(x+2)(x-3)}$ **11.** The LCD is $(x+3)(x-2)$. $\frac{1}{x-2}$
13. $-\frac{3x^2-16x+17}{3(x+2)(x-2)}$ **15.** The LCD is $(x+1)(x-1)(x+5)$. $\frac{4}{(x+1)(x+5)}$ **17.** $\frac{x^2+13x-21}{(x+1)(x-5)(x-2)}$ **19.** $-\frac{1}{5}$

5-6-1

1. a **3.** $\frac{1}{xy}$ **5.** $\frac{1}{ab}$ **7.** $\frac{a+1}{a^2b}$ **9.** $\frac{1}{x}$ **11.** $\frac{a-b}{ab}$ **13.** y^2-x^2 **15.** $\frac{x+2}{x-3}$ **17.** $\frac{2a+b}{a^2+2a+ab}$ **19.** $\frac{xy-x^2}{y}$
21. $-\frac{1}{x+5}$ **23.** $\frac{x^2+5x+1}{x^2+1}$

5-7-1

1. 10 **3.** $-\frac{9}{8}$ **5.** No solution **7.** 4 **9.** -2 **11.** No solution **13.** 3 **15.** 0 **17.** -2 **19.** 5 **21.** 9
23. $\frac{3}{7}$ **25.** 18 cm **27.** 4 hr **29.** 20 hr

Chapter 5 Review

The number in brackets following each answer indicates the section that contains that type of problem.

1. $\frac{x+15}{4}$ [5-1] **3.** $\frac{x+3}{x+8}$ [5-1] **5.** $\frac{3}{5(x-1)}$ [5-1] **7.** $\frac{x-1}{x-4}$ [5-2] **9.** $\frac{2x+1}{2x-1}$ [5-2] **11.** $\frac{2}{3}$ [5-2]
13. $\frac{x+2}{x+4}$ [5-2] **15.** $\frac{x-1}{x-4}$ [5-2] **17.** $x(x+3)$ [5-3] **19.** $(x+1)(x+2)(x-2)$ [5-3] **21.** $\frac{5x+6}{x(x+2)}$ [5-4]
23. $\frac{x^2+2x+3}{x(x-5)(x+1)}$ [5-4] **25.** $\frac{2x-3}{(x-1)(x-2)}$ [5-4] **27.** $\frac{x-8}{(x+1)(x-2)}$ [5-5] **29.** $\frac{x^2+6x-8}{(x+2)(x-2)(x+6)}$ [5-5]
31. $\frac{1}{ab}$ [5-6] **33.** $\frac{x^2+2xy+y^2}{xy}$ [5-6] **35.** $-\frac{1}{x-7}$ [5-6] **37.** $\frac{7}{6}$ [5-7] **39.** $\frac{1}{3}$ [5-7] **41.** 20 ohms [5-7]

Chapter 5 Practice Test

The number in brackets following each answer indicates the section that contains that type of problem.

1. $\frac{x+3}{4}$ [5-1] **2.** $\frac{x+3}{x+5}$ [5-1] **3.** $\frac{x+5}{x-1}$ [5-2] **4.** $\frac{2}{3}$ [5-2] **5.** $x(x-7)$ [5-3] **6.** $(x+6)(x-6)(x+4)$ [5-3]
7. $\frac{5x+2}{x(x+1)}$ [5-4] **8.** $\frac{3x^2+11x-3}{(x+2)(x+4)}$ [5-5] **9.** $\frac{1}{a}$ [5-6] **10.** x^2-3x [5-6] **11.** $x=2$ [5-7] **12.** $1\frac{1}{3}$ min [5-7]

Chapter 6 Survey

The number in brackets following each answer indicates the section that contains that type of problem.

1. x^4y^8 [6–1] **2.** 2 [6–1] **3.** $\frac{x^2}{y^6}$ [6–2] **4.** $\frac{1}{\left(\frac{1}{a}+\frac{1}{b}\right)^2}$ or $\frac{a^2b^2}{(b+a)^2}$ [6–2] **5.** 3.5×10^6 [6–2] **6.** -4 [6–3] **7.** 9 [6–4]

8. x^3y^2 [6–5] **9.** $-2xy^3\sqrt[5]{2x^3y}$ [6–5] **10.** $\sqrt{15}+\sqrt{10}+\sqrt{6}$ [6–6] **11.** -163 [6–6] **12.** $\frac{3\sqrt[3]{x^2}}{x}$ [6–7]

13. $\frac{4(\sqrt{5}-\sqrt{2})}{3}$ [6–7]

6–1–1

1. x^8 **3.** x^{10} **5.** x^4y^3 **7.** x^3 **9.** $\frac{1}{x}$

6–1–2

1. x^{10} **3.** x^8y^4 **5.** $\frac{x^6}{y^9}$ **7.** $16x^8$ **9.** $\frac{8}{x^6y^3}$ **11.** $\frac{y^2}{x^2}$ **13.** $8x^6y^6$ **15.** $64x^{18}y^6$ **17.** $-3x^2$ **19.** $-\frac{y^3}{5x}$ **21.** $2x^8y^4$

23. $-4x^8y^9$ **25.** $288x^{22}y^{16}$ **27.** $-\frac{2}{x}$ **29.** $-\frac{1}{10xy^4}$ **31.** $120x^6y^5$

6–2–1

1. $\frac{1}{x^3}$ **3.** $\frac{1}{x^3y^5}$ **5.** $\left(\frac{b}{a}\right)^5$ **7.** $\frac{1}{ab}$ **9.** $\frac{1}{a}+\frac{1}{b}$ **11.** $\frac{1}{x^6}$ **13.** $\frac{1}{x^8}$ **15.** $\frac{1}{x^{15}}$ **17.** $\frac{1}{8x}$ **19.** $\frac{1}{8}$ **21.** $\frac{1}{y^7}$

23. $-\frac{9}{8x^{11}}$ **25.** $\frac{x^{25}}{y^{10}}$ **27.** $\frac{x+y}{x^2y^2}$ **29.** $\frac{x^2y^2}{(x+y)^2}$

6–2–2

1. Yes **3.** Yes **5.** Yes **7.** No **9.** Yes **11.** 5×10^3 **13.** 2.35×10^{-7} **15.** 5.2×10^{-9} **17.** 6.8×10^1
19. 7.28×10^5 **21.** 320,100 **23.** 623,000,000,000,000,000,000,000 **25.** 50,200,000,000 **27.** 537.62 **29.** 3.6
31. 4.9×10^7 miles **33.** 6,000,000,000,000 miles **35.** .00000001 cm **37.** 1×10^{-3} cm **39.** 4.3×10^{19}

6–3–1

1. 2 **3.** 3 **5.** -1 **7.** 10 **9.** 2 **11.** -2 **13.** 15 **15.** 12 **17.** 25 **19.** -1

6–3–2

1. 3.873 **3.** 2.881 **5.** 4.626 **7.** 16.371 **9.** 21.726 **11.** 32.787

6–4–1

1. $\sqrt[3]{x}$ **3.** $\sqrt[4]{a^3}$ **5.** $\frac{1}{\sqrt[3]{x^2}}$ **7.** $\sqrt[7]{(-3)^2}$ **9.** $\frac{1}{\sqrt[3]{(ab)^2}}$ **11.** $x^{1/2}$ **13.** $a^{3/4}$ **15.** $x^{-1/2}$ **17.** $(ab)^{2/3}$ **19.** $x^{4/7}$ **21.** 2

23. 9 **25.** 16 **27.** 4 **29.** 1 **31.** 9 **33.** 1 **35.** 125 **37.** -2 **39.** -21

6–5–1

1. $2\sqrt{2}$ **3.** $12\sqrt{3}$ **5.** $2\sqrt[3]{2}$ **7.** $x\sqrt{x}$ **9.** x^2y^3 **11.** $3x^2\sqrt{2x}$ **13.** $4y^3\sqrt{2x}$ **15.** $10x\sqrt[3]{2x^2}$ **17.** $2x^2\sqrt[5]{2x}$ **19.** $3x^2y\sqrt{2xy}$
21. $3x^2y\sqrt[4]{2xy}$ **23.** $-2x^2\sqrt[5]{2y^2}$ **25.** $3x^4y^2\sqrt[3]{3x^2y}$

6–6–1

1. $3\sqrt{3}$ **3.** $7\sqrt{5}-3\sqrt{3}$ **5.** $-3\sqrt[3]{2}$ **7.** $3\sqrt{2}-3$ **9.** $\sqrt{2}-\sqrt[3]{2}$ **11.** $2\sqrt{3}$

6–6–2

1. $\sqrt{10}$ **3.** $6\sqrt{14}$ **5.** $8a^2$ **7.** $x\sqrt[4]{x}$ **9.** $-40x^2\sqrt{10}$

6–6–3

1. $\sqrt{6}+\sqrt{10}$ **3.** $6\sqrt{15}-9\sqrt{21}$ **5.** $30\sqrt{2}+15\sqrt{5}$ **7.** $12\sqrt{2}-12+12\sqrt{6}$ **9.** $\sqrt{21}+\sqrt{10}$ **11.** $6\sqrt{3}-8\sqrt{5}-24$ **13.** $4\sqrt[3]{6}-12$

6–6–4

1. $\sqrt{6}+\sqrt{15}+2+\sqrt{10}$ **3.** $6\sqrt{15}+\sqrt{10}-15-2\sqrt{6}$ **5.** -2 **7.** -167 **9.** $6\sqrt{2}-3\sqrt{30}+2\sqrt{6}\sqrt[3]{2}-3\sqrt{10}\sqrt[3]{2}$ **11.** $47+6\sqrt{10}$ **13.** $187-20\sqrt{21}$

6–7–1

1. $\frac{\sqrt{2}}{2}$ **3.** $\frac{\sqrt{7}}{7}$ **5.** $\frac{\sqrt{3}}{6}$ **7.** $\frac{\sqrt[3]{4}}{2}$ **9.** $\frac{\sqrt[5]{4}}{2}$ **11.** $\frac{\sqrt[3]{x}}{x}$ **13.** $\frac{3\sqrt[3]{4x^2}}{2x}$ **15.** $\frac{\sqrt[3]{2xy^2}}{x}$ **17.** $\frac{\sqrt{2}}{2}$ **19.** $\frac{\sqrt{2}}{4}$ **21.** $\frac{\sqrt{x+3}}{x+3}$ **23.** $\frac{2\sqrt{2x-4}}{x-2}$ **25.** $\sqrt{x+y}$

6–7–2

1. $\sqrt{2}-1$ **3.** $\frac{\sqrt{5}-1}{2}$ **5.** $\frac{6\sqrt{x}-6y}{x-y^2}$ **7.** $\frac{\sqrt{3}+\sqrt{5}}{-2}$ **9.** $\sqrt{3}+\sqrt{2}$ **11.** $\sqrt{a}+1$ **13.** $\sqrt{7}+\sqrt{2}$ **15.** $\frac{12+7\sqrt{6}}{-10}$ **17.** $\frac{6x-5y\sqrt{x}+y^2}{9x-y^2}$

Chapter 6 Review

The number in brackets following each answer indicates the section that contains that type of problem.

1. $32x^{10}y^{15}$ [6–1] **3.** $\frac{8x^9}{27y^3}$ [6–1] **5.** $-\frac{1}{2x^4}$ [6–1] **7.** x^3+x [6–2] **9.** $\frac{x^7}{4y^4}$ [6–2] **11.** 5.42×10^7 [6–2] **13.** .000032 [6–2]

15. 1.18×10^9 watts [6–2] **17.** -5 [6–3] **19.** 13.229 [6–3] **21.** 9 [6–4] **23.** $\frac{4}{25}$ [6–4] **25.** 36 [6–4]

27. $2\sqrt[3]{4}$ [6–5] **29.** $5xy^3\sqrt{3x}$ [6–5] **31.** $6\sqrt{6}-7\sqrt{3}$ [6–6] **33.** $144x^4$ [6–6] **35.** $26-8\sqrt{3}$ [6–6] **37.** $\frac{3\sqrt[3]{2}}{2}$ [6–7] **39.** $\sqrt{x}+\sqrt{2}$ [6–7]

Chapter 6 Practice Test

The number in brackets following each answer indicates the section that contains that type of problem.

1. $x^{24}y^{36}$ [6–1] **2.** $\frac{1}{4x^7y}$ [6–1] **3.** x^4 [6–2] **4.** x^6+x^2 [6–2] **5.** $\frac{1}{a^{16}}$ [6–2] **6.** $7\sqrt{2}$ [6–5] **7.** $xy^2\sqrt[5]{xy^2}$ [6–5]

8. $-3b^2\sqrt[3]{ab^2}$ [6–5] **9.** $20\sqrt{2}$ [6–6] **10.** $28-\sqrt{14}$ [6–6] **11.** $\frac{\sqrt{5}}{5}$ [6–7] **12.** $\frac{5\sqrt{2}+5\sqrt{x}}{2-x}$ [6–7] **13.** $\frac{\sqrt[3]{4}}{2}$ [6–7]

14. $8+2\sqrt{15}$ [6–6] **15.** 5.61×10^{-6} [6–2] **16.** 380,000,000 [6–2] **17.** 4 [6–4] **18.** $\frac{5}{7}$ [6–4] **19.** -6 [6–3]

20. 25 [6–4]

Chapter 7 Survey

The number in brackets following each answer indicates the section that contains that type of problem.

1. $x=\frac{1}{3}, -4$ [7–1] **2.** $x=0, \frac{5}{3}$ [7–2] **3.** $x=\pm\sqrt{2}$ [7–2] **4.** $\frac{10\pm\sqrt{85}}{5}$ [7–3] **5.** $\frac{-4\pm\sqrt{31}}{3}$ [7–4]

6. -11, no real roots [7–5] **7.** $\frac{-1+13i}{5}$ [7–6] **8.** $\frac{\pm\sqrt{6}}{3}, \pm\sqrt{2}$ [7–7] **9.** -1 [7–8] **10.** 4 and 9 *or* -4 and -9 [7–9]

7-1-1

1. $x^2 - 3x + 2 = 0$ **3.** $6x^2 - 5x + 1 = 0$ **5.** $2x^2 - 5x + 1 = 0$ **7.** $5x^2 + 5x - 2 = 0$

7-1-2

1. $\{-2,-1\}$ **3.** $\{-7,-1\}$ **5.** $\{-4,2\}$ **7.** $\{3,6\}$ **9.** $\{-7,3\}$ **11.** $\{-3,1\}$ **13.** $\{-1\}$ **15.** $\{-4,0\}$ **17.** $\left\{0,\frac{1}{2}\right\}$ **19.** $\{-5,-2\}$ **21.** $\left\{-\frac{3}{2},-1\right\}$ **23.** $\left\{-\frac{3}{2},-\frac{2}{5}\right\}$ **25.** $\left\{-\frac{1}{3},\frac{5}{2}\right\}$ **27.** $\left\{-\frac{5}{2},\frac{3}{5}\right\}$ **29.** $\left\{-2,-\frac{5}{6}\right\}$

7-2-1

1. $\{-3,0\}$ **3.** $\left\{0,\frac{3}{2}\right\}$ **5.** $\{0,8\}$ **7.** $\left\{0,\frac{5}{3}\right\}$ **9.** $\{0,2\}$ **11.** $\{\pm 2\}$ **13.** $\{\pm\sqrt{5}\}$ **15.** $\{\pm 2\sqrt{5}\}$ **17.** $\{\pm\sqrt{14}\}$ **19.** $\{\pm 6\}$ **21.** $\{\pm 4\}$ **23.** $\{\pm\sqrt{2}\}$ **25.** $\{\pm 2\sqrt{2}\}$ **27.** $\{\pm\sqrt{3}\}$ **29.** No real solution

7-3-1

1. 12 **3.** 20 **5.** 4 **7.** 24

7-3-2

1. 16 **3.** 121 **5.** 144 **7.** $\frac{25}{4}$

7-3-3

1. $\{-5,1\}$ **3.** $\{-1,3\}$ **5.** No real solution **7.** $\left\{\frac{5 \pm \sqrt{37}}{2}\right\}$ **9.** No real solution **11.** $\left\{\frac{1}{2},2\right\}$ **13.** $\left\{\frac{10 \pm \sqrt{85}}{5}\right\}$ **15.** $\left\{\frac{1}{2}\right\}$

7-4-1

1. $\{-5,3\}$ **3.** $\left\{-\frac{3}{5},2\right\}$ **5.** $\left\{\frac{-3 \pm \sqrt{5}}{2}\right\}$ **7.** $\left\{\frac{1}{2},1\right\}$ **9.** $\left\{-\frac{1}{3},1\right\}$ **11.** $\left\{-1,-\frac{2}{3}\right\}$ **13.** $\left\{\frac{3 \pm \sqrt{3}}{2}\right\}$ **15.** $\{5\}$ **17.** $\left\{-\frac{2}{3}\right\}$ **19.** No real solution

7-5-1

1. 1, unequal, rational **3.** -11, no real roots **5.** 0, real, equal **7.** 72, unequal, irrational **9.** 0, real, equal **11.** 25, unequal, rational **13.** 0, real, equal **15.** 1, unequal, rational **17.** -3, no real roots **19.** 17, unequal, irrational

7-6-1

1. $4i$ **3.** $i\sqrt{30}$ **5.** $5i\sqrt{2}$ **7.** $7i$ **9.** $3i\sqrt{2}$

7-6-2

1. $7 + 7i$ **3.** 13 **5.** $8 + 3i$ **7.** $-10 + 11i$ **9.** $9 + 40i$ **11.** $\frac{-4 + 7i}{5}$ **13.** $\frac{2 + 12i}{37}$ **15.** $\frac{14 - 5i}{13}$

7-6-3

1. $\{2 \pm 2i\}$ **3.** $\left\{\frac{1 \pm i}{4}\right\}$ **5.** $\left\{\frac{1 \pm i\sqrt{3}}{2}\right\}$ **7.** $\left\{\frac{-1 \pm i\sqrt{11}}{6}\right\}$ **9.** $\left\{\frac{1 \pm i\sqrt{14}}{5}\right\}$

7–7–1

1. $\{\pm 1, \pm 2\}$ **3.** $\{\pm\sqrt{2}, \pm 2\}$ **5.** $\{\pm\sqrt{2}, \pm\sqrt{5}\}$ **7.** $\left\{\frac{\pm\sqrt{6}}{3}, \pm\sqrt{2}\right\}$ **9.** $\{\pm 3, \pm i\sqrt{2}\}$ **11.** $\{16,81\}$ **13.** $\{4,16\}$ **15.** $\{-27,1\}$

7–8–1

1. $\{4\}$ **3.** $\{5\}$ **5.** $\{7\}$ **7.** $\{0,4\}$ **9.** $\{5\}$ **11.** $\{1\}$ **13.** $\{-3\}$ **15.** $\{2\}$

7–9–1

1. $\{3,8\}$ **3.** $\{11,13\}$ **5.** $\{4,5,6\}$ **7.** {6 cm,8 cm} **9.** {5 and 9 *or* −9 and −5} **11.** $\{9\}$ **13.** $\left\{5 \text{ cm}, \frac{13}{3} \text{ cm}\right\}$ **15.** $\{5,10\}$
17. {3 sec} **19.** {2 km/hr} **21.** Pump A = 12 hours, pump B = 4 hours

Chapter 7 Review

The number in brackets following each answer indicates the section that contains that type of problem.

1. $\{-7,-3\}$ [7–1] **3.** $\{3,8\}$ [7–1] **5.** $\left\{-\frac{3}{2}, \frac{1}{3}\right\}$ [7–1] **7.** $\{\pm\sqrt{17}\}$ [7–2] **9.** $\{\pm 4i\}$ [7–2] **11.** $\{-5,-3\}$ [7–3]
13. $\left\{\frac{-3 \pm \sqrt{13}}{2}\right\}$ [7–3] **15.** $\left\{\frac{-5 \pm i\sqrt{7}}{4}\right\}$ [7–3] **17.** $\{1,3\}$ [7–4] **19.** $\left\{\frac{-4 \pm \sqrt{2}}{2}\right\}$ [7–4]
21. 17, irrational, unequal [7–5] **23.** 0, real, equal [7–5] **25.** 1, rational, unequal [7–5] **27.** $3 + 3i$ [7–6] **29.** $3 - i$ [7–6]
31. $\left\{-\frac{1}{2}, \frac{4}{3}\right\}$ [7–1] **33.** $\left\{0, \frac{2}{3}\right\}$ [7–2] **35.** $\left\{\frac{3 \pm i\sqrt{31}}{4}\right\}$ [7–6] **37.** $\{\pm\sqrt{2}, \pm i\sqrt{5}\}$ [7–7] **39.** $\{16,25\}$ [7–7]
41. $\{9\}$ [7–8] **43.** $\{8\}$ [7–8] **45.** $\{-6\}$ [7–8] **47.** {−7 and −5 *or* 5 and 7} [7–9] **49.** $\left\{\frac{2}{5}, 5\right\}$ [7–9]

Chapter 7 Practice Test

The number in brackets following each answer indicates the section that contains that type of problem.

1. 17, c [7–5]; 0, b [7–5]; −7, a [7–5] **2.** 14 [7–3] **3.** $c = 25$ [7–3] **4.** $\{\pm\sqrt{10}\}$ [7–2] **5.** $\{\pm 2\}$ [7–2] **6.** $\{0,7\}$ [7–1]
7. $\left\{\frac{3 \pm i\sqrt{7}}{2}\right\}$ [7–5] **8.** $\{-3,1\}$ [7–1] **9.** $\{2 \pm \sqrt{3}\}$ [7–4] **10.** $\left\{-3, \frac{1}{2}\right\}$ [7–1] **11.** $\{\pm\sqrt{5}, \pm i\sqrt{2}\}$ [7–7] **12.** $\{18\}$ [7–8]
13. $\{1\}$ [7–8] **14.** {7 m} [7–9]

Chapter 8 Survey

The number in brackets following each answer indicates the section that contains that type of problem.

1. a. Yes, all real numbers $x > 3$ [8–1] **b.** No [8–1]

2.

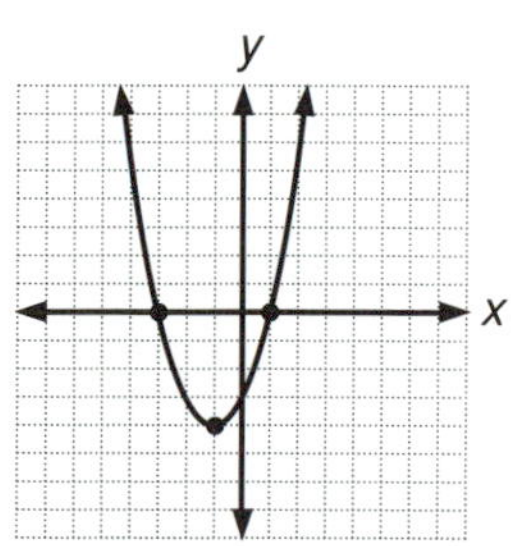

[8–2]

3.

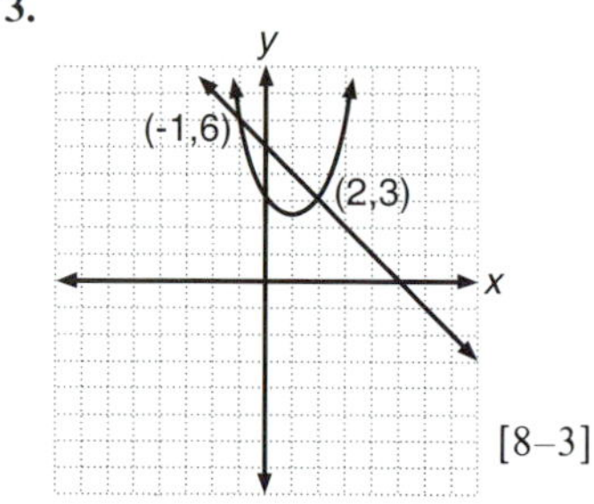

[8–3]

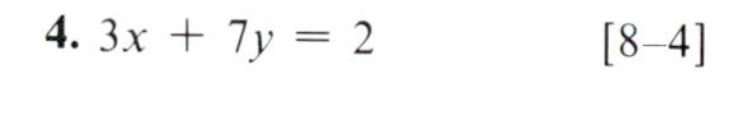

4. $3x + 7y = 2$ [8–4]

5.

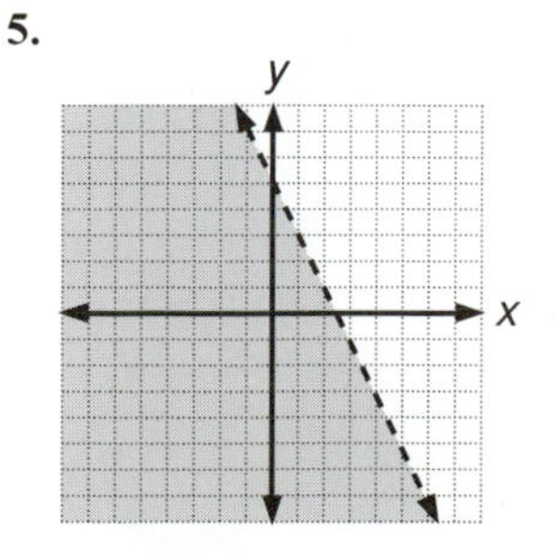

[8–5]

6.

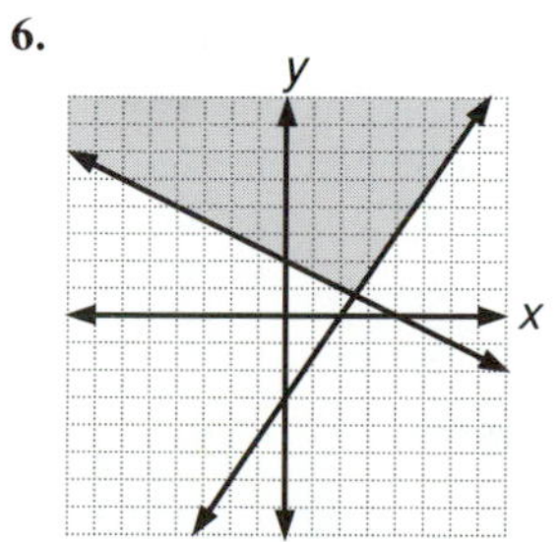

[8–6]

7. $(-3,4)$ [8–7] **8.** $(-2,1,0)$ [8–8]

8-1-1

1. Yes, all real numbers **3.** Yes, all real numbers **5.** Yes, all real numbers except 5 **7.** No
9. Yes, all real numbers except -2 and 2

8-2-1

1.

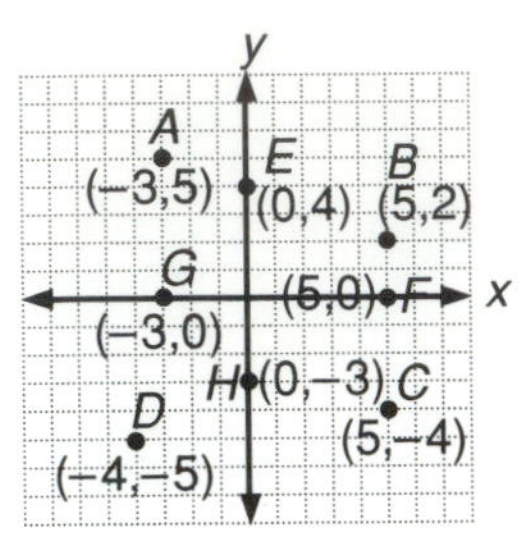

A: $(-3,5)$ *E:* $(0,4)$ *B:* $(5,2)$ *F:* $(5,0)$ *G:* $(-3,0)$ *H:* $(0,-3)$ *D:* $(-4,-5)$ *C:* $(5,-4)$

8-2-2

1. Function

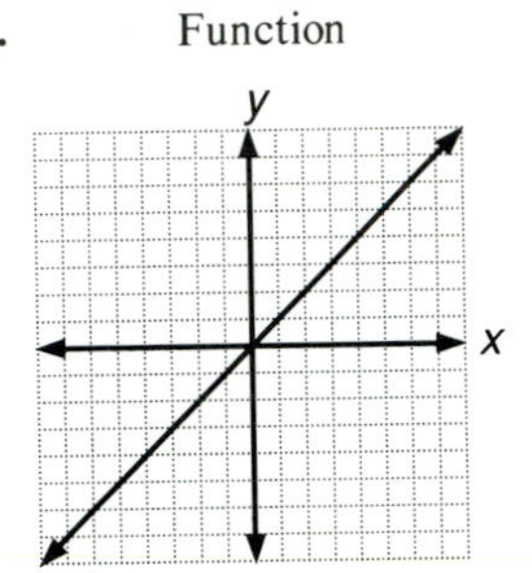

3. Function

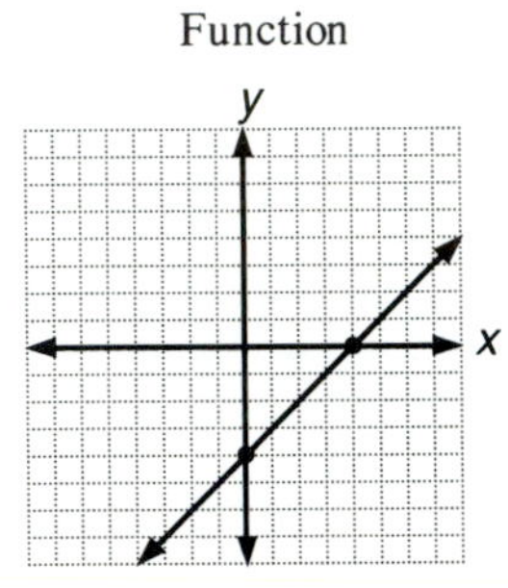

5. Function

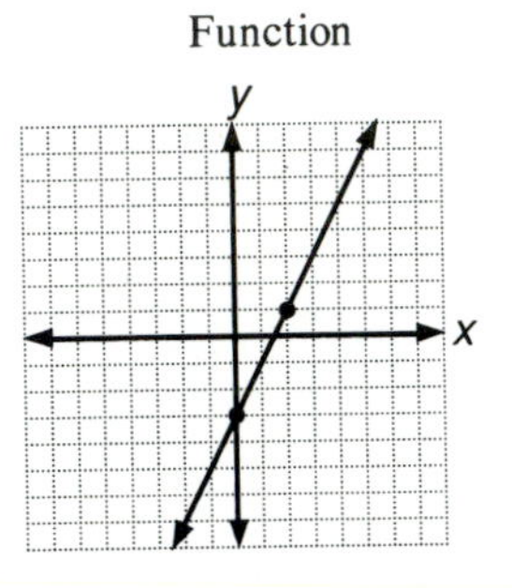

7. Function

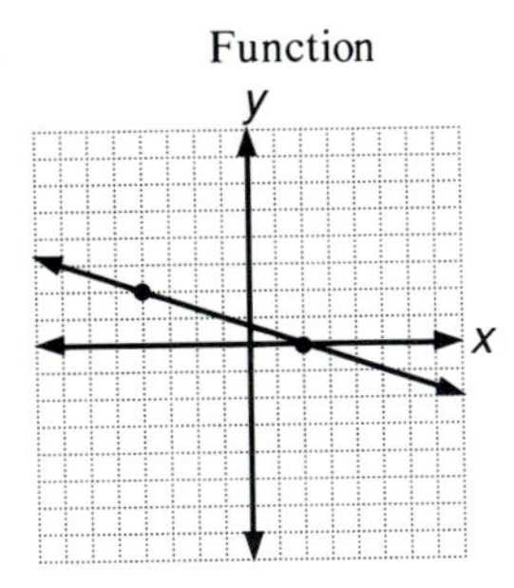

9. Function

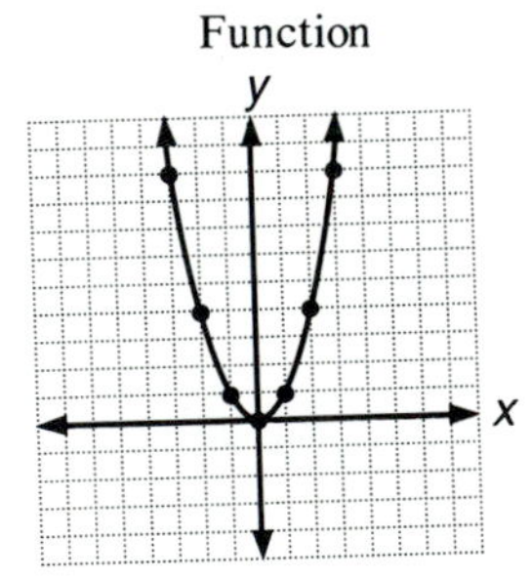

11. Function

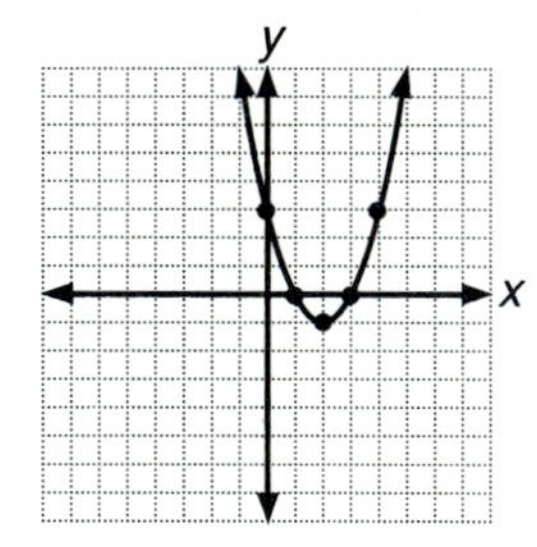

13. Not a function

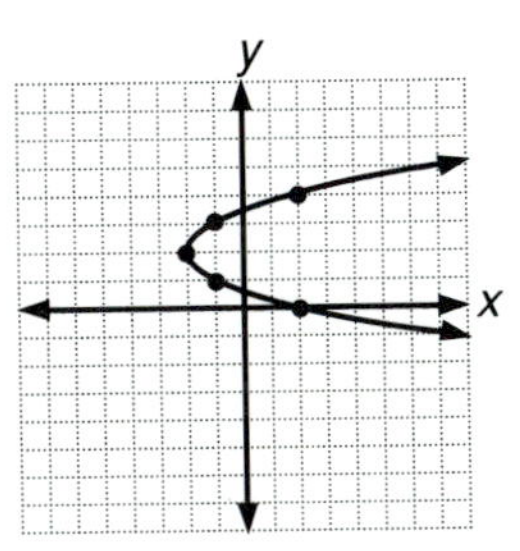

15. Function

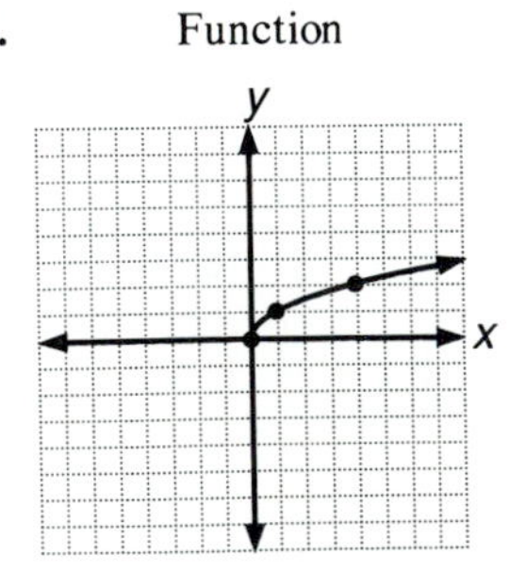

17. Any line parallel to the y-axis will intersect the graph of a function no more than once.

8-3-1

1.

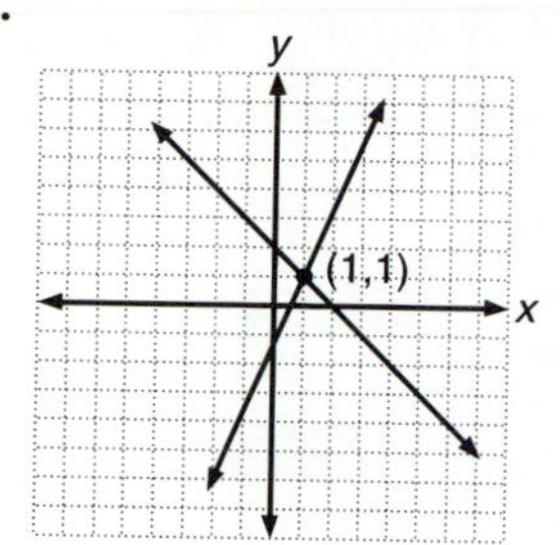

3.

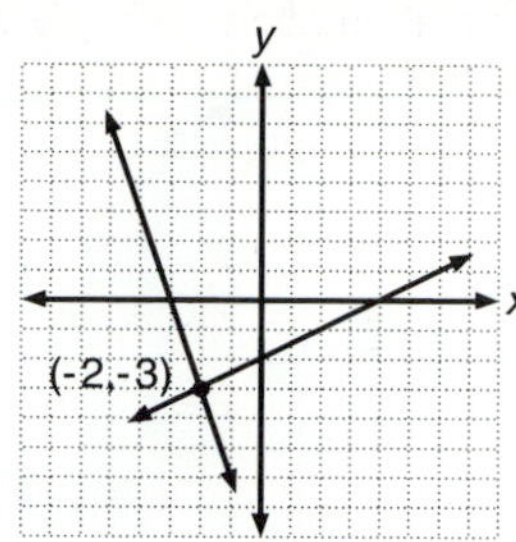

5.

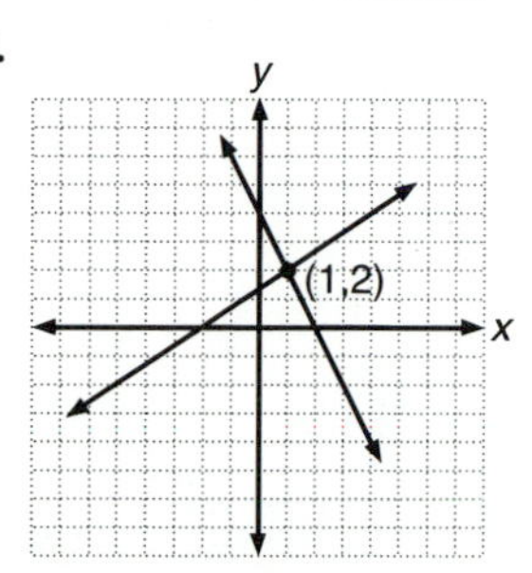

7.

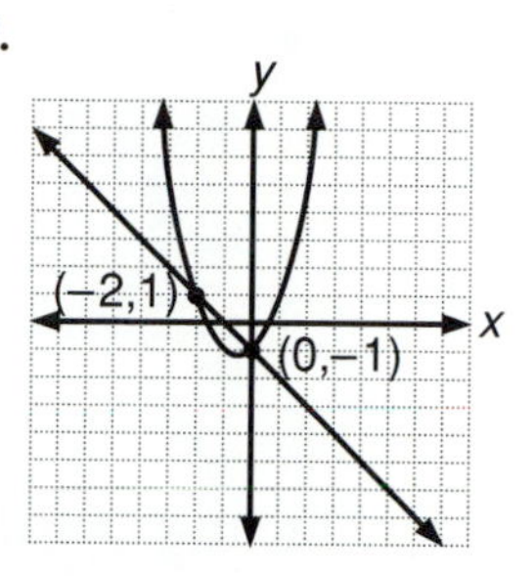

9. One point, no points, infinitely many points

8-4-1

1. $8x + 5y = 21$ **3.** $5x + 12y = 20$ **5.** $4x + 11y = -43$

7.

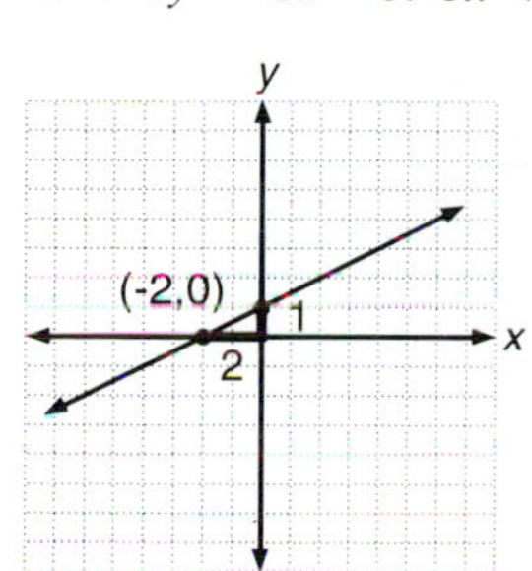

$x - 2y = -2$

9.

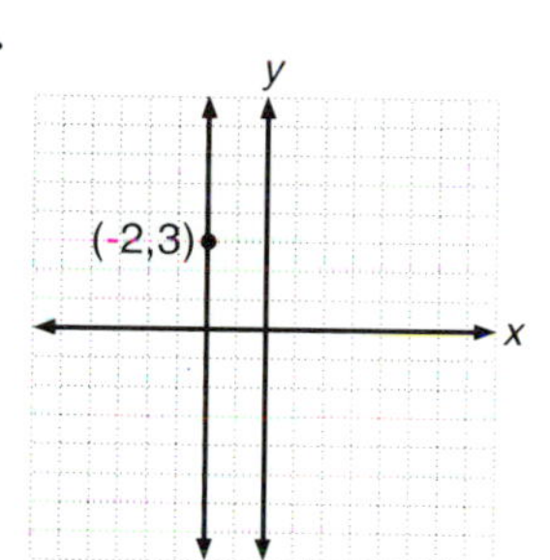

$x = -2$

11. $y = \frac{1}{4}x - 5$; $m = \frac{1}{4}$; $b = -5$ **13.** $y = \frac{2}{5}x - 3$; $m = \frac{2}{5}$; $b = -3$

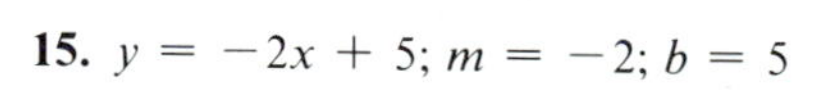
15. $y = -2x + 5$; $m = -2$; $b = 5$

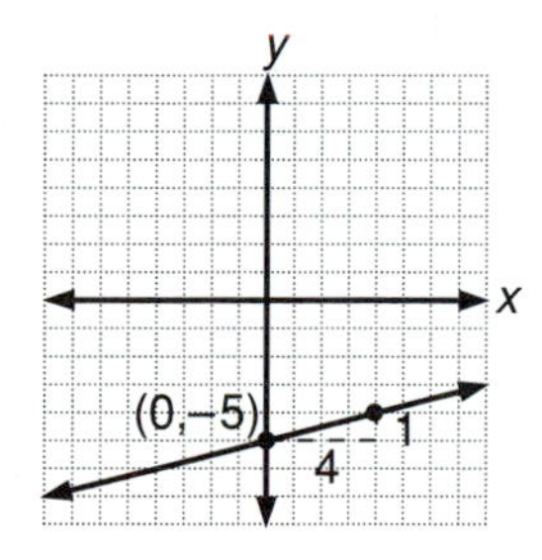

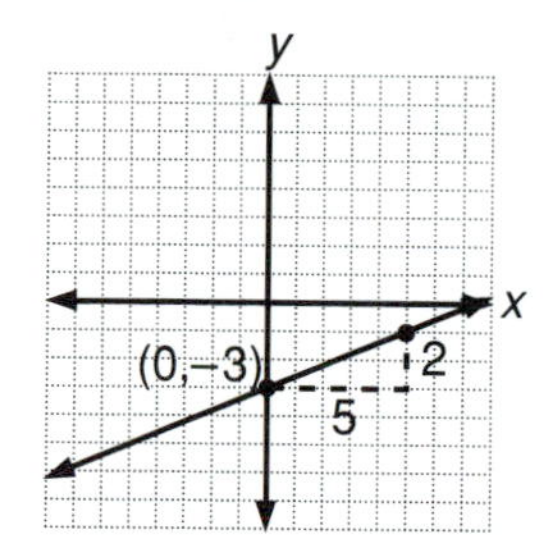

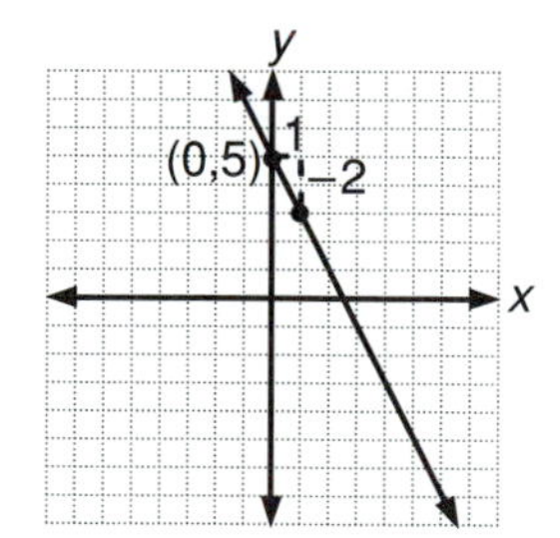

17. $2x - y = 5$ **19.** $y = 1$ **21.** $3x + 2y = 13$ **23.** $5x - y = -9$

8–5–1

1.

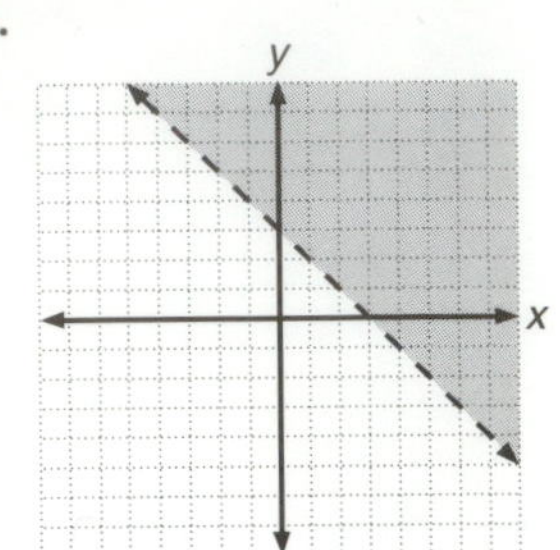

3.

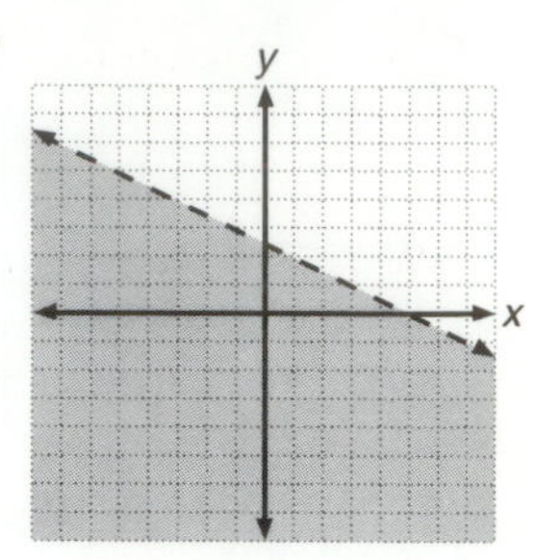

5.

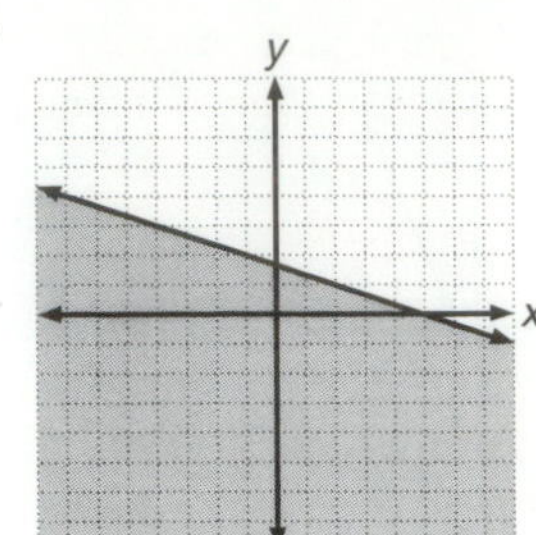

7.

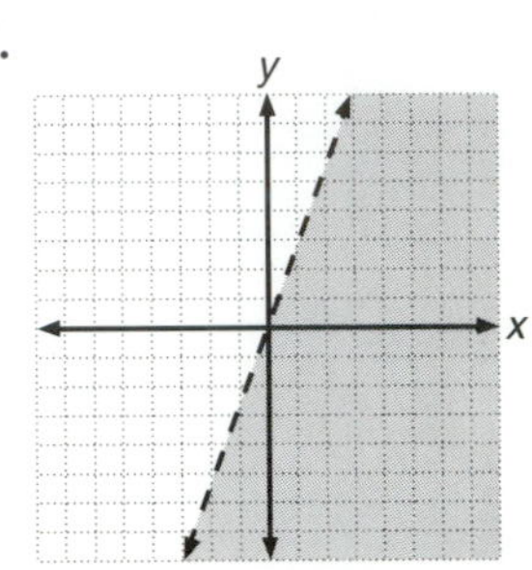

9.

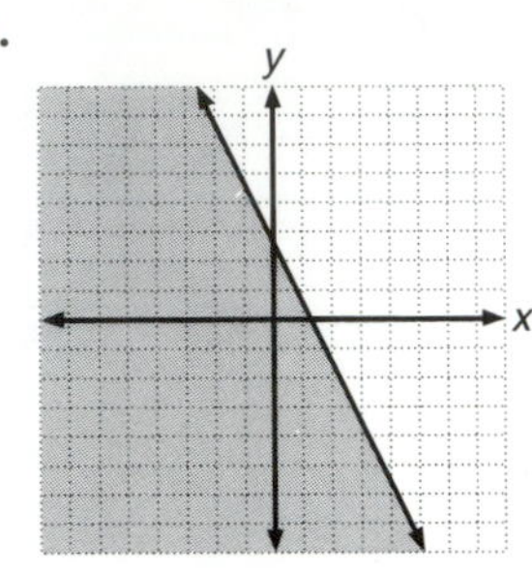

8–6–1

1.

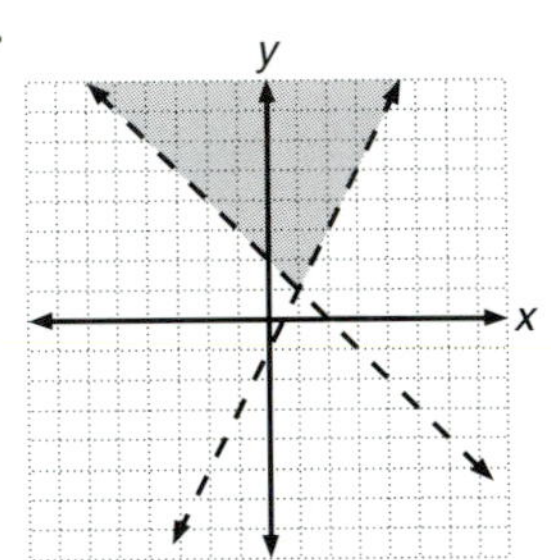

3.

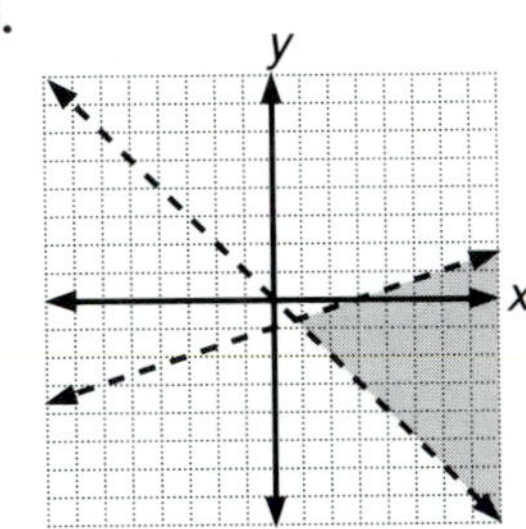

5.

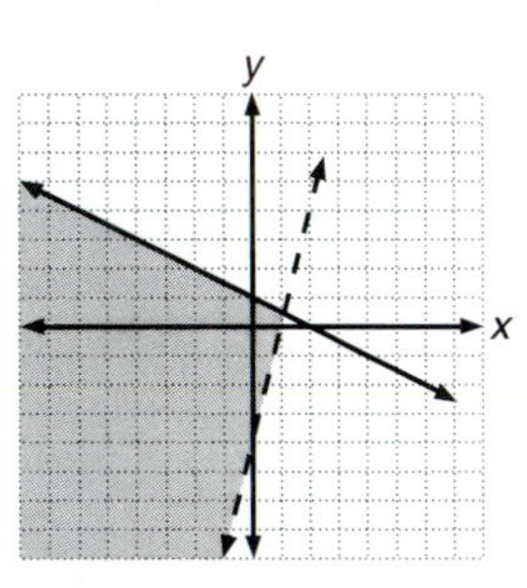

7.

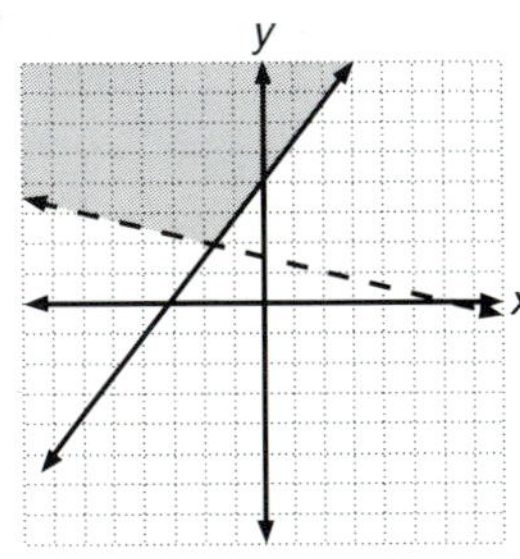

9.

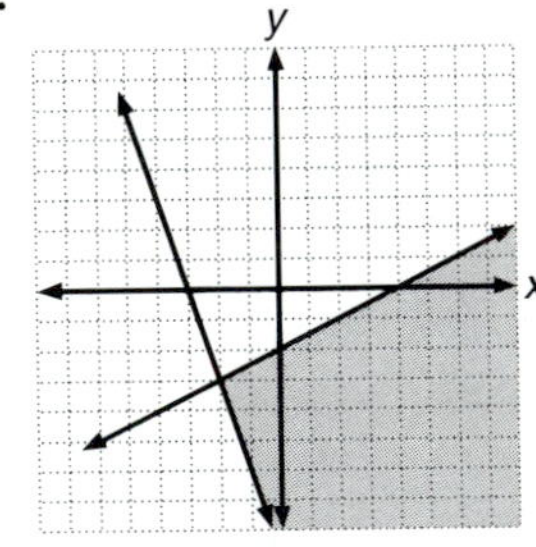

8–7–1

1. (2,1) **3.** (1,−1) **5.** $\left(\frac{1}{2},4\right)$ **7.** $\left(8,3\frac{1}{2}\right)$ **9.** (3,−10) **11.** 27 men, 21 women

8–7–2

1. (3,−1) **3.** (−1,4) **5.** (2,1) **7.** (2,−8) **9.** (−2,5) **11.** Golden Gate Bridge = 4,200 ft; Brooklyn Bridge = 1,595 ft **13.** 64, 83 **15.** 39, 46 **17.** Rate of boat: 9 km/hr, rate of current: 3 km/hr **19.** 16 nickels, 18 dimes

8–7–3

1. (2,−3) **3.** (1,4) **5.** (0,−2) **7.** Dependent **9.** (−3,7) **11.** $\left(\frac{1}{2},3\right)$ **13.** Dependent **15.** (3,−8)

8-8-1

1. (1,2,3) **3.** (−2,1,3) **5.** (2,1,−3) **7.** (−2,1,0) **9.** 5 nickels, 10 dimes, 13 quarters
11. 2 liters 5%, 3 liters 20%, 4 liters 50%

Chapter 8 Review

The number in brackets following each answer indicates the section that contains that type of problem.

1. Yes, all real numbers [8–1] **3.** No [8–1] **5.** Yes, all nonnegative real numbers except $x = 9$ [8–1]

7.

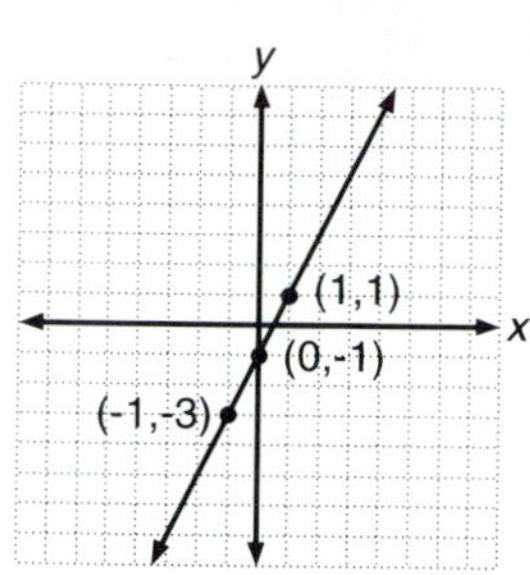

[8–2]

9.

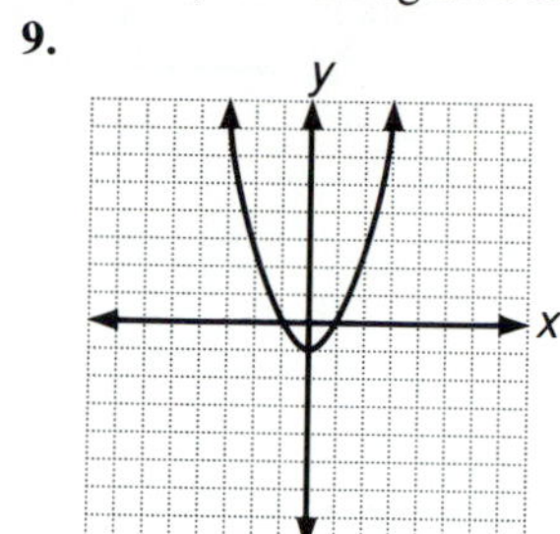

[8–2]

11.

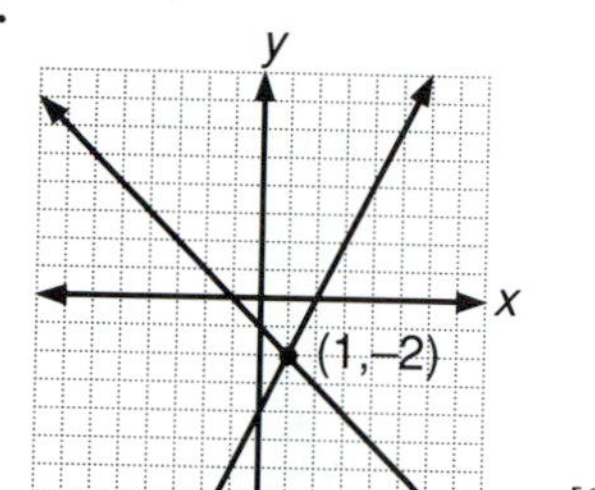

[8–3]

13.

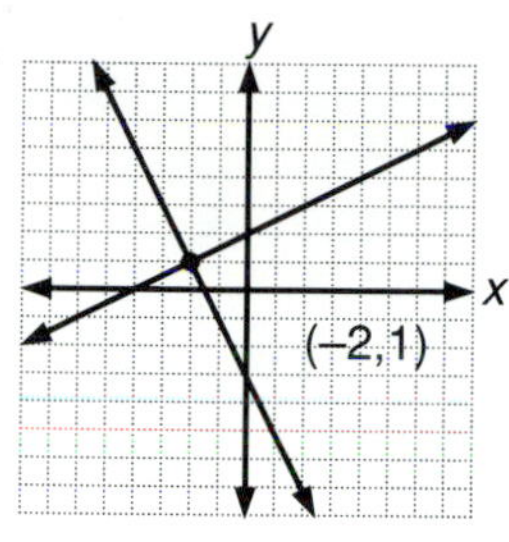

[8–3]

15.

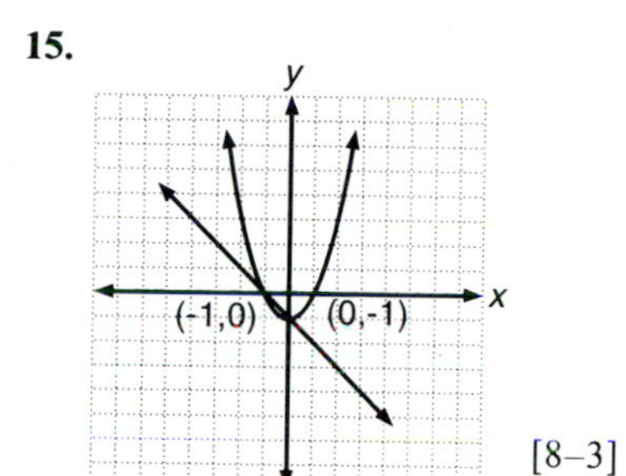

[8–3]

17.

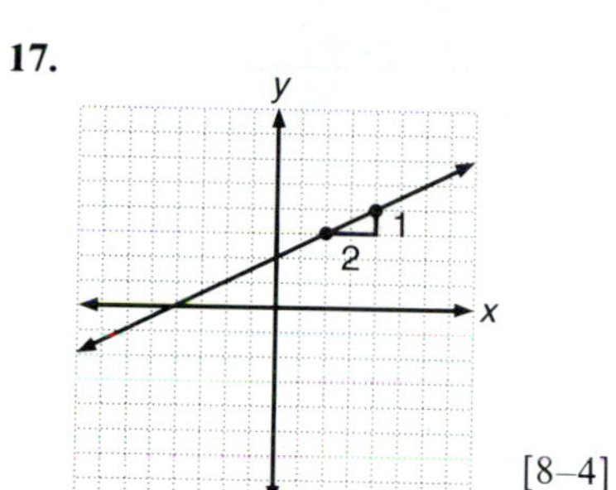

[8–4]

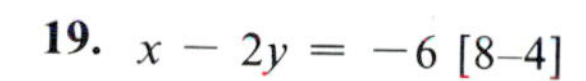

19. $x - 2y = -6$ [8–4]

21.

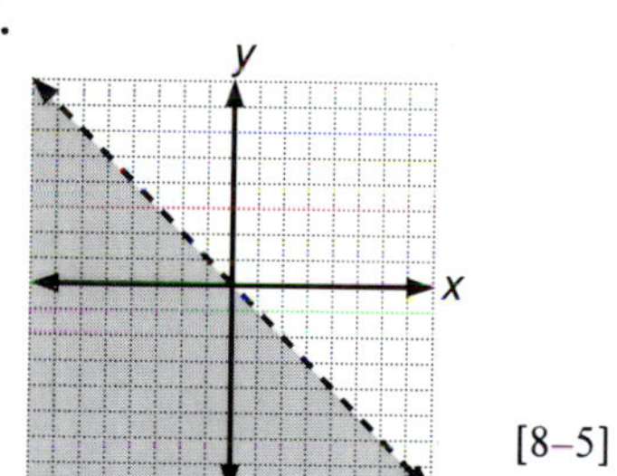

[8–5]

23.

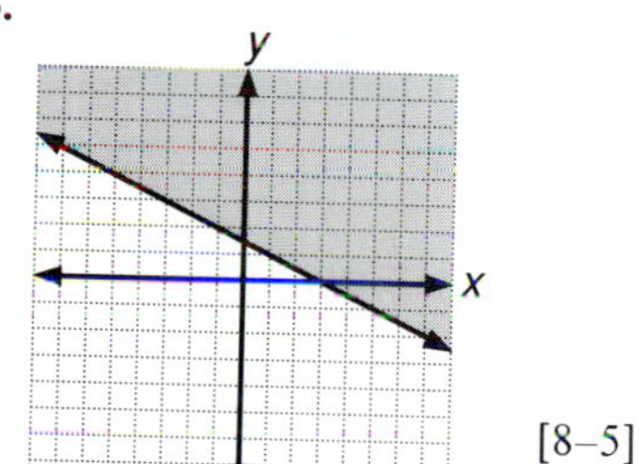

[8–5]

25.

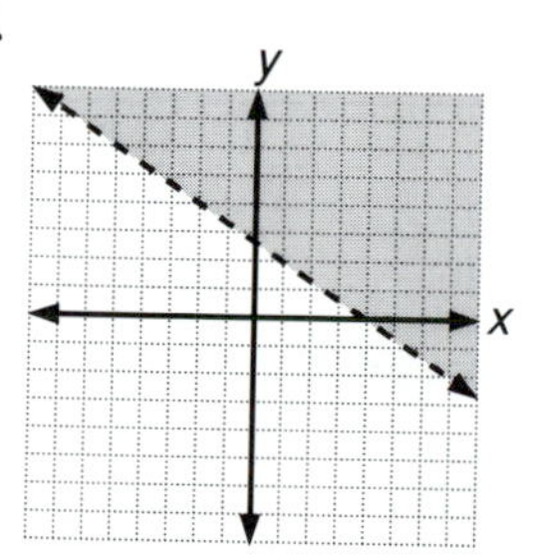

[8–5]

27.

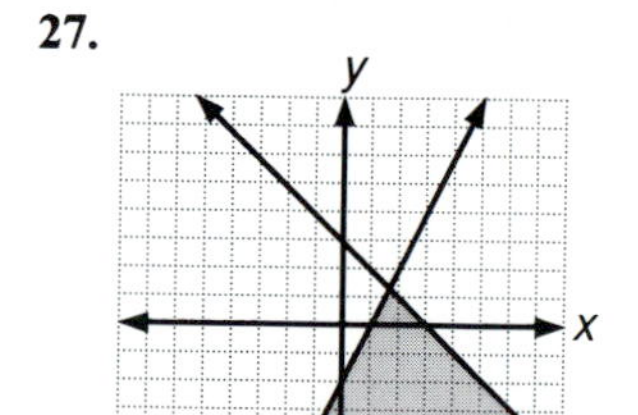

[8–6]

29.

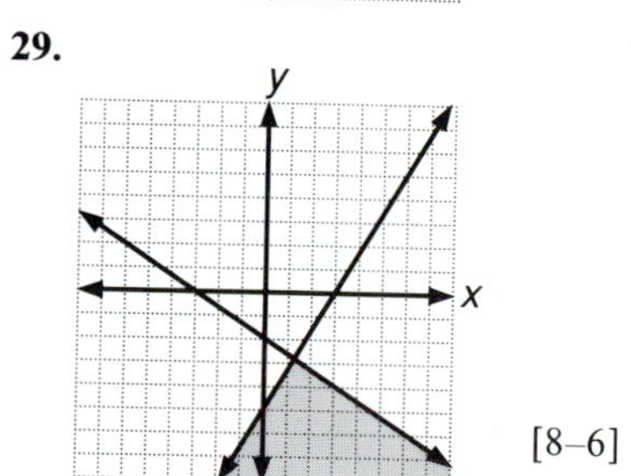

[8–6]

31. (2,−1) [8–7] **33.** (0,−5) [8–7] **35.** Speed of the boat = 19 km/hr, speed of the current = 4 km/hr [8–7]
37. (1,−1,3) [8–8] **39.** (0,−2,5) [8–8]

Chapter 8 Practice Test

The number in brackets following each answer indicates the section that contains that type of problem.

1. Yes, all real numbers $x > 1$ [8–1]

2. a.

x	-2	-1	0	1	2
y	3	0	-1	0	3

[8–2]

b.

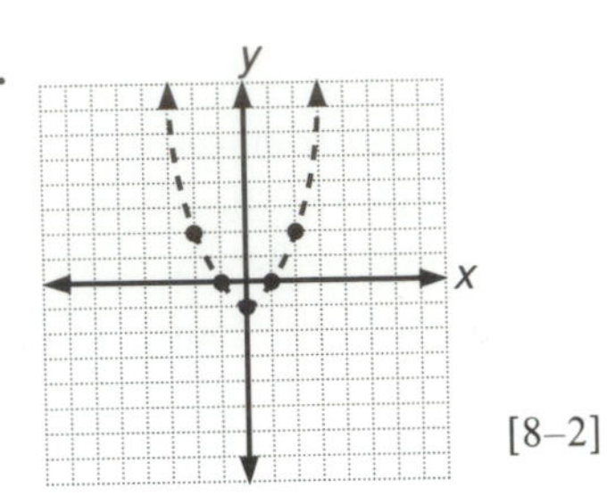

[8–2]

3.

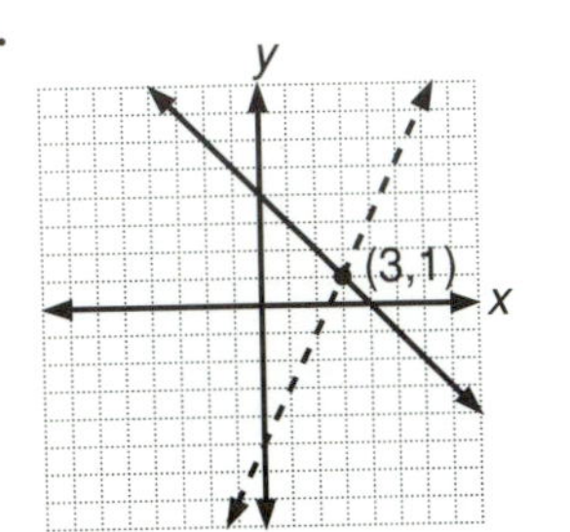

[8–3]

4. $3x - 5y = 10$ [8–4]

5.

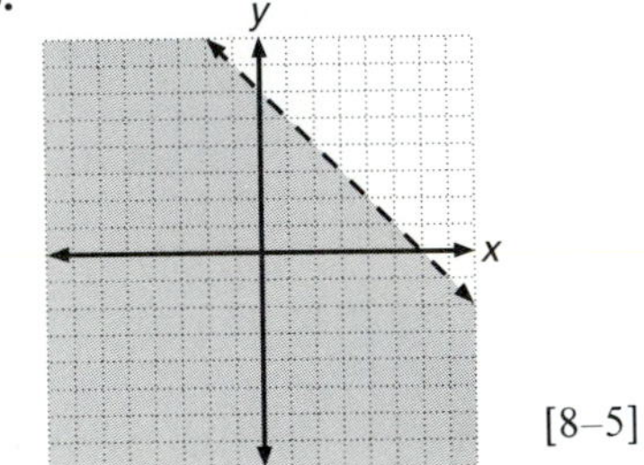

[8–5]

6.

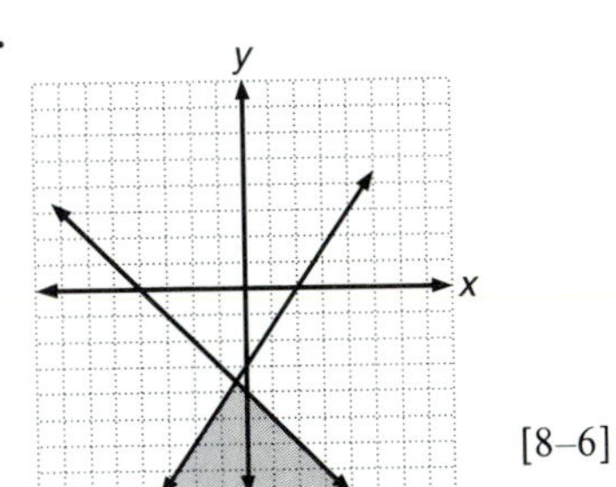

[8–6]

7. $(3,-4)$ [8–7] **8.** $(5,-4)$ [8–7] **9.** $(2,-1,1)$ [8–8] **10.** First number $= 1$, second number $= -7$, third number $= 6$ [8–8]

End-of-Book Test

The number in brackets refers to the chapter and section that discusses that type of problem.

1. -2 [1–2] **2.** $-8a^2$ [1–4] **3.** $4a^5b^4 - 6a^2b^4$ [1–6] **4.** $17x^2 - 4x$ [1–7] **5.** 5 [2–3] **6.** 4 [2–3] **7.** 4 [2–3]

8. $-\dfrac{9a}{4}$ [2–4] **9.** $x < 3$ [2–5] **10.** $-4 < x < 10$ [2–6] **11.** $3ab^2(3a - 2b - 4a^2)$ [4–1] **12.** $7(x - 1)(x + 1)$ [4–6]

13. $(2x - 7)^2$ [4–5] **14.** $4(2x - 1)(x + 3)$ [4–6] **15.** $\dfrac{x + 7}{x + 5}$ [5–2] **16.** $\dfrac{3}{x - 4}$ [5–2] **17.** $\dfrac{9x + 15}{x(x + 3)}$ [5–4]

18. $\dfrac{2x^2 + 5x - 2}{(x + 5)(x - 1)}$ [5–5] **19.** $x^2 - 3x$ [5–6] **20.** -2 [5–7] **21.** $\dfrac{1}{64}$ [6–4] **22.** $-27x^3y^{12}$ [6–1] **23.** $-72x^7y^{11}$ [6–1]

24. $\dfrac{25x^8}{9y^4}$ [6–2] **25.** $\dfrac{\sqrt{x - 2}}{x - 2}$ [6–7] **26.** $\dfrac{\sqrt{x} - 2}{x - 4}$ [6–7] **27.** $\left\{\pm\dfrac{2\sqrt{3}}{3}\right\}$ [7–2] **28.** $\left\{-\dfrac{1}{2},5\right\}$ [7–1] **29.** $\left\{\dfrac{7}{2}\right\}$ [7–1]

30. $\{2\}$ [7–8] **31.** $(-3,4)$ [8–7] **32.** $(3,6)$ [8–7] **33.** $5x - 4y = 35$ [8–4]

34. 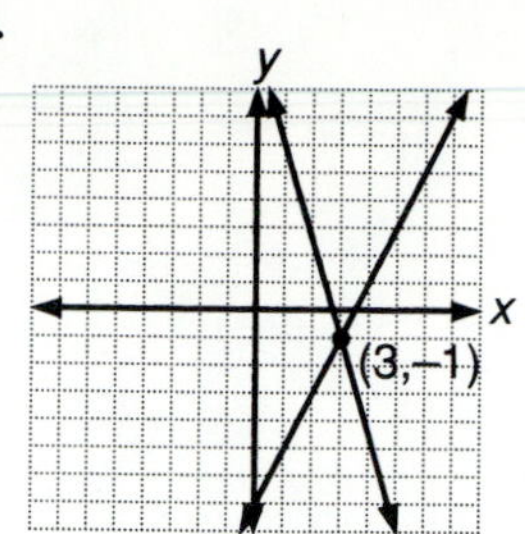

[8–3]

35. 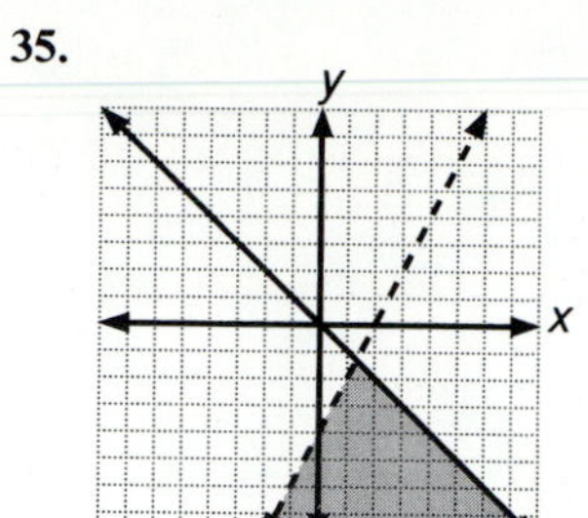

[8–6]

36. (2,1,−2) [8–8] **37.** Salesperson A: \$906; salesperson B: \$883; salesperson C: \$1,766 [3–3] **38.** $17\frac{1}{2}$ hours [3–4] **39.** 40 lb of \$2.50, 60 lb of \$2.00 [3–5] **40.** 60 m by 40 m [7–9]

Index

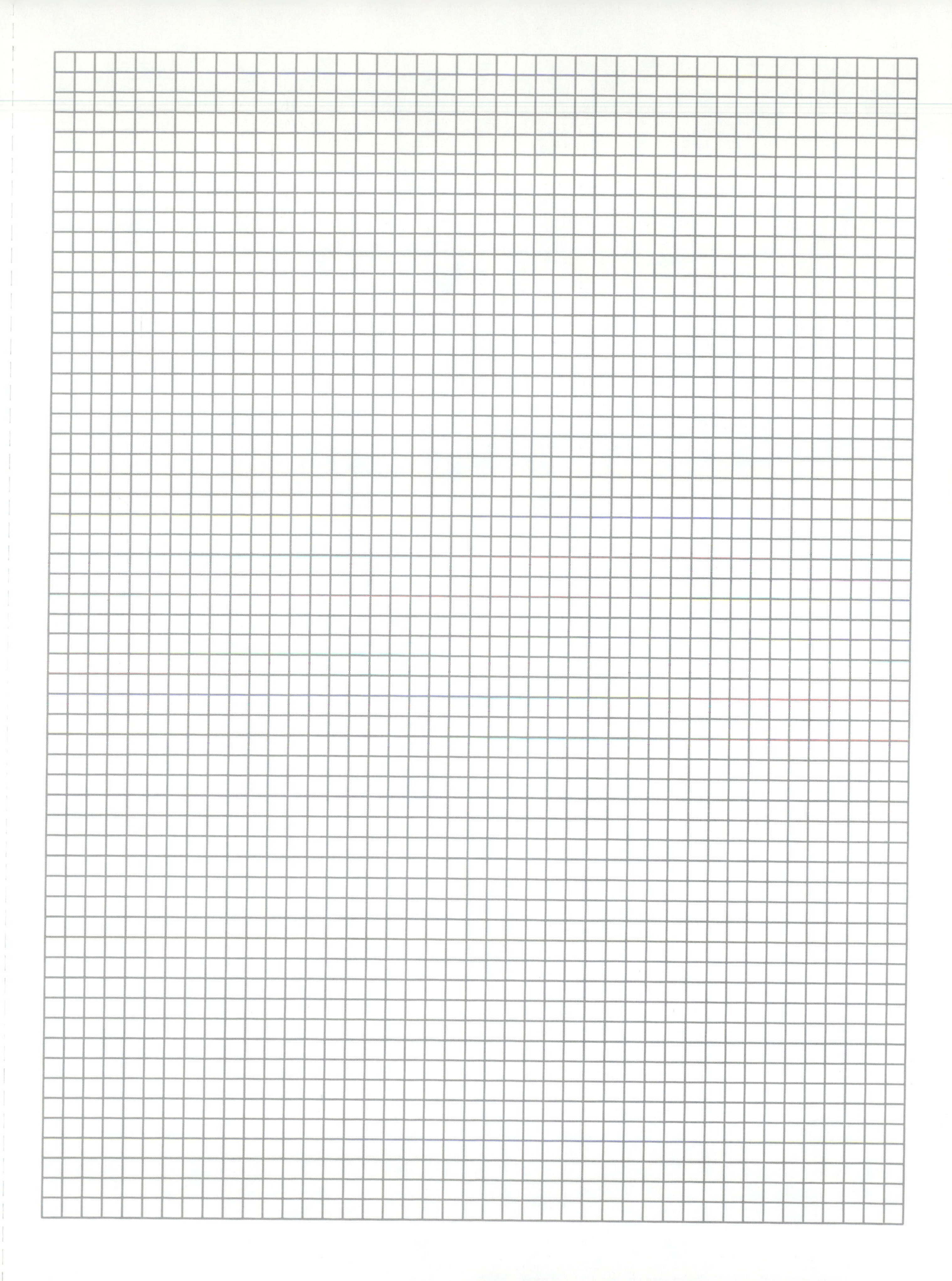

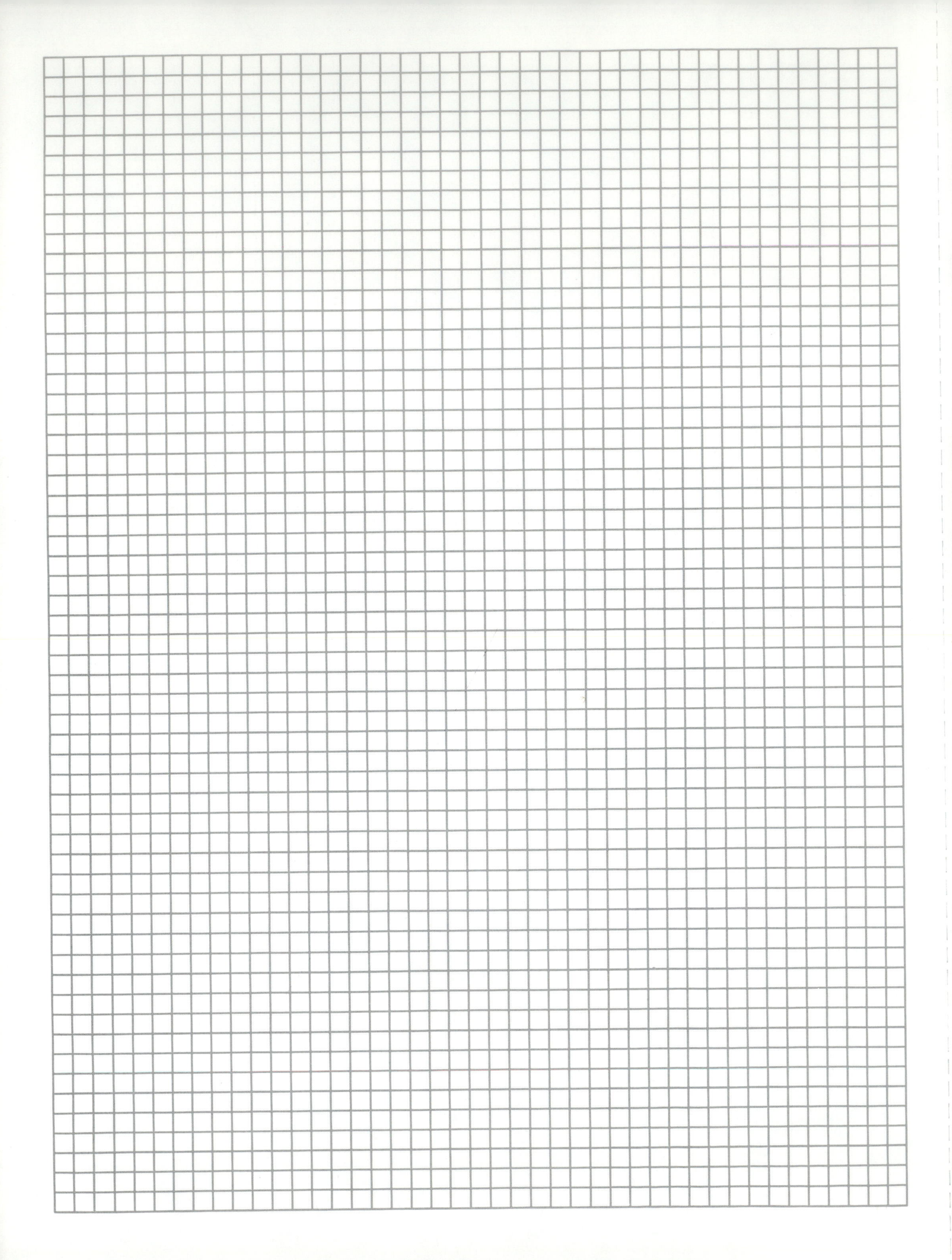

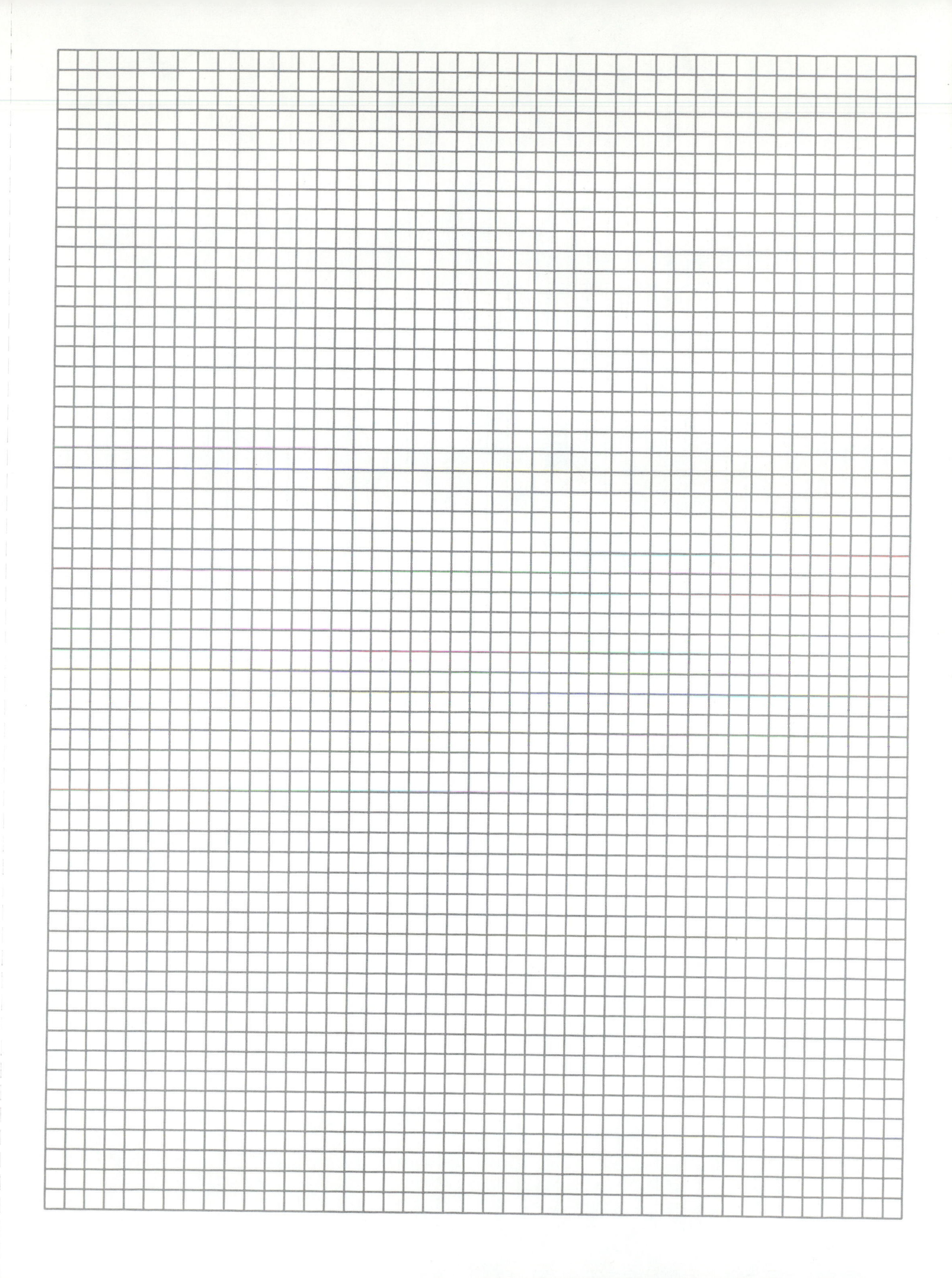

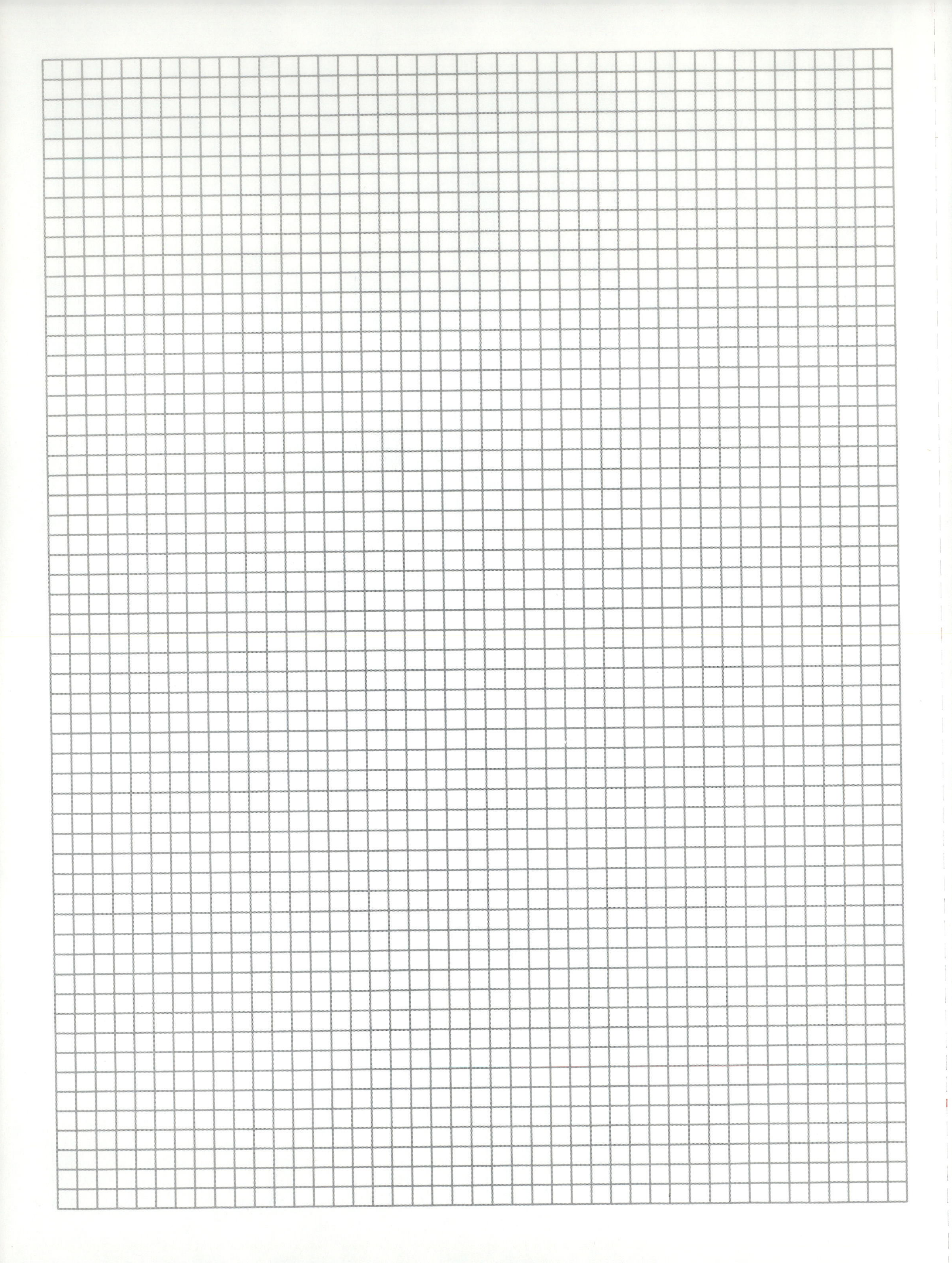

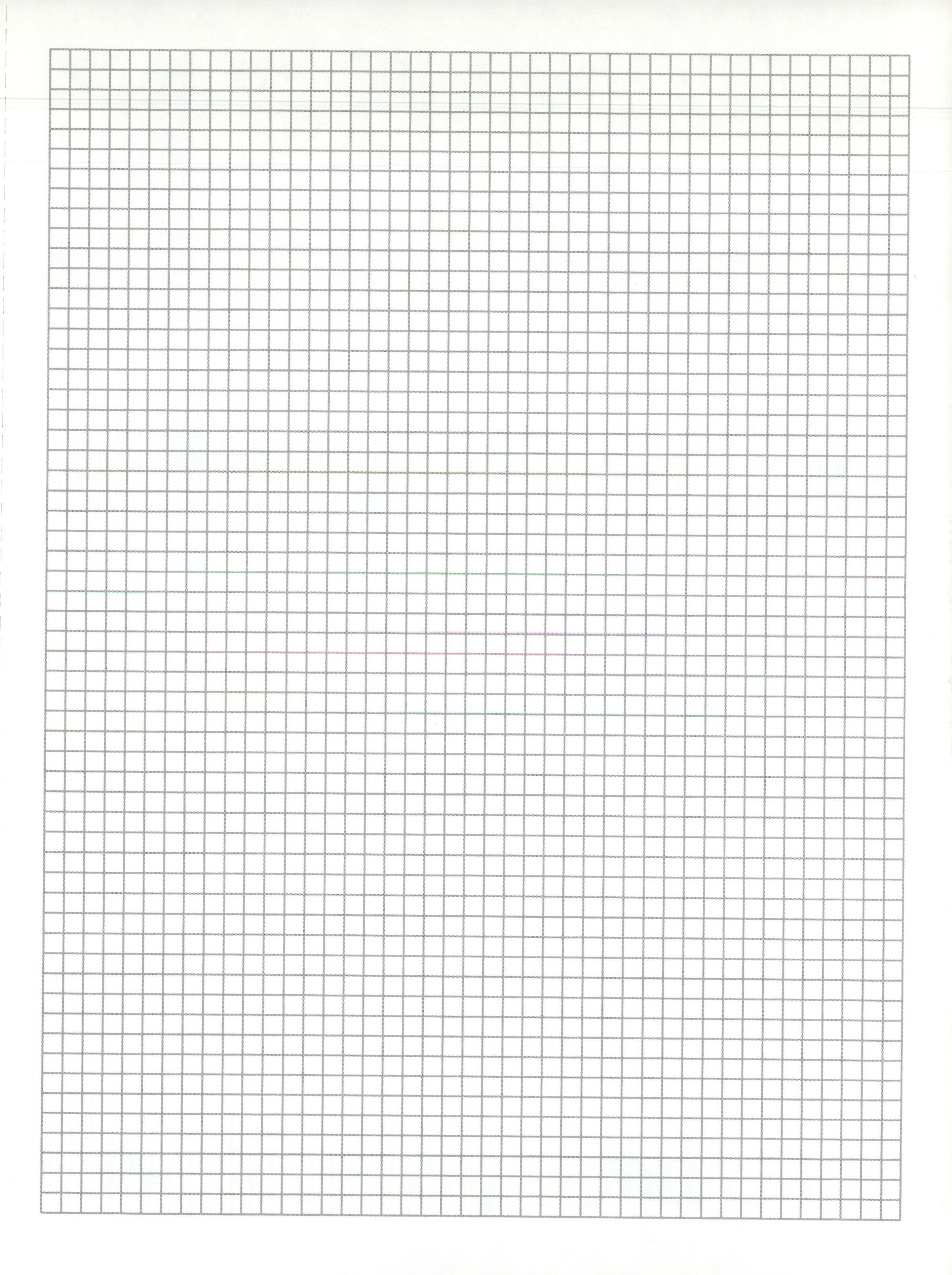

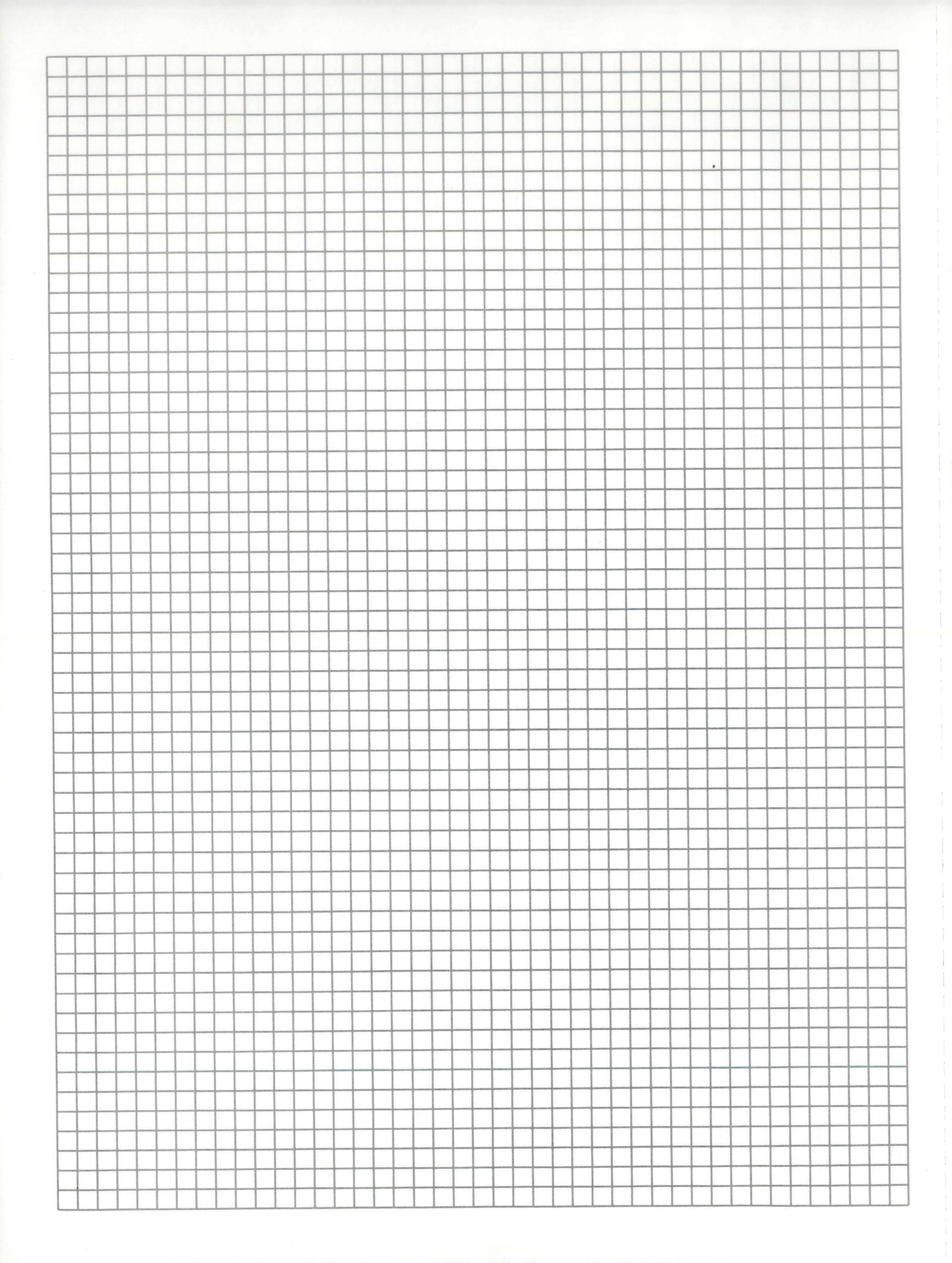

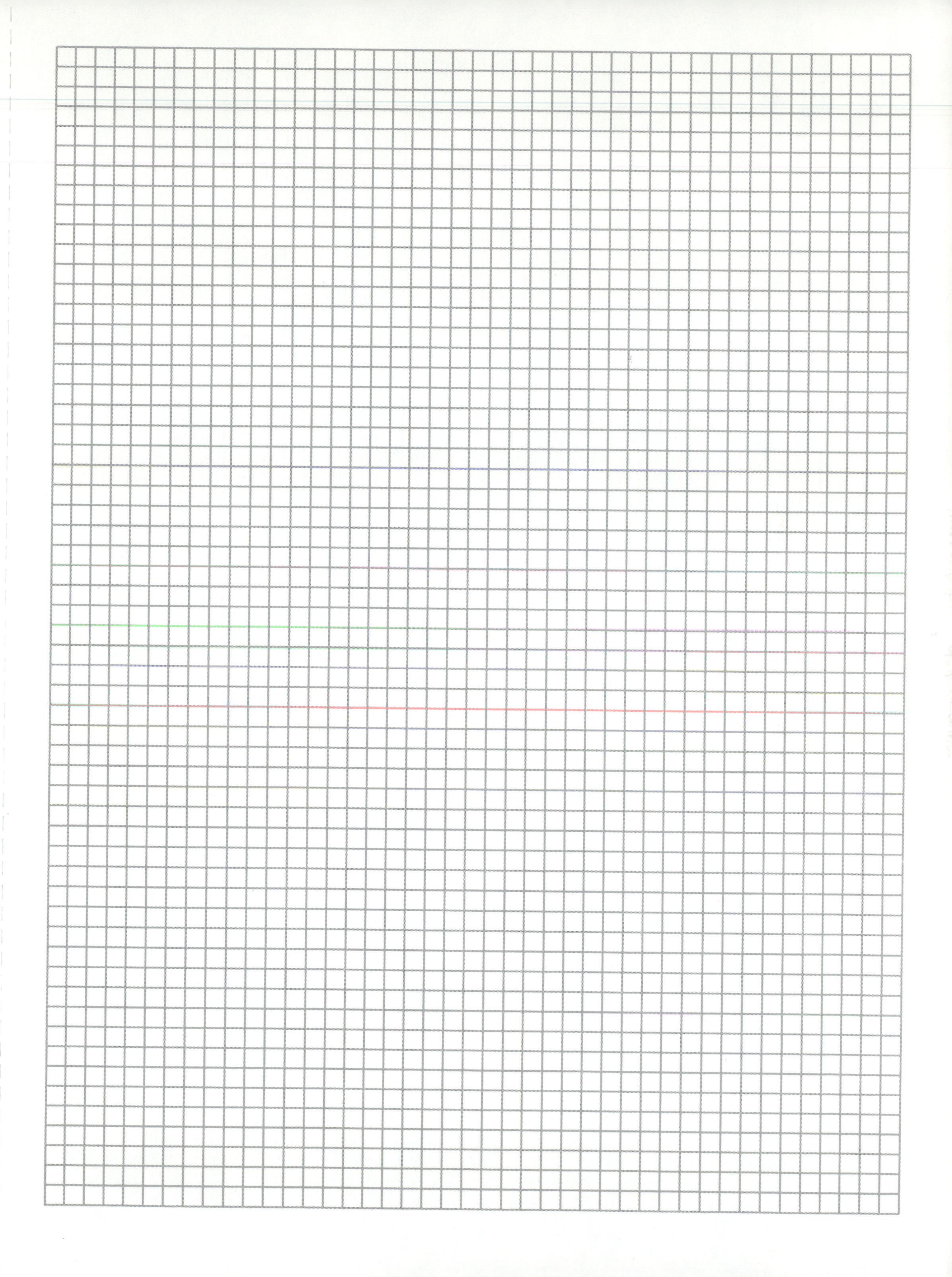

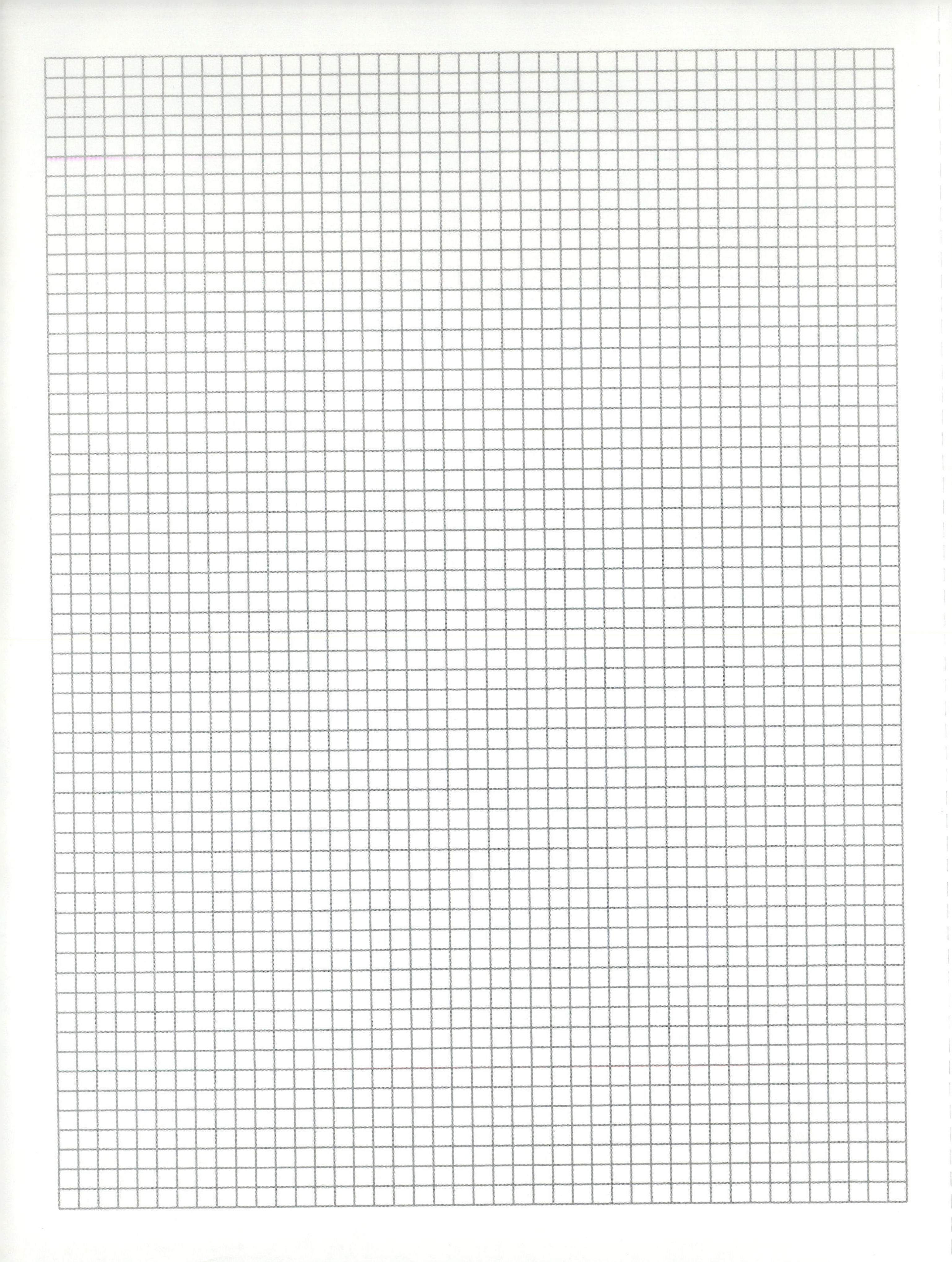

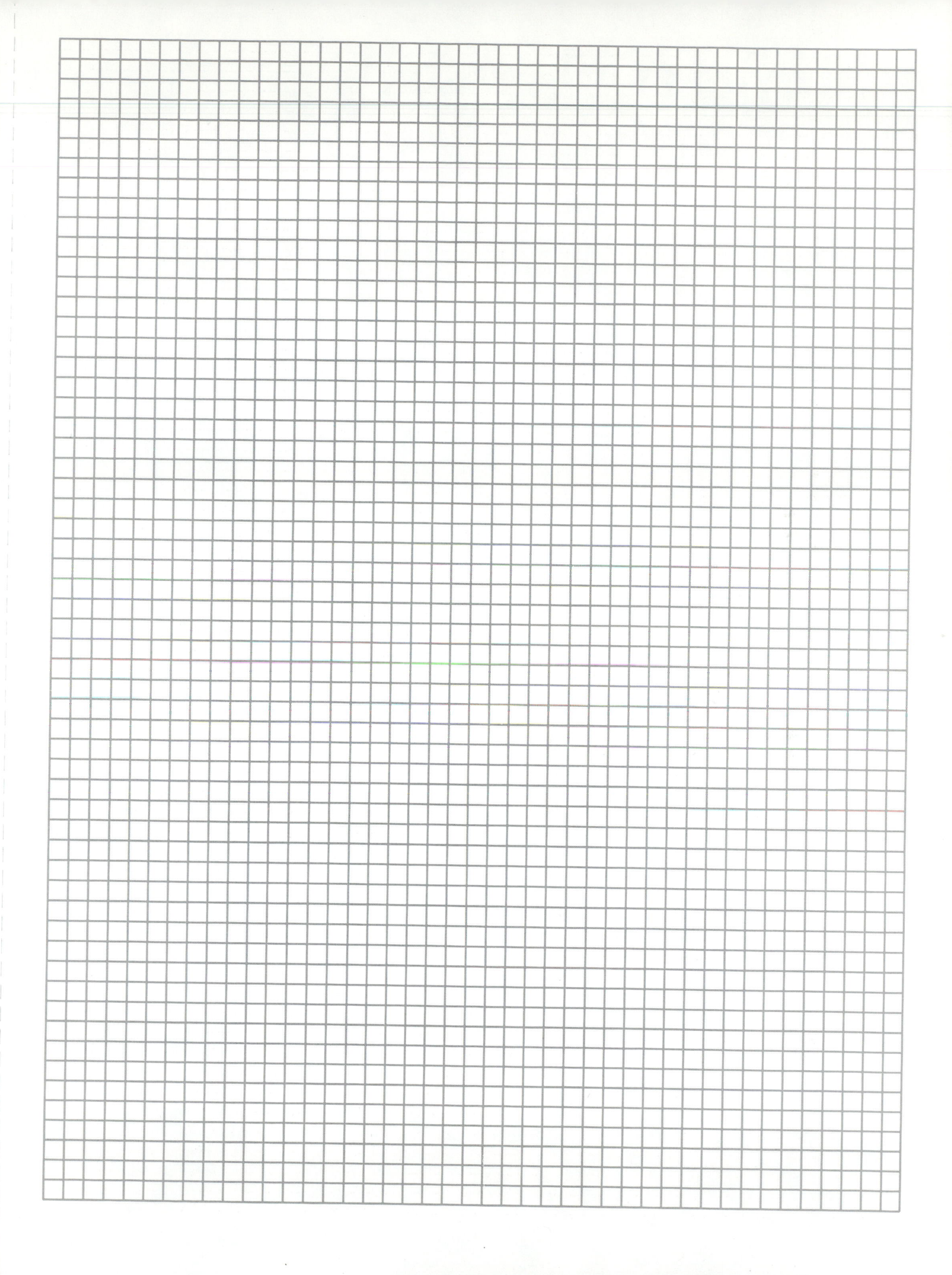

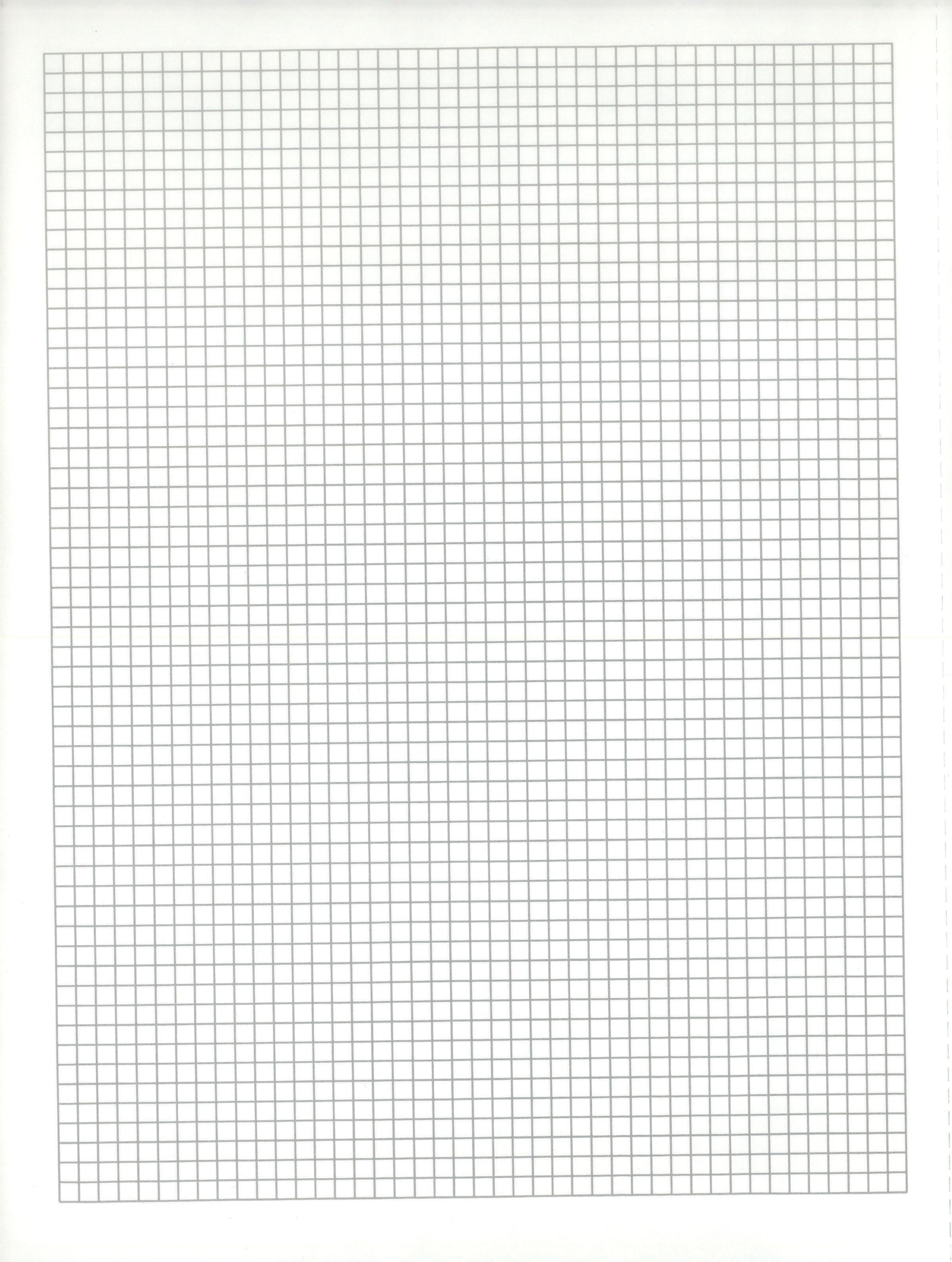

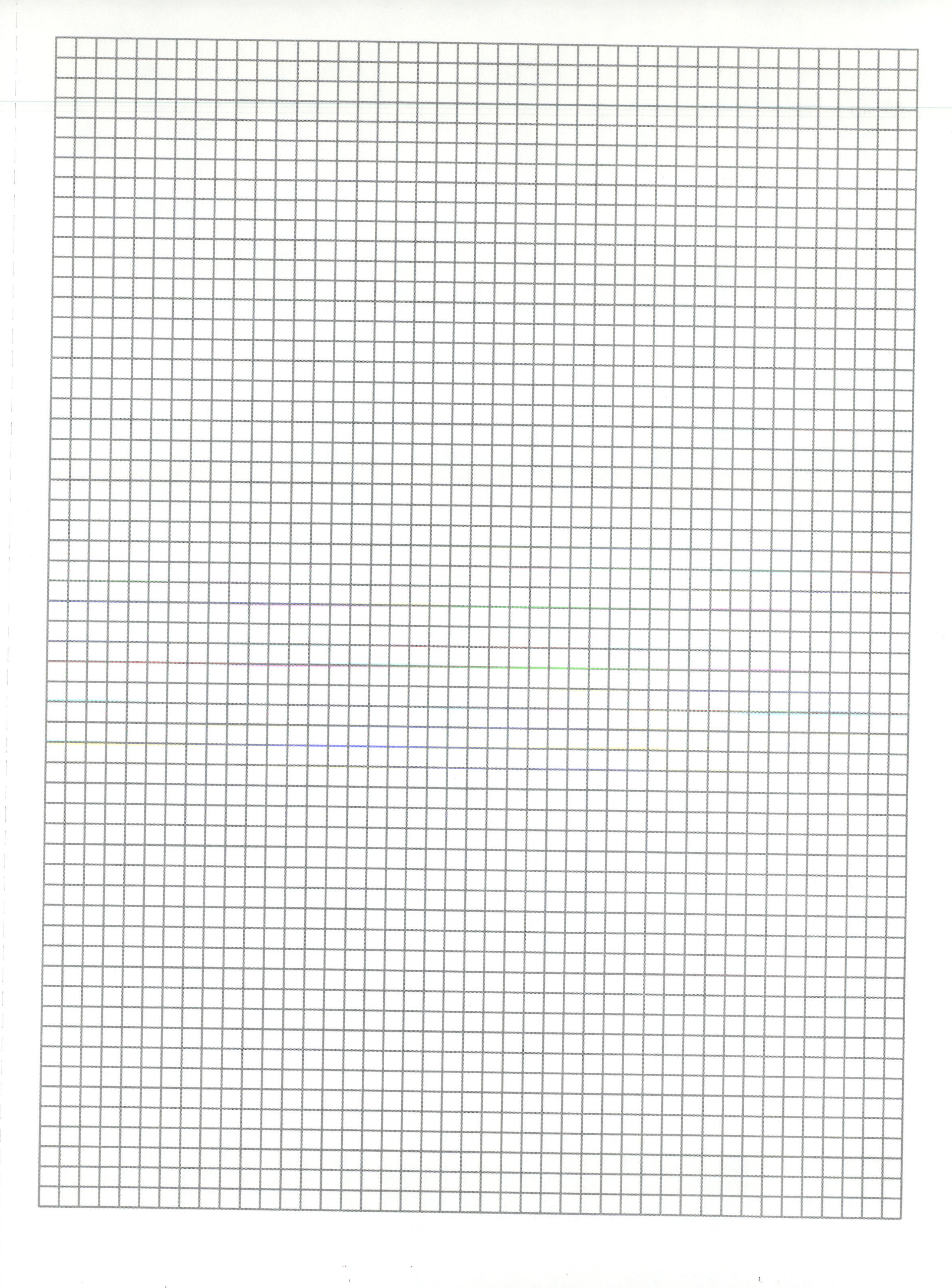

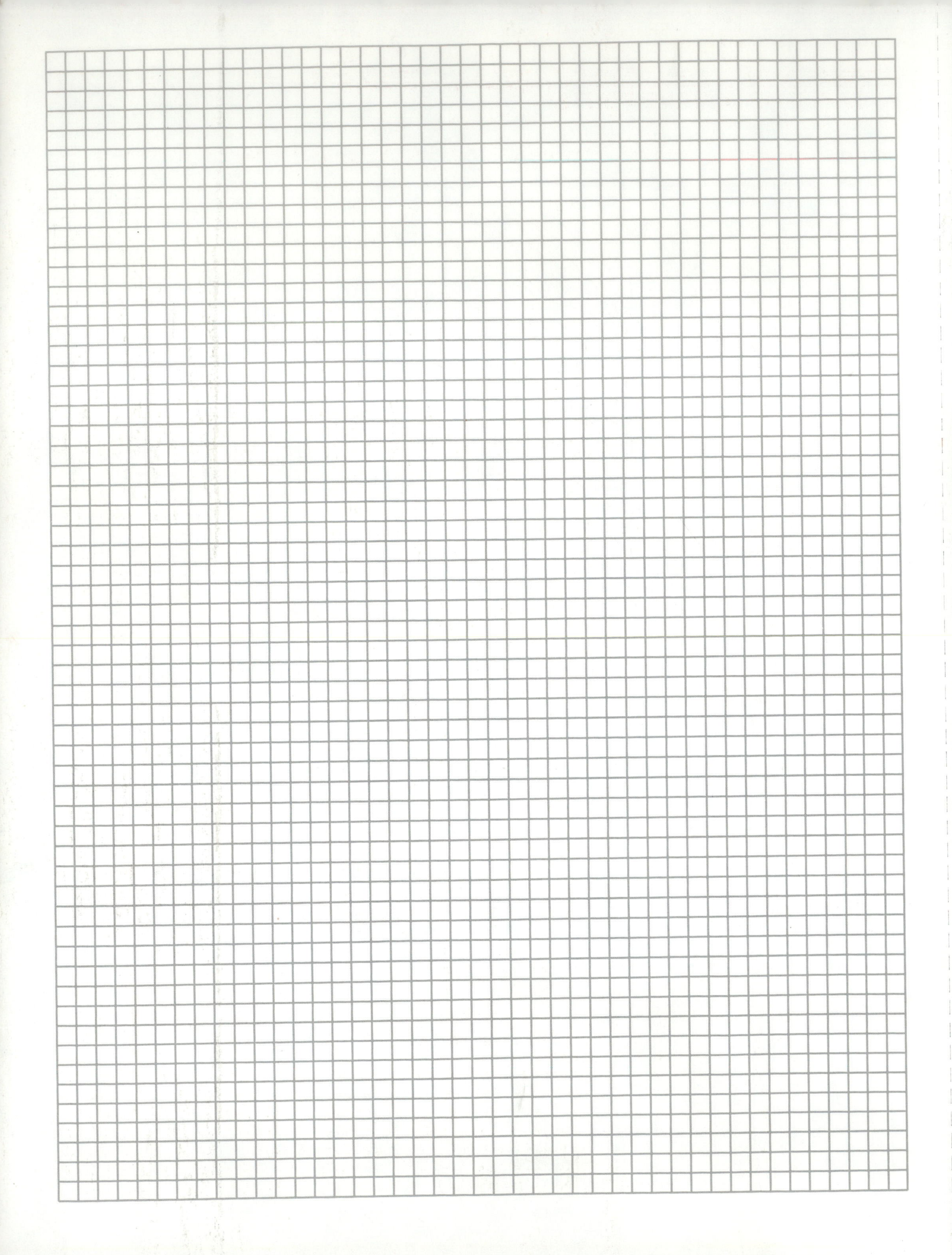